T0348580

VOLUME THIRTY ONE

THE ENZYMES

Eukaryotic RNases and their Partners in
RNA Degradation and Biogenesis, Part A

VOLUME THIRTY ONE

THE ENZYMES

Eukaryotic RNases and their Partners in
RNA Degradation and Biogenesis, Part A

Edited by

GUILLAUME F. CHANFREAU
Department of Chemistry and Biochemistry
Molecular Biology Institute
University of California
Los Angeles, CA 90095, USA

FUYUHIKO TAMANOI
Department of Microbiology, Immunology,
and Molecular Genetics, Molecular Biology Institute
University of California
Los Angeles, CA 90095, USA

AMSTERDAM · BOSTON · HEIDELBERG · LONDON
NEW YORK · OXFORD · PARIS · SAN DIEGO
SAN FRANCISCO · SINGAPORE · SYDNEY · TOKYO
Academic Press is an imprint of Elsevier

Academic Press is an imprint of Elsevier
525 B Street, Suite 1900, San Diego, CA 92101-4495, USA
225 Wyman Street, Waltham, MA 02451, USA
32 Jamestown Road, London NW1 7BY, UK
Radarweg 29, PO Box 211, 1000 AE Amsterdam, The Netherlands

First edition 2012

ISBN: 978-0-12-404740-2
ISSN: 1874-6047

For information on all Academic Press publications
visit our website at store.elsevier.com

Printed and bound by CPI Group (UK) Ltd, Croydon, CR0 4YY

Transferred to Digital Printing, 2013

**Working together to grow
libraries in developing countries**

www.elsevier.com | www.bookaid.org | www.sabre.org

ELSEVIER   BOOK AID
             International   Sabre Foundation

CONTENTS

Preface   *xi*

1. Biochemistry and Function of RNA Exosomes   1
Michal Lubas, Aleksander Chlebowski, Andrzej Dziembowski,
and Torben Heick Jensen

1. Introduction   2
2. Cellular Functions of the Exosome   3
3. Composition of Exosomes and Their Related Complexes   6
4. Exosome Cofactors   15
5. Concluding Remarks   21
Acknowledgments   22
References   22

2. Plant Exosomes and Cofactors   31
Heike Lange and Dominique Gagliardi

1. Introduction   32
2. Composition of the Plant Exosome Core Complex   32
3. Exosome-Related Activities and Cofactors in Plants   39
4. Impact of Exosome-Mediated Degradation in Plants   44
5. Final Remarks   48
References   48

3. Structure and Activities of the Eukaryotic RNA Exosome   53
Elizabeth V. Wasmuth and Christopher D. Lima

1. Introduction   53
2. Global Architecture of the Eukaryotic Exosome Core   56
3. RNase PH-Like Domains Comprise a PH-Like Ring in Eukaryotic Exosomes   56
4. S1 and KH Domains Cap the PH-Like Ring   59
5. Rrp44, a Hydrolytic Endoribonuclease and Processive Exoribonuclease   61
6. Rrp44 and the 10-Component Exosome   65
7. Rrp6, a Eukaryotic Exosome Subunit with Distributive Hydrolytic Activities 67
8. Rrp44, Rrp6, and the 11-Component Nuclear Exosome   70
9. Conclusions   71
Acknowledgments   71
References   71

v

4. TRAMP Stimulation of Exosome   **77**

Peter Holub and Stepanka Vanacova

1. TRAMP Is a Cofactor of Nuclear Exosome   78
2. RNA Substrate Repertoire of TRAMP   79
3. TRAMP Biochemistry and Structure   84
4. TRAMP Complex in Activation of the Exosome   88
5. TRAMP Complexes in Different Organisms   89
Acknowledgments   90
References   90

5. XRN1: A Major 5′ to 3′ Exoribonuclease in Eukaryotic Cells   **97**

Sarah Geisler and Jeff Coller

1. Introduction   98
2. XRN1 in mRNA Decay   98
3. XRN1 in mRNA Quality Control   100
4. XRN1 in miRNA-Mediated Decay   101
5. XRN1 in siRNA-Mediated Decay   103
6. Localization of XRN1 in Cells   104
7. XRN1 in lncRNA Decay   105
8. XRN1 in tRNA Quality Control   108
9. XRN1 in rRNA and snoRNA Processing   108
10. Regulation of XRN1 Activity   109
11. Summary   110
Acknowledgments   110
References   110

6. Structures of 5′–3′ Exoribonucleases   **115**

Jeong Ho Chang, Song Xiang, and Liang Tong

1. Introduction   116
2. Crystal Structure of Rat1/Xrn2   116
3. Crystal Structures of Xrn1   118
4. Structural Homologs of XRNs   121
5. Active Site of XRNs   122
6. Structure of the Rat1–Rai1 Complex   124
7. Structure of Rai1/Dom3Z   125
8. Perspectives   126
Acknowledgments   127
References   127

7. Rat1 and Xrn2: The Diverse Functions of the Nuclear Rat1/Xrn2 Exonuclease   131

Michal Krzyszton, Monika Zakrzewska-Placzek, Michal Koper, and Joanna Kufel

1. Introduction   132
2. The Life Story of Rat1/Xrn2   133
3. Sequence and Structural Data   135
4. One Protein, Many Functions   138
5. Overlapping Functions of Rat1 and Xrn1   154
6. Xrn2 as a Silencing Suppressor   155
7. Perspectives   156
Acknowledgments   157
References   157

8. Normal and Aberrantly Capped mRNA Decapping   165

Megerditch Kiledjian, Mi Zhou, and Xinfu Jiao

1. Introduction   166
2. mRNA-Decapping Proteins in the Exonucleolytic Pathway of mRNA Decay   167
3. Presence of an Aberrant Cap-Decapping Protein in *S. cerevisiae*   170
4. Additional Potential Functions of Rai1   174
5. Future Directions   176
Acknowledgment   177
References   177

9. Activity and Function of Deadenylases   181

Christiane Harnisch, Bodo Moritz, Christiane Rammelt, Claudia Temme, and Elmar Wahle

1. Introduction   182
2. The Poly(A) Nuclease (PAN)   183
3. The Poly(A)-Specific Ribonuclease   185
4. The CCR4–NOT Complex   190
Acknowledgments   203
References   203

10. The Diverse Functions of Fungal RNase III Enzymes in RNA Metabolism   213

Kevin Roy and Guillaume F. Chanfreau

1. Introduction   214
2. Phylogenetic Distribution and Conservation of RNase III Enzymes in Fungi   214

3. Ribosomal RNA Processing and RNA Polymerase I Transcriptional
   Termination   217
4. Small Nuclear RNAs Processing   219
5. Functions in Small Nucleolar RNAs Processing   221
6. (Pre)-mRNA Surveillance, Degradation, and Regulation   226
7. RNA Polymerase II Termination   228
8. Conclusions and Perspectives   230
Acknowledgments   231
References   231

Author Index   237
Subject Index   259

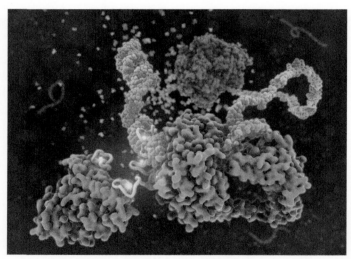

The TRAMP complex is in front and the gray structure in the back is the exosome. In yellow are zinc ions coordinated by Air2. Courtesy of Stepanka Vanacova. (See color plate section in the back of the book.)

PREFACE

Enzymes involved in RNA processing and RNA turnover play important functions in gene expression pathways. Many messenger RNAs and noncoding RNAs are functional only after proper processing, highlighting the essential role of enzymes involved in the biogenesis of cellular RNAs. In addition, the steady-state level of a specific RNA is determined not only by its rate of transcription but also by how efficiently this RNA molecule is degraded, making RNA degradation an important step in gene expression control. Although ribonucleases involved in eukaryotic RNA processing and degradation have been characterized for several decades, the field has made important advances in the recent years.

First, high-resolution structures of many eukaryotic RNA processing and degradation enzymes have become available. These structures have allowed us to understand at the atomic level the mechanism of these enzymes, their substrate specificity, and regulation by cofactors. In addition, genomic technologies adapted to the profiling of RNAs in systems where the function of these enzymes has been perturbed or to the dentification of RNAs that crosslink to these enzymes have provided extensive maps of the repertoire of eukaryotic ribonucleases. Finally, ribonucleases have been shown to be involved in quality control processes, as they promote discard pathways of RNA molecules that are defective, either because of mutations or because of errors in their biogenesis pathways. This concept of quality control has been demonstrated in a growing number of RNA biogenesis pathways.

We designed this volume to highlight these recent advances in our understanding of the structure and functions of eukaryotic ribonucleases. This volume is focused mostly on ribonucleases involved in the processing of noncoding RNAs and the degradation of mRNAs, with a special emphasis on exonucleases and their partners. The topics covered include structure and function of the RNA exosomes, the XRN family of exonucleases, deadenylases, and fungal RNase III. Some important enzymes involved in eukaryotic RNA metabolism could not be included in this volume. In the next volume, enzymes involved in the biogenesis of small RNAs such as siRNA and miRNA will be presented in detail. We believe that this first volume provides an important survey of the structure and function of some of the most important players in eukaryotic RNA processing and degradation.

We would like to thank the authors for their efforts in providing insightful contributions to this volume. We would also like to thank Sarah Latham from Elsevier for help and guidance during the preparation of this volume.

GUILLAUME CHANFREAU

FUYUHIKO TAMANOI

June 28, 2012

CHAPTER ONE

Biochemistry and Function of RNA Exosomes

Michal Lubas[*,†,‡], Aleksander Chlebowski[†,‡], Andrzej Dziembowski[†,‡,1], Torben Heick Jensen[*,1]

*Department of Molecular Biology and Genetics, Centre for mRNP Biogenesis and Metabolism, Aarhus University, C.F. Møllers Allé 3, Aarhus C, Denmark
†Institute of Biochemistry and Biophysics, Polish Academy of Sciences, ul. Pawińskiego 5a, Warsaw, Poland
‡Institute of Genetics and Biotechnology, Faculty of Biology, University of Warsaw, ul. Pawińskiego 5a, Warsaw, Poland
1Corresponding authors: e-mail address: andrzejd@ibb.waw.pl; thj@mb.au.dk

Contents

1. Introduction 2
2. Cellular Functions of the Exosome 3
   2.1 Nuclear substrates 3
   2.2 RNA QC by the exosome 3
   2.3 mRNA turnover 5
   2.4 Regulation of chromatin activity by the exosome 5
3. Composition of Exosomes and Their Related Complexes 6
   3.1 Ancestors of the eukaryotic exosome 6
   3.2 The eukaryotic exosome 7
   3.3 Enzymatic activities of the eukaryotic exosome 12
4. Exosome Cofactors 15
   4.1 Mtr4: A central helicase in nuclear RNA metabolism 15
   4.2 The Saccharomyces cerevisiae TRAMP complex and its human counterpart 17
   4.3 Rrp47p/C1D 18
   4.4 Mpp6 18
   4.5 Nrd1p and Nab3p 19
   4.6 The NEXT complex 20
   4.7 The SKI complex 20
5. Concluding Remarks 21
Acknowledgments 22
References 22

Abstract

Discovery of the evolutionary conserved RNA exosome was a milestone in RNA biology. First identified as an activity essential for the processing of ribosomal RNA, the exosome has since proved to be central for RNA processing and degradation in both the nucleus and the cytoplasm of eukaryotic cells. This multisubunit protein complex

The Enzymes, Volume 31
ISSN 1874-6047
http://dx.doi.org/10.1016/B978-0-12-404740-2.00001-X

1

consists of a catalytically inert 9-subunit core endowed with associated ribonucleolytic activities and further assisted by compartment-specific cofactors required for its activation and substrate targeting. Although many features of exosome biology are known, fundamental aspects are still under investigation. In this chapter, we review current biochemical and functional knowledge of eukaryotic exosomes. After introducing some of their nuclear and cytoplasmic functions, we discuss the structural organization and evolutionary aspects of exosome complexes. Finally, we describe catalytic properties of the complex and its regulation by cofactors.

1. INTRODUCTION

The discovery of the RNA exosome in 1997 came in the midst of an increasing interest in RNA metabolism and posttranscriptional mechanisms of gene regulation. Research into the biogenesis of ribosomal RNA (rRNA) in *Saccharomyces cerevisiae* demonstrated that maturation of 5.8S rRNA requires $3'$-$5'$ exoribonucleolytic trimming and depends on the *RRP4* gene [1]. Further investigation identified the Rrp41p, Rrp42p, Rrp43p, and Dis3p (*Dis*junction abnormal, also called Rrp44p) factors, which together with Rrp4p were shown to form a hetero-multimeric complex named the exosome [2]. At the time, catalytic activity was, erroneously, assigned to three of these factors: distributive hydrolytic activity to Rrp4p, processive hydrolytic activity to Rrp44p, and phosphorolytic activity to Rrp41p [2,3]. The exosome was recognized as a major *Saccharomyces cerevisiae* ribonuclease and—due to its strong evolutionary conservation— quickly became a major theme of research. What followed were many purification trials as well as comparative studies related to exosome-ancestral proteins (RNase PH, RNase II, and RNase R) [4,5]. While this has shaped a near-final view of the exosome's composition, next-generation techniques now serve to expose even further the cellular versatility of this amazing complex.

Today, we know that the nuclear exosome of *Saccharomyces cerevisiae* consists of 11 components, 10 of which are essential, even though 9 of these have no catalytic activity. The 10th and 11th subunits possess three enzymatic activities between them: two $3'$-$5'$ hydrolytic exoribonucleases (one processive and one distributive) and one endoribonuclease. Armed with these, the exosome is not limited to processing of stable RNAs, but engages in pathways of degradation and maturation of virtually all transcript classes.

2. CELLULAR FUNCTIONS OF THE EXOSOME

2.1. Nuclear substrates

The nuclear exosome acts on a multitude of substrates (Fig. 1.1, left). The abundant and stable rRNAs, small nuclear RNAs (snRNAs), and small nucleolar RNAs (snoRNAs) all require the nuclear exosome for the 3′end trimming of their precursors during biogenesis (Fig. 1.1A) [6,7]. This is a precise process, in which exonucleolytic decay must be stopped at particular positions. On the other hand, production of most, if not all, functional RNA is also subjected to strict exosome-dependent quality control (QC) mechanisms, detecting and removing faulty molecules entirely (Fig. 1.1B) [8,9]. Complete decay is also the fate of RNA maturation by-products, such as excised introns and the external transcribed spacer (ETS) regions of rRNA transcripts, the removal of which constitutes a major task for the nuclear exosome activity. Finally, the nuclear exosome also acts on a diverse group of noncoding RNAs (ncRNAs) arising from transcription of intergenic regions and exemplified by so-called cryptic unstable transcripts (CUTs) in *Saccharomyces cerevisiae* and PROMoter uPstream Transcripts (PROMPTs) in humans [10–15]. Some CUTs have been implicated in regulation of gene expression. Such regulation can be achieved by interference mechanisms involving sense or antisense RNAs [16–19]. However, in most cases, it remains to be determined whether the RNA itself, or the transcription event producing it, provides the actual activity.

A major challenge for the exosome is to distinguish whether an RNA molecule should undergo only limited exonucleolytic trimming or complete removal. One level of discrimination is the ability of "healthy" RNAs to fold into ribonuclease-immune assemblies bound by protective proteins (RNA-protein particles (RNPs)). Another takes advantage of the exosome's employment of cofactors harboring structure and/or sequence specificity. This allows the exosome to identify the majority of ncRNAs, which lack stabilizing features, and target these for rapid decay (Fig. 1.1B) [20].

2.2. RNA QC by the exosome

Biogenesis of mRNA is a multistep process including 5′end capping, splicing, 3′end formation, and packaging of mRNA with protein into mRNP. These reactions are error-prone and subject to QC, so that contamination of the

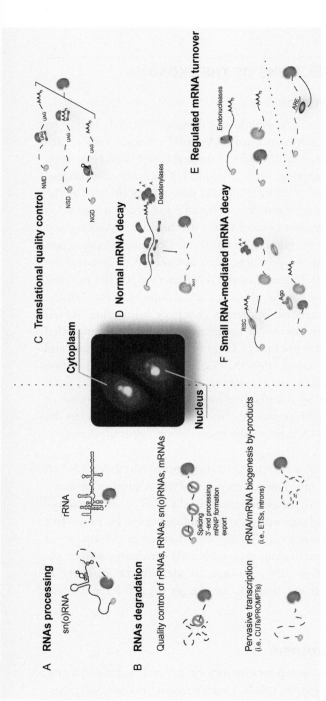

Figure 1.1 Substrates of the eukaryotic exosome. In eukaryotes, the exosome (green pacman) is present in both the nucleus (left) and the cytoplasm (right). The nuclear complex displays a nucleolar accumulation as indicated by the localization of a RRP45–EGFP fusion construct expressed at a near-endogenous level (unpublished result from the authors). Nuclear exosome functions encompass the processing of stable RNAs (sn(o)RNA and rRNA) (A), and the degradation of aberrant molecules as well as RNA maturation by-products (B). The latter comprises (i) QC at multiple steps along the biogenesis of pre-tRNAs, pre-mRNAs, pre-rRNAs, and pre-sn(o)RNAs (indicated by red signs), (ii) elimination of pervasive and unstable transcripts (i.e., CUTs/PROMPTs), and (iii) removal of introns and pre-rRNA ETSs. In the cytoplasm, the exosome functions in the degradation of mRNA engaged with stalled ribosomes (red shapes, right) during nonsense-mediated decay (NMD), nonstop decay (NSD), and no-go decay (NGD) (C). It partakes in normal cytoplasmic mRNA decay (D) as well as in regulated mRNA turnover pathways initiated by endonucleases (violet) and/or *cis*-acting AU-rich elements and associated proteins (dark blue) (E). Finally, the exosome assists the RNAi machinery, RNA-induced silencing complex (RISC) and Argonaute proteins (Ago), in removing spare RNA intermediates (F). See text for more details. Yellow and brown balls signify the 5′cap structure of relevant RNAs and sn(o)RNA-binding proteins, respectively. (See color plate section in the back of the book.)

cell by faulty molecules can be prevented [21]. In the nucleus, the exosome "monitors" the splicing process and ensures elimination of retained and unspliced pre-mRNA as well as splicing intermediates (Fig. 1.1B) [22–24]. Moreover, mRNA 3′end processing and mRNP formation are monitored by the exosome; errors result in the removal of transcripts lacking a protective poly(A) tail [25–30]. In the cytoplasm, the RNA exosome also engages in specialized QC pathways that remove aberrant mRNAs to prevent their translation: nonsense-mediated decay (NMD), no-go decay (NGD), and nonstop decay (NSD), where ribosomes stall at premature termination codons, strong secondary structures or 3′termini of mRNAs that lack stop codons altogether, respectively (Fig. 1.1C) [31–35].

2.3. mRNA turnover

All mRNAs have a finite lifetime and are eventually degraded. In the cytoplasm, exonucleolytic decay of the mRNA body is preceded by removal of the poly(A) tail (Fig. 1.1D) [36]. Subsequent degradation is then carried out in the 3′-5′ direction by the exosome or in the 5′-3′ direction by Xrn1 after removal of the mRNA 5′cap structure [37,38]. Consistently, in the cytoplasm of *Saccharomyces cerevisiae*, the exosome and Xrn1p exhibit a high level of redundancy: cells become inviable only when both exonucleolytic activities are blocked simultaneously [39,40].

Besides constitutive turnover, many mRNAs are degraded in a regulated fashion, often involving special destabilizing *cis*-elements or an initial cleavage by endonucleases prior to exosomal decay (Fig. 1.1E and F) [41]. In mammals, this is often achieved by AU-rich elements (AREs)—*cis*-acting regulatory sequences usually located in mRNA 3′ untranslated regions (3′UTRs), where they serve as binding sites for factors that either decrease (e.g., TTP, KSRP, AUF1) or increase (e.g., HuR) RNA stability. In the former cases, the exosome can be directly recruited by ARE-binding proteins or by the RNA helicase RHAU [42–44]. Binding sites for microRNAs in mRNA 3′UTRs can also be considered destabilizing *cis*-elements in RNA interference (RNAi). Here, the mRNA is cleaved by an endonuclease and the products are degraded by the exosome and Xrn1 RNases (Fig. 1.1F) [9].

2.4. Regulation of chromatin activity by the exosome

Major parts of eukaryotic genomes are transcriptionally active [20]. Such "pervasive" transcription produces transcripts from diverse genomic regions and with highly variable half-lives. For example, in *Schizosaccharomyces pombe*, transcription is detected from highly compact chromatin regions [45,46], which are regulated by the RNAi machinery generating small interfering (si)RNAs that act back on DNA and maintain transcriptional repression. The *Schizosaccharomyces pombe* RNA exosome contributes to such maintenance of chromatin (in)activity by its degradation of heterochromatic transcripts [45–47]. In *Saccharomyces cerevisiae*, where the RNAi system is absent, the nuclear exosome also reduces levels of heterochromatic transcripts that appear to be a subgroup of the previously mentioned CUTs [48]. Depletion of the nuclear exosome and its cofactors causes a decrease in chromatin compaction, including in rDNA and telomeric regions [49].

3. COMPOSITION OF EXOSOMES AND THEIR RELATED COMPLEXES

3.1. Ancestors of the eukaryotic exosome

Part of our knowledge about the evolution of exosome complexes is based on studies of their prokaryotic homologs: the homo–hexameric bacterial RNase PH (Fig. 1.2A) and the homo–trimeric polynucleotide phosphorylase (PNPase) (Fig. 1.2B) [51–54]. In RNase PH, the six subunits form a ring-like structure, in which every subunit is endowed with phosphorolytic exonuclease activity. The polypeptide chain of each monomer in PNPase contains two RNase PH-like domains, PH1 and PH2. In the trimer, these assume the same spatial arrangement as in RNase PH, but only the PH1 domains remain active. Unlike RNase PH, PNPase also contains the RNA-binding domains S1 and KH, three of each in the trimer, which "cap" the hexameric ring on one side. This cap is also ring-shaped, so that a channel traverses the whole structure. RNase PH and PNPase processively degrade RNA substrates in the $3'$-$5'$ direction using inorganic phosphate to attack the RNA backbone. The reaction releases nucleoside $5'$diphosphates and is readily reversible [55]. In *Escherichia coli*, PNPase is part of a larger complex, the degradosome, that participates in most RNA degradation pathways. In addition to PNPase, the degradosome is composed of the endonuclease RNase E, the RNA helicase RhlB, and the enolase protein of unknown function [56–58]. PNPase is also found in mitochondria and chloroplasts of higher eukaryotes [59–61], where there is no exosome.

A protein complex like the exosome is found in most *Archaea*. This archaebacterial exosome is composed of three to four independently encoded

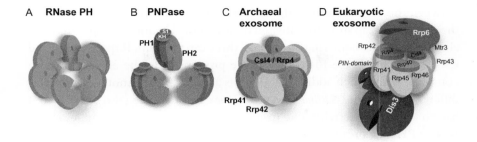

Figure 1.2 Architecture of the eukaryotic RNA exosome complex and its phosphorolytic ancestors. Schematic representations of RNase PH, PNPase, and the archaeal and eukaryotic exosomes with phosphorolytically active and inactive subunits indicated in green and orange, respectively. RNase PH forms a homo-hexameric ring composed of PH subunits (A). In PNPase, each monomer consists of a PH1 and a PH2 domain with two RNA-binding domains, S1 and KH (B). The archaeal exosome forms a six-subunit heteromeric RNase PH-like ring with active, Rrp41, and inactive, Rrp42, components as well as three KH/S1 domain-containing subunits, Rrp4, Csl4, or both (light blue), on the top of the ring (C). The eukaryotic exosome core comprises nine different and catalytically inert subunits, three of which (Rrp40, Rrp4, and Csl4) form an RNA-binding cap structure (light blue) (D). Hydrolytic RNases associate with the core: the Dis3 (dark blue) endo/exoribonuclease at the bottom and the Rrp6 distributive exoribonuclease (purple) presumably at the top. Dis3 interacts with Rrp41–Rrp45 and its N-terminal PIN domain is crucial to maintain this interaction [50]. (See color plate section in the back of the book.)

proteins and shows close functional and structural similarity to PNPase (Fig. 1.2C) [54,62–65]. However, the four domains of PNPase, PH1, PH2, KH, and S1 (Fig. 1.2B), have been separated out into individual proteins: aRrp42 (PH1), aRrp41 (PH2), and aRrp4 (KH and S1 domains) and/or aCsl4 (S1 domain) (Fig. 1.2C). Together, these factors form a structure like that of PNPase: a hexameric ring capped by aRrp4 and/or aCsl4. The similarity between the exosome of *Archaea* and PNPase of *E. coli* extends to their biochemical properties: the phosphorolytic archaeal complex also reversibly degrades RNA thanks to its three catalytic sites localized in the aRrp41 subunits inside the bottom part of the channel. Unstructured RNA substrates can be threaded through the central channel between the RNA-binding cap proteins situated on top of the core structure [66].

3.2. The eukaryotic exosome

The architecture of the eukaryotic exosome resembles that of the archaeal complex. The six RNase PH domains can be identified in distinct eukaryotic components related to archaeal aRrp41 (Rrp41, Rrp46, and Mtr3) or aRrp42 (Rrp42, Rrp43, and Rrp45) (Fig. 1.2D). The main difference between the organization of the archaeal and eukaryotic core complexes is the eukaryotic cap structure situated on top of the hexameric ring. Unlike in *Archaea*, the hexameric ring requires all three cap proteins, Rrp4, Rrp40, and Csl4, to remain stable [64]. These nine proteins form the eukaryotic core complex, with which additional subunits are associated. Another important difference is that the *Saccharomyces cerevisiae* exosome core (also called Exo9) has no enzymatic activity, due to mutations in key catalytic residues in the RNase PH–like proteins, and instead associates with two hydrolytic ribonucleases: Dis3p and Rrp6p (Fig. 1.2D) [2–4]. Despite their lack of catalytic activity, all core exosome subunits are essential for growth [3].

All exosome core subunits are present in the cytosol and the nucleus. Dis3p has the same localization pattern; however, Rrp6p is absent from the cytoplasm (Fig. 1.3 and Table 1.1) [4,67]. Thus two types of exosome complexes exist in *Saccharomyces cerevisiae*, a cytoplasmic one, composed of 10 subunits (Exo10), and a nuclear one, composed of all 11 subunits (Exo11) (Fig. 1.3, left). The Exo9 complex is highly preserved in human cells [68], yet human exosomes are more multifaceted in terms of their associated catalytic subunits and at least four different variants

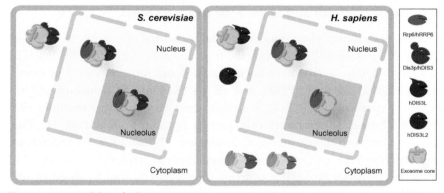

Figure 1.3 Models of the composition and localization of exosome complexes in *Saccharomyces cerevisiae* and *H. sapiens*. The exosome core (Exo9) is present throughout the cell, leaving variability to the differential localization of the associated nucleases. In *Saccharomyces cerevisiae*, nuclear Exo9 associates with Dis3p and Rrp6p, whereas cytoplasmic Exo9 only binds Dis3p (left). In human cells, hRRP6 associates with Exo9 in the nucleus (with nucleolar accumulation) and to a minor extent in the cytoplasm (right). hDIS3 is mainly nuclear localized and limited to the non-nucleolar part of the nucleus. Moreover, it exhibits a minor cytoplasmic presence. In contrast, hDIS3L is strictly cytoplasmic. In addition, DIS3L2 is cytoplasmic, but does not associate with the exosome core. Factor symbols are shown to the right and low opacity indicates less prominent cytoplasmic exosomes. See main text for references. (See color plate section in the back of the book.)

Table 1.1 Human and yeast exosome components and auxiliary proteins

Complex/protein	Protein name		Localization		Domains
	Saccharomyces cerevisiae	H. sapiens	Saccharomyces cerevisiae	H. sapiens	
Exosome core components	Csl4p	hCSL4/EXOSC1	no, n, c	no, n, c	S1, KH
	Rrp4p	hRRP4/EXOSC2	no, n, c	no, n, c	S1, KH
	Rrp40p	hRRP40/EXOSC3	no, n, c	no, n, c	S1, KH
	Rrp41p	hRRP41/EXOSC4	no, n, c	no, n, c	RNase PH
	Rrp42p	hRRP42/EXOSC7	no, n, c	no, n, c	RNase PH
	Rrp43p	hRRP43/EXOSC8	no, n, c	no, n, c	RNase PH
	Rrp45p	hRRP45/EXOSC9	no, n, c	no, n, c	RNase PH
	Rrp46p	hRRP46/EXOSC5	no, n, c	no, n, c	RNase PH
	Mtr3p	hMTR3/EXOSC6	no, n, c	no, n, c	RNase PH
Catalytic subunits	Rrp44p	hDIS3/RRP44	no, n, c	n, c	RNB, PIN
	–	hDIS3L/RRP44L	–	c	RNB, PIN[a]
	Rrp6p	hRRP6/EXOSC10	no, n	no, n, c	RNase D
Single-protein activators	Rrp47p/Lrp1p	C1D	no, n	no, n	Noncanonical RNA-binding motif
	Mpp6p	hMPP6	no, n	no, n	
	Ski7p	HBS1L	c	c[b]	EF-Tu/EF-1A

Continued

Table 1.1 Human and yeast exosome components and auxiliary proteins—cont'd

Complex/protein	Protein name		Localization		Domains
	Saccharomyces cerevisiae	H. sapiens	Saccharomyces cerevisiae	H. sapiens	
TRAMP complex	Trf4p/Trf5p	hTRF4-2 (PAPD5)/hTRF4-1 (PAPD7)	no, n	no, n	NTP, PAP_assoc
	Air1p/2p	ZCCHC7	no, n	no	Zf-CCHC
	Mtr4p	hMTR4/SKIV2L2	no, n	no, n	DEAD/DExH
Nrd1–Nab3–Sen1 complex	Nrd1p	—	no, n	—	CID, RRM
	Nab3p	—	no, n	—	RRM
	Sen1p	SETX			helicase SEN1
NEXT complex	—	RBM7	—	n	RRM
	—	ZCCHC8	—	n	Zf-CCHC
	Mtr4p	hMTR4/SKIV2L2	no, n	no, n	DEAD/DExH
SKI complex	Ski2p	SKIV2L	c	n, c	DEAD/DExH
	Ski3p	TTC37/hSKI3	c	n, c	TPR
	Ski8p	WDR61/hSKI8	c	n, c	WD40

[a] Catalytically inactive.
[b] Unpublished date. (S. Lykke–Andersen, T.H. Jensen)
No, nucleolus; n, nucleus; c, cytoplasm.
Domain names abbreviations: EF-Tu/EF-1A, translation elongation factor domains; NTD, nucleotidyltransferase domain; PAP_assoc, PAP/25A-associated domain—noncanonical poly(A) polymerase; DEAD/DExH, box helicase domain; Zf-CCHC, zinc finger domain CysCysHisCys (CCHC) type; CID-RNAPII, CTD-interacting domain; RRM, RNA recognition motif; TPR, tetratricopeptide repeat region; WD40, motif terminating in a tryptophan–aspartic acids (W-D); see text for information about domains.

were proposed based on recent coimmunoprecipitation (co-IP) and localization studies (Fig. 1.3) [69–72]. Like in *Saccharomyces cerevisiae*, the human cell nucleus contains hDIS3 (homolog of Dis3p) and hRRP6/ PM/Scl-100 (Rrp6p) [4,70,72]; however, the nuclear distributions of hDIS3 and hRRP6 differ. Whereas hRRP6 is present all over the nucleus with nucleolar accumulation, nuclear hDIS3 is excluded from nucleoli [70,72,73]. Also, the relative cytoplasmic presence of hDIS3 is clearly diminished relative to *Saccharomyces cerevisiae*; most human Exo9 complexes in this compartment associate with paralog of hDIS3, hDIS3L (see below). Moreover, hRRP6 is also present in the cytoplasm, as shown by intracellular localization analyses [70,74–77]. Notably, in co-IP experiments the obtained ratio of hRRP6 to core subunits is significantly lower when DIS3L, as opposed to hDIS3, is used as the bait [69,74]. Moreover, when using hRRP6 as the bait, DIS3L is not detected at all, but hDIS3 is [69,72]. This suggests that cytoplasmic hRRP6 levels are low and that exosomes containing both DIS3L and hRRP6 are scarce. Taken together, these data imply that an Exo10 complex, harboring hDIS3L as its nucleolytic component, is the predominant exosome in the cytoplasm of human cells (Fig. 1.3, right).

Although the RNA exosomes from *Saccharomyces cerevisiae* and human cells are the best described homologs have also been characterized in protists, insects, and plants [78–80]. Of these data, electron microscopy (EM) suggested that RRP6 binds to the heterotrimeric cap of the *Leishmania tarentolae* exosome core [81]. Although being in line with Dis3p binding to RNase PH subunits at the bottom of the exosome core as demonstrated in *Saccharomyces cerevisiae* [50], it still awaits further verification whether such Rrp6 localization extends to *Saccharomyces cerevisiae* and human exosomes. Unexpectedly, the *Drosophila* exosome was purified without the Rrp43 and Rrp45 subunits, leading the authors to suggest the possible existence of exosome subcomplexes, a scenario also awaiting further investigation [75,80]. Finally, interesting results were obtained concerning the composition of plant exosomes [82]. Driven by multiple gene duplications, plant genomes encode several copies of core components (RRP40A and RRP40B as well as RRP45A and RRP45B) and catalytic subunits (AtRRP6-like 1–3). Moreover, the plant Exo9 complex appears to have retained phosphorolytic activity [79]. The impact of these differences on the organization and function of plant exosomes will be interesting to learn more about.

3.3. Enzymatic activities of the eukaryotic exosome

As mentioned above, the eukaryotic exosome harbors three ribonuclease activities contained in two of its subunits. Rrp6p/hRRP6 is a distributive exoribonuclease, while Dis3p/hDIS3 is a processive exo- as well as an endoribonuclease [83–86]. The Exo9 complex is catalytically inert and may serve as a docking platform to get the nucleases in proximity to substrate RNAs and regulatory factors [3,62-63,74,87].

3.3.1 The endo/exonuclease Dis3

The exonuclease activity of *Saccharomyces cerevisiae* Dis3p is located in a large region of homology to bacterial RNase II/R enzymes, consisting of two cold shock domains (CSDs), an RNB domain and a C-terminal S1 domain (Fig. 1.4). It degrades its RNA substrates down to 3–4 nt products and releases nucleoside 5'monophosphates [3,83]. As in RNase II, the RNB domain contains a catalytic center with four acidic amino acid residues coordinating two magnesium ions and a single amino acid mutation of any of these residues completely abolishes Dis3p exonucleolytic activity [3,65,88].

Dis3p is also an endonuclease [65,84,85]. The endonucleolytic activity is embedded within its N-terminal PilT N (PIN) domain, which is also responsible for Dis3p association with the exosome core. The active site consists of four acidic amino acid residues that coordinate two divalent metal cations [50,83–85]. The PIN domain of Dis3p exhibits high similarity to the RNase E component of the bacterial degradosome and consistently displays high specificity toward RNA substrates with phosphorylated 5'ends [85,89]. Dis3p endonuclease activity is hypothesized to cleave highly structured regions of substrates, thereby providing access for exonucleolytic decay and increasing the overall degradation efficiency of the exosome.

Figure 1.4 Domain composition of *Saccharomyces cerevisiae* Dis3p and its human homologs and *E. coli* RNase R. *Saccharomyces cerevisiae* Dis3p harbors exo-(RNB) and endo-(PIN) nucleolytic domains as well as three RNA-binding domains (CSD1, CSD2, and S1). Positions of active sites' amino acid residues responsible for endo- and exonucleolytic activity are shown in red. The RNB domain possesses one highly evolutionary conserved active site. The hDIS3L PIN domain is inactive due to the lack of two of four conserved amino acid residues. (See color plate section in the back of the book.)

The human genome contains three homologs of the *Saccharomyces cerevisiae DIS3* gene: *hDIS3*, *hDIS3L*, and *hDIS3L2* [70,74,90] (Figs. 1.3 and 1.4). The PIN domain of hDIS3L has two of the four amino acids of the hDIS3 active site mutated, which predictably disables its endonucleolytic activity (Fig. 1.4) [70,74], as shown by experiments failing to detect any endonucleolytic activity of hDIS3L [72]. Interestingly, the third *DIS3* homolog, hDIS3L2, holds an extended CSD1 domain and a significantly shorter N-terminal region with no PIN domain at all. Consistently, it is insufficient to support interaction with the exosome core (M. Lubas, T.H. Jensen and A. Dziembowski, unpublished) and DIS3L2 therefore acts independently of the exosome. Of the three Dis3p homologs in human, only the *hDIS3* gene can partly complement a *dis3Δ Saccharomyces cerevisiae* strain [74]. Consistently, hDIS3 is the only protein with both endo- and exonucleolytic activities preserved.

3.3.2 The exonuclease Rrp6

The second 3′-5′ exoribonuclease of the *Saccharomyces cerevisiae* and human exosomes, Rrp6p/hRRP6, is homologous to bacterial RNase D and a member of the DEDD nuclease superfamily [2,91,92]. Notably, Rrp6p is the only nonessential exosome subunit in *Saccharomyces cerevisiae*; however, an *rrp6Δ* strain is heat-sensitive [91]. Rrp6p/hRRP6 comprises a helicase/RNase D C-terminal (HRDC)-, a DEDD-, and a polycystin 2 N-terminal (PMC2NT) domain. The DEDD domain is required for the enzyme's distributive 3′-5′ exonucleolytic activity, which also requires divalent metal ions, ideally magnesium [93,94]. The PMC2NT domain serves as a binding site for the Rrp6p cofactor and RNA-binding protein Rrp47p [95]. The function of the HRDC domain remains unclear. In *Saccharomyces cerevisiae*, the DEDD and HRDC domains are sufficient to retain catalytic activity *in vitro*, while *in vivo* the presence of the HRDC and PMC2NT domains are necessary for the maturation of 5.8S rRNA and some sn(o)RNAs [96,97]. Human hRRP6 is strongly inhibited by HRDC deletion, yet this hRRP6 variant still preserves a minor exonucleolytic activity and ability to interact with RNA *in vitro* [94]. Both *Saccharomyces cerevisiae* and human Rrp6 prefer single-stranded substrates, although secondary structures can also be degraded if they have a single-stranded region to their 5′end. This is in agreement with endogenous substrates of Rrp6 including 7S pre-rRNA, small nuclear, and snoRNAs.

3.3.3 Activities of Exo10 and Exo11

Structural studies of Dis3p demonstrate that the active site of the RNB domain lies at the bottom of a narrow channel that can encompass about 10 nt of single-stranded (ss) RNA [3,64,65]. Dis3p can only initiate exonucleolytic decay of structured substrates if a 5 nt stretch of ssRNA is available at the substrate 3'end. The unwinding ability of Dis3p was found to determine its RNA degradation rate [98]. Dis3p, as well as its homolog RNase R, accumulates energy released from substrate hydrolysis until 4 nt are unwound at once in a so-called spring mechanism. When Dis3p associates with the exosome core in the Exo10 complex, its length requirement for a ssRNA overhang increases to 30 nt [50]. This is because the substrate must be threaded through the internal channel of Exo9 before it reaches the Dis3p active site, and now the additional 25 nt of ssRNA are needed to cover the distance from the Exo9 cap to the bottom of its core [50]. This 30 nt ssRNA requirement for Exo10 activity may serve to avoid the undesired degradation of properly folded mature RNAs, which only have to keep their unstructured termini short [99]. Experimental support that RNA threading also occurs *in vivo* comes from the appearance of a 30 nt 3'extended 5.8S rRNA processing intermediate when Rrp6p is absent [4]. RNA threading has also been confirmed by EM imaging [91,99]; however, a channel-independent path to the Dis3p active site has also been suggested [99]. Here, the RNA may contact Exo10 via the PIN domain of Dis3p, which is exposed to the solvent. This hypothetical pathway would then require some reversible structural changes in Dis3p to proceed from endo- to exonucleolytic decay. As the position of Rrp6 on the exosome core has not been definitively determined less is known about any potential functional crosstalk between this enzyme and the rest of the exosome. Provided that Rrp6p is situated on top of the Exo9 cap and removes the last 30 nt from the 5.8S rRNA precursor the enzyme may work independently of the central channel. Whether this is true or Exo9 cap proteins modulate Rrp6p activity remains to be investigated. Similarly, the *in vivo* indications that Rrp6 activity may cooperate with the endonucleolytic activity of the Dis3p PIN domain [83] await further analyses.

Unexpectedly, the three nucleolytic activities of the *Saccharomyces cerevisiae* exosome have quite different ionic requirements for their optimal activity *in vitro*. The Dis3p endonuclease is manganese-dependent, with an optimal Mn^{2+} concentration of 3–5 mM, whereas its exonuclease activity is inhibited in the presence of Mn^{2+} [83–85]. Additionally, Rrp6p displays optimal activity at millimolar concentrations of Mg^{2+}, where Dis3p is barely active at all [3,96]. The rationale behind these different requirements, including how the activities collaborate in the cell, is presently unclear.

4. EXOSOME COFACTORS

Cellular RNAs are usually well protected from spurious nucleolytic degradation. The RNA exosome gains access to its substrates with the help of a set of compartment-specific cofactors, which "boost" exosome activity. Some cofactors serve as adaptors between the exosome and specific RNA sequences or structures. Others are enzymes, like RNA helicases or poly (A) polymerases, providing exosome access by resolving double-stranded/ protein-bound regions or by adding unstructured tails to the 3′ends of substrates. Overall, exosome activators are believed to provide substrate specificity and discrimination of RNA processing and degradation pathways.

4.1. Mtr4: A central helicase in nuclear RNA metabolism

In the nucleus, Mtr4p(Dob1p)/hMTR4(SKIV2L2) participates in all known functions of the exosome, often linking its activities with additional compartment-specific RNA-binding adaptors (Fig. 1.5). The *MTR4* gene was initially described in a *Saccharomyces cerevisiae* screen, that is, its mutation results in the nuclear accumulation of poly(A)$^+$ RNA as well as in rRNA processing defects [100,101]. Indeed, the helicase contributes to the exosomal maturation of rRNAs, sno/snRNAs as well as the degradation of unfolded RNAs and pervasive transcripts [6,7,26,97,101–104] (Fig. 1.5A–C). In many of these pathways, Mtr4p functions in the context of the TRAMP complex, connecting polyadenylation with exosomal decay, which explains the nuclear accumulation of poly(A)$^+$ RNA observed upon Mtr4p mutation [6,7,105,106]. Mtr4p is a member of the Ski2-like DExH-box family of helicases, possessing ATPase and 3′-5′ directional helicase activity [104,107]. Recent structural data on Mtr4p have disclosed the presence of a so-called Kyrpides–Ouzounis–Woese (KOW) domain, which is unique for this helicase family and indispensable for Mtr4p function [108–110]. The KOW domain is a flexible arch-shaped region located close to the helicase core. Interestingly, a *Saccharomyces cerevisiae* strain harboring a KOW-less Mtr4p variant accumulates 5.8S rRNA with 30 nt 3′extensions, similar to cells lacking Rrp6p but different

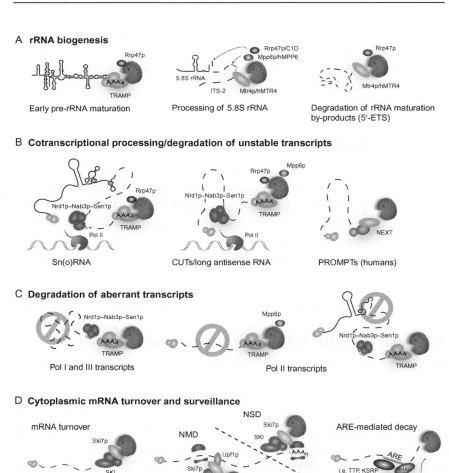

Figure 1.5 Models of RNA exosome activation. The RNA exosome requires the aid of cofactors to efficiently process and degrade its substrates (labeled as in Fig. 1.1). Most of the known cellular activities of the exosome require RNA helicases. During rRNA biogenesis, the nuclear exosome is activated by Mtr4p/hMTR4 alone or in the context of the TRAMP complex (A, left and middle). In addition, the two RNA adaptors, Rrp47p/C1D and Mpp6p/hMPP6, recruit the exosome in structure- and sequence-specific manners, respectively (A). In *Saccharomyces cerevisiae*, the processing of sn(o)RNAs relies on the Nrd1p–Nab3p–Sen1p complex facilitating the cotranscriptional recruitment of the exosome/TRAMP to well-defined RNA motifs (B, left). Similar mechanisms lead to the complete degradation of CUTs or long antisense RNAs, and perhaps also other transcripts lacking protective features (B, middle). In human cells, unstable PROMPTs are targeted for exosomal decay by the NEXT complex (B, right). Defective molecules generated by all three RNA polymerases are efficiently eliminated from the cellular pool of RNA (C). Exosome cofactors, for example, TRAMP, "sense" these faulty RNAs (marked

from exosome core mutants [6,108] (Fig. 1.5A). Thus, the Mtr4p KOW domain may be required for substrate degradation/processing by Rrp6p, for example, by displacing substrates from the exosome core and thereby providing access for Rrp6p [109].

4.2. The *Saccharomyces cerevisiae* TRAMP complex and its human counterpart

Mtr4p functions in the context of activator complexes. The best understood exosome activator is the *Saccharomyces cerevisiae* TRAMP complex, which stimulates exosomal activity by the addition of unstructured oligo(A) tails to substrate 3′ends [105,106]. These marks are introduced by the noncanonical poly(A) polymerases Trf4p or Trf5p and aided by the RNA-binding proteins Air1p or Air2p. Such adenylation–assisted RNA degradation is reminiscent of the prokaryotic world and an additional similarity between bacterial and eukaryotic 3′-5′ exonucleolytic decay [111–114]. TRAMP substrates do not share any obvious similarities. They are transcribed by all three RNA polymerases and include sn/snoRNAs, pre-rRNAs, aberrant tRNAs, and mRNAs as well as ncRNAs (e.g., CUTs and antisense RNAs) [12,49,103,106,112,115,116] (Fig. 1.5). Interestingly, TRAMP adenylation and unwinding activities are carefully coordinated in that the helicase activity of Mtr4p is stimulated by the presence of Trf4p–Air2p, while Mtr4p in turn diminishes the polyadenylation activity of Trf4p [112,114,117]. This presumably avoids the addition of too long poly(A) tails, predicted to bind poly(A)-binding proteins and inhibit rather than favor RNA decay.

A putative human TRAMP complex consists of the noncanonical poly (A) polymerase hTRF4-2 (PAPD5), the zinc-knuckle protein ZCCHC7, and the Mtr4p homologous RNA helicase hMTR4. These three proteins all appear in hRRP6 immunoprecipitates and localization studies revealed their nucleolar accumulation, although hTRF4-2 and hMTR4 were also detected in the non–nucleolar part of the nucleus [69]. Consistent with a

by red signs) and promote their exosomal degradation (C, left, middle, and right). In the cytoplasm, the SKI complex and the GTPase Ski7p serve as major exosome cofactors (D, left). Ski7p directly interacts with nonsense-mediated decay factor, Upf1, and participates in removal of faulty mRNAs. In human, sequence-specific recruitment of the cytoplasmic exosome requires RNA-binding factors (i.e., TTP, KSRP) that recognize ARE instability elements and involve helicase activity of RHAU protein (D, middle and right). (See color plate section in the back of the book.)

nucleolar function, hTRF4-2 requires ZCCHC7 for the adenylation of rRNA degradation intermediates arising as a result of cell treatment with Actinomycin D [69]. This resembles the nucleolar functions of *Saccharomyces cerevisiae* TRAMP and the data taken together imply that a TRAMP-like mode of action may vary between nuclear compartments in mammals [118,119]. Perhaps more structured nucleolar RNAs require the cooperative function of hMTR4, hTRF4-2, and ZCCHC7, whereas hTRF4-2 polyadenylation coordinated with unwinding by hMTR4 may suffice to deal with non-nucleolar RNAs. Finally, as described below, hMTR4 interacts with other factors in the Nuclear EXosome Targeting (NEXT) complex present in human nuclei.

4.3. Rrp47p/C1D

Sequence-unspecific recruitment of the nuclear exosome relies on the structural properties of substrates recognized by adaptors. This is illustrated by the nuclear RNA-binding protein Rrp47p/Lrp1p (C1D in human), which shows a strong binding preference for structured RNAs *in vitro* (Fig. 1.5A) [73,95,120]. Rrp47p depletion largely resembles the effects observed in *rrp6Δ* cells and to some extent in exosome core-mutated strains. These include alterations in the early stages of pre-rRNA maturation, in the final trimming of 5.8S rRNA, and in the degradation of the rRNA 5′ETS sequence (Fig. 1.5A) as well as in sn(o)RNA processing and the degradation of the rather poorly structured CUTs, which may form RNA duplexes with antisense RNAs (Fig. 1.5B) [83,121]. As mentioned above, Rrp47p interacts directly with the PMC2NT domain of Rrp6p and the cellular levels and nucleolar localizations of Rrp47p/C1D depend on Rrp6p/hRRP6 presence [73,120]. Consistently, Rrp47p and C1D are tightly associated with yeast and human exosomes, although in substoichiometric amounts [69,95,120,122]. The C1D protein was first characterized as a DNA-binding transcriptional repressor, partnering with a DNA-dependent protein kinase in DNA double-stranded break repair; a function later confirmed for *Saccharomyces cerevisiae* Rrp47p [123–125]. Thus, Rrp47p/C1D may hold additional exosome-independent functions.

4.4. Mpp6

In contrast to Rrp47p/C1D, the exosome cofactor M-phase phospho-protein 6 (Mpp6p/hMPP6) is a sequence-specific adaptor with high affinity to pyrimidine tracts [121,126]. In *Saccharomyces cerevisiae*, Mpp6p is involved in 5.8S rRNAs maturation (Fig. 1.5A), the degradation of CUTs (Fig. 1.5B), and the surveillance of pre-mRNA (Fig. 1.5C). However, reported effects in *mpp6Δ* cells are usually rather mild, which is perhaps due to its functional redundancy with other exosome cofactors—for example, codepletion of Mpp6p and Rrp47p displays a synergistic effect in the stabilization of CUTs [121]. Consistently, absence of Mpp6p is synthetically lethal with the absence of either Rrp6p or Rrp47p [120,121]. In human cells, hMPP6 accumulates in nucleoli where it favors the interaction with poly-pyrimidine stretches [126]. Similar to C1D, hMPP6 copurifies with the exosome in an hRRP6-dependent manner [69,72,126].

4.5. Nrd1p and Nab3p

Sequence-specific recruitment of the nuclear exosome is perhaps best exemplified by the *Saccharomyces cerevisiae* RNA-binding proteins Nrd1p and Nab3p [10,127]. Both these factors harbor RNA recognition motifs (RRMs) that bind GUAA/G and GNUCUUGU consensus sequences, respectively [11,112,128]. In complex with the RNA helicase Sen1p, Nrd1p and Nab3p physically interact with RNA polymerase II (RNAPII) to contribute to transcription termination of sno/snRNA, CUT, long antisense ncRNA, and some mRNA genes in a polyadenylation-site-independent manner (Fig. 1.5B) [10,112,129–132]. The Nrd1p–Nab3p–Sen1p complex recognizes binding sites on nascent transcripts and provides an inroad for exosome trimming of sn(o)RNAs down to protective features provided by protein-bound C/D- or H/ACA-boxes [6,7,128,133,134]. A similar transcription termination mechanism applies to CUTs, these however lack stabilizing features and are thus completely removed by the exosome (Fig. 1.5B) [10,11,112,132]. Interestingly, recent data suggest that the Nrd1p–Nab3p–Sen1p complex is not only restricted to RNAPII-derived transcripts but also recruits the TRAMP/exosome for posttranscriptional degradation of aberrant pre-tRNAs (Fig. 1.5C) [112]. *In vivo*, many Nrd1p/Nab3p-bound transcripts are enriched with nontemplated oligo(A)-tails in agreement with the recruitment of TRAMP by these factors [112]. Curiously, higher eukaryotes lack any obvious homologs of Nrd1p and Nab3p. This loss is perhaps explained by the different biogenic requirements for human snoRNAs, which, unlike in *Saccharomyces cerevisiae*, are mainly intron-encoded and processed from longer precursors.

4.6. The NEXT complex

In human cells, the profiling of nuclear exosome-associated factors has un-covered a set of putative RNA-binding proteins robustly copurifying with hRRP6 and hMTR4 [69,72]. These are exosome adaptor candidates. A remarkably stable and stoichiometric interaction was identified between hMTR4, RBM7, and ZCCHC8, described as the trimeric NEXT complex [69]. ZCCHC8 and RBM7 localize strictly to the nucleoplasm associating with the non-nucleolar pool of hMTR4 and the RNA exosome. Both RBM7 and ZCCHC8 are only conserved among vertebrates.

Functionally, NEXT has been reported to contribute to the exosomal degradation of PROMPTs (Fig. 1.5B) [69]. However, additional targets are expected. The exact mechanism by which NEXT targets PROMPTs is presently unknown, but RBM7 possesses a canonical RRM motif and consistently binds to RNA *in vivo* (M. Lubas, T.H. Jensen and A. Dziembowski, unpublished). ZCCHC8 holds CCHC-type zinc-finger and proline-rich domains, which may also assist in RNA binding.

Intriguingly, although PROMPTs share many similarities with *Saccharomyces cerevisiae* CUTs, their mechanisms of decay diverge. First, human homologs of *Saccharomyces cerevisiae* TRAMP components, ZCCHC7 and hTRF4-2, are dispensable for PROMPT degradation, though hTRF4-2 contributes to their adenylation [69,135]. Second, PROMPT decay is manageable without a Nrd1p–Nab3p–Sen1p homologous complex. Besides the exosome and its accessory factors, the NEXT complex copurifies with a previously reported complex of ARS2 and the cap-binding complex [69,136,137]. Thus, NEXT may be preferentially recruited to RNAPII transcripts. A better understanding of exosomal decay in human cells is needed to uncover to what extent *Saccharomyces cerevisiae* RNA surveillance pathways and exosome activation mechanisms are preserved in higher eukaryotes.

4.7. The SKI complex

In the cytoplasm of *Saccharomyces cerevisiae*, the exosome is activated by the superkiller (SKI) complex and the GTPase Ski7p [39,138,139]. SKI components were initially identified in a screen where their depletion caused the accumulation of viral dsRNA and production of toxins [40,140,141]. The SKI complex is composed of the DExH-box RNA helicase Ski2p, a tetratricopeptide protein Ski3p and two molecules of the WD40 protein Ski8p [142,143]. Existence of insect and human SKI complexes has also been confirmed, though their functionalities have not yet been extensively investigated [144–146]. Notably, both *Saccharomyces cerevisiae* exosomal activators, the nuclear TRAMP and the cytoplasmic SKI complex, involve RNA helicases, Mtr4p and Ski2p, respectively. Interestingly, structural studies revealed differences between these two proteins: a flexible β-barrel domain attached to the DExH core of Ski2p, which interacts with both ss- and dsRNAs [147]. In Mtr4p, this region contains the KOW motif that directs the binding of structured RNAs instead [108,109]. This structural divergence hints at different modes of substrate recognition by the two complexes.

The SKI complex interacts with the exosome through Ski7p [139], which contains a domain with similarity to regions of the translation factors eEF1A and eRF3. This allows Ski7p to, for example, recognize and help dissociate ribosomes stalled at 3′ ends of nonstop mRNAs, thereby providing access for the cytoplasmic exosome [35,148]. *Saccharomyces cerevisiae* Ski7p has a paralog, Hbs1p, which forms a complex with the Dom34p protein and uses similar mimicry to interact with ribosomes and stimulate degradation of NSD and NGD intermediates [31,149–152]. In most eukaryotes, HBS1L is the only homolog of Ski7p. HBS1L has also been found to coimmunoprecipitate with the cytoplasmic exosome [72,144,146,153].

5. CONCLUDING REMARKS

The eukaryotic RNA exosome exemplifies the evolutionary progress in RNA metabolism achieved in this domain of life. Based on present data, it is likely that early eukaryotic ancestors possessed phosphorolytic PNPase-like or archaeal exosome-like complexes as well as free hydrolytic RNases. Evolution of the exosome enzymes was probably not dictated by energy demands, so the phosphorolytic activities that utilize significantly less energy were lost in favor of more efficient hydrolytic nucleases. Perhaps the irreversible character of hydrolysis itself provided an additional evolutionary advantage.

Remarkably, the modern eukaryotic exosome retained its Exo9 core, which is crucial in the regulation of nucleolytic decay and serves as a docking platform for catalytic subunits. In light of the increasing variety of substrates and pathways involving the exosome, it must have been of advantage to integrate various activities into one multitasking enzyme complex. This could be due to the ability of the core-associated nucleases, Dis3 and Rrp6, to cooperate in the degradation/processing of some substrates, as indicated in this chapter.

Dis3 and Rrp6 have different properties, for example, (i) processive versus distributive nucleolytic activities; (ii) exo-only versus exo- and endonucleolytic decay; and (iii) variable preferences for substrates. Moreover, the nature and compartment-specific localization of these catalytic subunits vary among species, and are probably adjusted to deal with specific RNA targets.

The intrinsic activity and specificity of the eukaryotic exosome in its cellular environment is probably relatively low, which prevents unintended RNA degradation. However, various nuclear and cytoplasmic cofactor proteins supplement the exosome by enhancing the recognition and accessibility of RNA substrates. Perhaps a rapid change in the requirements for RNA metabolism has driven the formation of such an additional level of regulation/activation, which may possess an even higher adaptive potential. As exemplified above, the selection of RNA targets and exosome recruitment mechanisms vary depending on substrates, cellular compartments and hosts. Some exosome cofactors are evolutionarily conserved up to humans, that is, the RNA helicases and the RNA-binding proteins Rrp47 and Mpp6. Others, like Nrd1p and Nab3p, although essential for lower eukaryotes, are not preserved in metazoans. Conversely, many of the human factors that copurify with the exosome, for example, RBM7, ZCCHC8, ARS2, etc., are maintained only among higher eukaryotes. Of these, putative RNA-binding adaptors with yet unknown targets may undergo rapid evolutionary changes. Future analyses should further characterize the variety of exosome auxiliary proteins to precisely describe their functions.

ACKNOWLEDGMENTS

Manfred Schmid, Søren Lykke-Andersen, and Evgenia Ntini are thanked for critical reading of the manuscript. Work in the authors laboratories is supported by grants from the Danish National Research Foundation (T. H. J.), the NOVO Nordisk foundation (T. H. J.), the Danish Cancer Society (T. H. J.), National Science Centre: 2011/02/A/NZ1/00001 (A. D.), and Foundation for Polish Science: TEAM | 2008-2 | 1 (A. D.). M. L. is supported by a Boehringer Ingelheim Ph.D. fellowship.

REFERENCES

[1] Mitchell P, Petfalski E, Tollervey D. The 3' end of yeast 5.8S rRNA is generated by an exonuclease processing mechanism. Genes Dev 1996;10:502–13.
[2] Mitchell P, Petfalski E, Shevchenko A, Mann M, Tollervey D. The exosome: a conserved eukaryotic RNA processing complex containing multiple 3'→5' exoribonucleases. Cell 1997;91:457–66.
[3] Dziembowski A, Lorentzen E, Conti E, Seraphin B. A single subunit, Dis3, is essentially responsible for yeast exosome core activity. Nat Struct Mol Biol 2007;14:15–22.

[4] Allmang C, Petfalski E, Podtelejnikov A, Mann M, Tollervey D, Mitchell P. The yeast exosome and human PM-Scl are related complexes of 3' → 5' exonucleases. Genes Dev 1999;13:2148–58.

[5] Mian IS. Comparative sequence analysis of ribonucleases HII, III, II PH and D. Nucleic Acids Res 1997;25:3187–95.

[6] Allmang C, Kufel J, Chanfreau G, Mitchell P, Petfalski E, Tollervey D. Functions of the exosome in rRNA, snoRNA and snRNA synthesis. EMBO J 1999;18:5399–410.

[7] van Hoof A, Lennertz P, Parker R. Yeast exosome mutants accumulate 3'-extended polyadenylated forms of U4 small nuclear RNA and small nucleolar RNAs. Mol Cell Biol 2000;20:441–52.

[8] Schmid M, Jensen TH. The exosome: a multipurpose RNA-decay machine. Trends Biochem Sci 2008;33:501–10.

[9] Houseley J, Tollervey D. The many pathways of RNA degradation. Cell 2009;136:763–76.

[10] Arigo JT, Eyler DE, Carroll KL, Corden JL. Termination of cryptic unstable transcripts is directed by yeast RNA-binding proteins Nrd1 and Nab3. Mol Cell 2006;23:841–51.

[11] Creamer TJ, et al. Transcriptome-wide binding sites for components of the Saccharomyces cerevisiae non-poly(A) termination pathway: Nrd1, Nab3, and Sen1. PLoS Genet 2011;7:e1002329.

[12] Wyers F, et al. Cryptic pol II transcripts are degraded by a nuclear quality control pathway involving a new poly(A) polymerase. Cell 2005;121:725–37.

[13] Davis CA, Ares Jr. M. Accumulation of unstable promoter-associated transcripts upon loss of the nuclear exosome subunit Rrp6p in Saccharomyces cerevisiae. Proc Natl Acad Sci USA 2006;103:3262–7.

[14] Chekanova JA, et al. Genome-wide high-resolution mapping of exosome substrates reveals hidden features in the Arabidopsis transcriptome. Cell 2007;131:1340–53.

[15] Preker P, et al. RNA exosome depletion reveals transcription upstream of active human promoters. Science 2008;322:1851–4.

[16] Camblong J, Iglesias N, Fickentscher C, Dieppois G, Stutz F. Antisense RNA stabilization induces transcriptional gene silencing via histone deacetylation in S. cerevisiae. Cell 2007;131:706–17.

[17] Hongay CF, Grisafi PL, Galitski T, Fink GR. Antisense transcription controls cell fate in Saccharomyces cerevisiae. Cell 2006;127:735–45.

[18] Berretta J, Morillon A. Pervasive transcription constitutes a new level of eukaryotic genome regulation. EMBO Rep 2009;10:973–82.

[19] San Paolo S, et al. Distinct roles of non-canonical poly(A) polymerases in RNA metabolism. PLoS Genet 2009;5:e1000555.

[20] Jacquier A. The complex eukaryotic transcriptome: unexpected pervasive transcription and novel small RNAs. Nat Rev Genet 2009;10:833–44.

[21] Muhlemann O, Jensen TH. mRNP quality control goes regulatory. Trends Genet— TIG 2012;28:70–7.

[22] Bousquet-Antonelli C, Presutti C, Tollervey D. Identification of a regulated pathway for nuclear pre-mRNA turnover. Cell 2000;102:765–75.

[23] Galy V, Gadal O, Fromont-Racine M, Romano A, Jacquier A, Nehrbass U. Nuclear retention of unspliced mRNAs in yeast is mediated by perinuclear Mlp1. Cell 2004;116:63–73.

[24] Schmid M, et al. Rrp6p controls mRNA poly(A) tail length and its decoration with poly(A) binding proteins. Mol Cell 2012;47:267–80.

[25] Libri D, Dower K, Boulay J, Thomsen R, Rosbash M, Jensen TH. Interactions between mRNA export commitment, 3'-end quality control, and nuclear degradation. Mol Cell Biol 2002;22:8254–66.

[26] Milligan L, Torchet C, Allmang C, Shipman T, Tollervey D. A nuclear surveillance pathway for mRNAs with defective polyadenylation. Mol Cell Biol 2005;25:9996–10004.

[27] Torchet C, Bousquet-Antonelli C, Milligan L, Thompson E, Kufel J, Tollervey D. Processing of 3'-extended read-through transcripts by the exosome can generate functional mRNAs. Mol Cell 2002;9:1285–96.

[28] Hilleren P, McCarthy T, Rosbash M, Parker R, Jensen TH. Quality control of mRNA 3'-end processing is linked to the nuclear exosome. Nature 2001;413:538–42.

[29] Saguez C, et al. Nuclear mRNA surveillance in THO/sub2 mutants is triggered by inefficient polyadenylation. Mol Cell 2008;31:91–103.

[30] Rougemaille M, et al. Dissecting mechanisms of nuclear mRNA surveillance in THO/sub2 complex mutants. EMBO J 2007;26:2317–26.

[31] Doma MK, Parker R. Endonucleolytic cleavage of eukaryotic mRNAs with stalls in translation elongation. Nature 2006;440:561–4.

[32] Conti E, Izaurralde E. Nonsense-mediated mRNA decay: molecular insights and mechanistic variations across species. Curr Opin Cell Biol 2005;17:316–25.

[33] Inada T, Aiba H. Translation of aberrant mRNAs lacking a termination codon or with a shortened 3'-UTR is repressed after initiation in yeast. EMBO J 2005;24:1584–95.

[34] Frischmeyer PA, van Hoof A, O'Donnell K, Guerrerio AL, Parker R, Dietz HC. An mRNA surveillance mechanism that eliminates transcripts lacking termination codons. Science 2002;295:2258–61.

[35] van Hoof A, Frischmeyer PA, Dietz HC, Parker R. Exosome-mediated recognition and degradation of mRNAs lacking a termination codon. Science 2002;295:2262–4.

[36] Chen CY, Shyu AB. Mechanisms of deadenylation-dependent decay. Wiley Interdiscip Rev RNA 2011;2:167–83.

[37] Hsu CL, Stevens A. Yeast cells lacking 5'→3' exoribonuclease 1 contain mRNA species that are poly(A) deficient and partially lack the 5' cap structure. Mol Cell Biol 1993;13:4826–35.

[38] Muhlrad D, Decker CJ, Parker R. Deadenylation of the unstable mRNA encoded by the yeast MFA2 gene leads to decapping followed by 5'→3' digestion of the transcript. Genes Dev 1994;8:855–66.

[39] Anderson JS, Parker RP. The 3' to 5' degradation of yeast mRNAs is a general mechanism for mRNA turnover that requires the SKI2 DEVH box protein and 3' to 5' exonucleases of the exosome complex. EMBO J 1998;17:1497–506.

[40] Johnson AW, Kolodner RD. Synthetic lethality of sep1 (xrn1) ski2 and sep1 (xrn1) ski3 mutants of Saccharomyces cerevisiae is independent of killer virus and suggests a general role for these genes in translation control. Mol Cell Biol 1995;15:2719–27.

[41] Schoenberg DR, Maquat LE. Regulation of cytoplasmic mRNA decay. Nat Rev Genet 2012;13:246–59.

[42] Chen CY, et al. AU binding proteins recruit the exosome to degrade ARE-containing mRNAs. Cell 2001;107:451–64.

[43] Gherzi R, et al. A KH domain RNA binding protein, KSRP, promotes ARE-directed mRNA turnover by recruiting the degradation machinery. Mol Cell 2004;14:571–83.

[44] Tran H, Schilling M, Wirbelauer C, Hess D, Nagamine Y. Facilitation of mRNA deadenylation and decay by the exosome-bound, DExH protein RHAU. Mol Cell 2004;13:101–11.

[45] Buhler M, Moazed D. Transcription and RNAi in heterochromatic gene silencing. Nat Struct Mol Biol 2007;14:1041–8.

[46] Wang SW, Stevenson AL, Kearsey SE, Watt S, Bahler J. Global role for polyadenylation-assisted nuclear RNA degradation in posttranscriptional gene silencing. Mol Cell Biol 2008;28:656–65.

[47] Buhler M, Spies N, Bartel DP, Moazed D. TRAMP-mediated RNA surveillance prevents spurious entry of RNAs into the Schizosaccharomyces pombe siRNA pathway. Nat Struct Mol Biol 2008;15:1015–23.

[48] Vasiljeva L, Kim M, Terzi N, Soares LM, Buratowski S. Transcription termination and RNA degradation contribute to silencing of RNA polymerase II transcription within heterochromatin. Mol Cell 2008;29:313–23.

[49] Houseley J, Kotovic K, El Hage A, Tollervey D. Trf4 targets ncRNAs from telomeric and rDNA spacer regions and functions in rDNA copy number control. EMBO J 2007;26:4996–5006.

[50] Bonneau F, Basquin J, Ebert J, Lorentzen E, Conti E. The yeast exosome functions as a macromolecular cage to channel RNA substrates for degradation. Cell 2009;139:547–59.

[51] Harlow LS, Kadziola A, Jensen KF, Larsen S. Crystal structure of the phosphorolytic exoribonuclease RNase PH from Bacillus subtilis and implications for its quaternary structure and tRNA binding. Protein Sci 2004;13:668–77.

[52] Choi JM, Park EY, Kim JH, Chang SK, Cho Y. Probing the functional importance of the hexameric ring structure of RNase PH. J Biol Chem 2004;279:755–64.

[53] Ishii R, Nureki O, Yokoyama S. Crystal structure of the tRNA processing enzyme RNase PH from Aquifex aeolicus. J Biol Chem 2003;278:32397–404.

[54] Shi Z, Yang WZ, Lin-Chao S, Chak KF, Yuan HS. Crystal structure of Escherichia coli PNPase: central channel residues are involved in processive RNA degradation. RNA 2008;14:2361–71.

[55] Deutscher MP, Marshall GT, Cudny H. RNase PH: an Escherichia coli phosphate-dependent nuclease distinct from polynucleotide phosphorylase. Proc Natl Acad Sci USA 1988;85:4710–4.

[56] Py B, Causton H, Mudd EA, Higgins CF. A protein complex mediating mRNA degradation in Escherichia coli. Mol Microbiol 1994;14:717–29.

[57] Py B, Higgins CF, Krisch HM, Carpousis AJ. A DEAD-box RNA helicase in the Escherichia coli RNA degradosome. Nature 1996;381:169–72.

[58] Miczak A, Kaberdin VR, Wei CL, Lin-Chao S. Proteins associated with RNase E in a multicomponent ribonucleolytic complex. Proc Natl Acad Sci USA 1996;93:3865–9.

[59] Piwowarski J, Grzechnik P, Dziembowski A, Dmochowska A, Minczuk M, Stepien PP. Human polynucleotide phosphorylase, hPNPase, is localized in mitochondria. J Mol Biol 2003;329:853–7.

[60] Yehudai-Resheff S, Hirsh M, Schuster G. Polynucleotide phosphorylase functions as both an exonuclease and a poly(A) polymerase in spinach chloroplasts. Mol Cell Biol 2001;21:5408–16.

[61] Leszczyniecka M, et al. Identification and cloning of human polynucleotide phosphorylase, hPNPase old-35, in the context of terminal differentiation and cellular senescence. Proc Natl Acad Sci USA 2002;99:16636–41.

[62] Lorentzen E, Walter P, Fribourg S, Evguenieva-Hackenberg E, Klug G, Conti E. The archaeal exosome core is a hexameric ring structure with three catalytic subunits. Nat Struct Mol Biol 2005;12:575–81.

[62a] Buttner K, Wenig K, Hopfner KP. Structural framework for the mechanism of archaeal exosomes in RNA processing. Mol Cell 2005;20:461–71.

[63] Lykke-Andersen S, Brodersen DE, Jensen TH. Origins and activities of the eukaryotic exosome. J Cell Sci 2009;122:1487–94.

[64] Liu Q, Greimann JC, Lima CD. Reconstitution, activities, and structure of the eukaryotic RNA exosome. Cell 2006;127:1223–37.

[65] Lorentzen E, Basquin J, Tomecki R, Dziembowski A, Conti E. Structure of the active subunit of the yeast exosome core, Rrp44: diverse modes of substrate recruitment in the RNase II nuclease family. Mol Cell 2008;29:717–28.

[66] Lorentzen E, Dziembowski A, Lindner D, Seraphin B, Conti E. RNA channelling by the archaeal exosome. EMBO Rep 2007;8:470–6.

[67] Huh WK, et al. Global analysis of protein localization in budding yeast. Nature 2003;425:686–91.

[68] Januszyk K, Lima CD. Structural components and architectures of RNA exosomes. Adv Exp Med Biol 2011;702:9–28.

[69] Lubas M, et al. Interaction profiling identifies the human nuclear exosome targeting complex. Mol Cell 2011;43:624–37.

[70] Staals RH, et al. Dis3-like 1: a novel exoribonuclease associated with the human exosome. EMBO J 2010;29:2358–67.

[71] Lykke-Andersen S, Tomecki R, Jensen TH, Dziembowski A. The eukaryotic RNA exosome: same scaffold but variable catalytic subunits. RNA Biol 2011;8:61–6.

[72] Tomecki R, et al. The human core exosome interacts with differentially localized processive RNases: hDIS3 and hDIS3L. EMBO J 2010;29:2342–57.

[73] Schilders G, van Dijk E, Pruijn GJ. C1D and hMtr4p associate with the human exosome subunit PM/Scl-100 and are involved in pre-rRNA processing. Nucleic Acids Res 2007;35:2564–72.

[74] Tomecki R, Drazkowska K, Dziembowski A. Mechanisms of RNA degradation by the eukaryotic exosome. Chembiochem 2010;11:938–45.

[75] Graham AC, Kiss DL, Andrulis ED. Differential distribution of exosome subunits at the nuclear lamina and in cytoplasmic foci. Mol Biol Cell 2006;17:1399–409.

[76] Haile S, Cristodero M, Clayton C, Estevez AM. The subcellular localisation of trypanosome RRP6 and its association with the exosome. Mol Biochem Parasitol 2007;151:52–8.

[77] Lejeune F, Li X, Maquat LE. Nonsense-mediated mRNA decay in mammalian cells involves decapping, deadenylating, and exonucleolytic activities. Mol Cell 2003;12:675–87.

[78] Estevez AM, Kempf T, Clayton C. The exosome of Trypanosoma brucei. EMBO J 2001;20:3831–9.

[79] Chekanova JA, Dutko JA, Mian IS, Belostotsky DA. Arabidopsis thaliana exosome subunit AtRrp4p is a hydrolytic 3'→5' exonuclease containing S1 and KH RNA-binding domains. Nucleic Acids Res 2002;30:695–700.

[80] Andrulis ED, Werner J, Nazarian A, Erdjument-Bromage H, Tempst P, Lis JT. The RNA processing exosome is linked to elongating RNA polymerase II in Drosophila. Nature 2002;420:837–41.

[81] Cristodero M, Bottcher B, Diepholz M, Scheffzek K, Clayton C. The Leishmania tarentolae exosome: purification and structural analysis by electron microscopy. Mol Biochem Parasitol 2008;159:24–9.

[82] Lange H, Gagliardi D. The exosome and 3'-5' RNA degradation in plants. Adv Exp Med Biol 2011;702:50–62.

[83] Lebreton A, Tomecki R, Dziembowski A, Seraphin B. Endonucleolytic RNA cleavage by a eukaryotic exosome. Nature 2008;456:993–6.

[84] Schneider C, Leung E, Brown J, Tollervey D. The N-terminal PIN domain of the exosome subunit Rrp44 harbors endonuclease activity and tethers Rrp44 to the yeast core exosome. Nucleic Acids Res 2009;37:1127–40.

[85] Schaeffer D, et al. The exosome contains domains with specific endoribonuclease, exoribonuclease and cytoplasmic mRNA decay activities. Nat Struct Mol Biol 2009;16:56–62.

[86] Chlebowski A, Tomecki R, Lopez ME, Seraphin B, Dziembowski A. Catalytic properties of the eukaryotic exosome. Adv Exp Med Biol 2011;702:63–78.

[87] Hernandez H, Dziembowski A, Taverner T, Seraphin B, Robinson CV. Subunit architecture of multimeric complexes isolated directly from cells. EMBO Rep 2006;7:605–10.

[88] Frazao C, et al. Unravelling the dynamics of RNA degradation by ribonuclease II and its RNA-bound complex. Nature 2006;443:110–4.

[89] Mackie GA. Ribonuclease E is a 5'-end-dependent endonuclease. Nature 1998;395: 720–3.

[90] Astuti D, et al. Germline mutations in DIS3L2 cause the Perlman syndrome of overgrowth and Wilms tumor susceptibility. Nat Genet 2012;44:277–84.

[91] Briggs MW, Burkard KT, Butler JS. Rrp6p, the yeast homologue of the human PM-Scl 100-kDa autoantigen, is essential for efficient 5.8 S rRNA 3' end formation. J Biol Chem 1998;273:13255–63.

[92] Zuo Y, Deutscher MP. Exoribonuclease superfamilies: structural analysis and phylogenetic distribution. Nucleic Acids Res 2001;29:1017–26.

[93] Midtgaard SF, Assenholt J, Jonstrup AT, Van LB, Jensen TH, Brodersen DE. Structure of the nuclear exosome component Rrp6p reveals an interplay between the active site and the HRDC domain. Proc Natl Acad Sci USA 2006;103:11898–903.

[94] Januszyk K, Liu Q, Lima CD. Activities of human RRP6 and structure of the human RRP6 catalytic domain. RNA 2011;17:1566–77.

[95] Stead JA, Costello JL, Livingstone MJ, Mitchell P. The PMC2NT domain of the catalytic exosome subunit Rrp6p provides the interface for binding with its cofactor Rrp47p, a nucleic acid-binding protein. Nucleic Acids Res 2007;35: 5556–67.

[96] Assenholt J, Mouaikel J, Andersen KR, Brodersen DE, Libri D, Jensen TH. Exonucleolysis is required for nuclear mRNA quality control in yeast THO mutants. RNA 2008;14:2305–13.

[97] Callahan KP, Butler JS. Evidence for core exosome independent function of the nuclear exoribonuclease Rrp6p. Nucleic Acids Res 2008;36:6645–55.

[98] Lee G, Bratkowski MA, Ding F, Ke A, Ha T. Elastic coupling between RNA degradation and unwinding by an exoribonuclease. Science 2012;336:1726–9.

[99] Malet H, et al. RNA channelling by the eukaryotic exosome. EMBO Rep 2010;11:936–42.

[100] Liang S, Hitomi M, Hu YH, Liu Y, Tartakoff AM. A DEAD-box-family protein is required for nucleocytoplasmic transport of yeast mRNA. Mol Cell Biol 1996;16:5139–46.

[101] de la Cruz J, Kressler D, Tollervey D, Linder P. Dob1p (Mtr4p) is a putative ATP-dependent RNA helicase required for the 3' end formation of 5.8S rRNA in Saccharomyces cerevisiae. EMBO J 1998;17:1128–40.

[102] Bernstein KA, Granneman S, Lee AV, Manickam S, Baserga SJ. Comprehensive mutational analysis of yeast DEXD/H box RNA helicases involved in large ribosomal subunit biogenesis. Mol Cell Biol 2006;26:1195–208.

[103] Kadaba S, Wang X, Anderson JT. Nuclear RNA surveillance in Saccharomyces cerevisiae: Trf4p-dependent polyadenylation of nascent hypomethylated tRNA and an aberrant form of 5S rRNA. RNA 2006;12:508–21.

[104] Wang X, Jia H, Jankowsky E, Anderson JT. Degradation of hypomodified tRNA (iMet) in vivo involves RNA-dependent ATPase activity of the DExH helicase Mtr4p. RNA 2008;14:107–16.

[105] Vanacova S, et al. A new yeast poly(A) polymerase complex involved in RNA quality control. PLoS Biol 2005;3:e189.

[106] LaCava J, et al. RNA degradation by the exosome is promoted by a nuclear polyadenylation complex. Cell 2005;121:713–24.

[107] Bernstein J, Patterson DN, Wilson GM, Toth EA. Characterization of the essential activities of Saccharomyces cerevisiae Mtr4p, a 3'→5' helicase partner of the nuclear exosome. J Biol Chem 2008;283:4930–42.

[108] Weir JR, Bonneau F, Hentschel J, Conti E. Structural analysis reveals the characteristic features of Mtr4, a DExH helicase involved in nuclear RNA processing and surveillance. Proc Natl Acad Sci USA 2010;107:12139–44.

[109] Jackson RN, Klauer AA, Hintze BJ, Robinson H, van Hoof A, Johnson SJ. The crystal structure of Mtr4 reveals a novel arch domain required for rRNA processing. EMBO J 2010;29:2205–16.

[110] Holub P, et al. Air2p is critical for the assembly and RNA-binding of the TRAMP complex and the KOW domain of Mtr4p is crucial for exosome activation. Nucleic Acids Res 2012;40:5679–93.

[111] Spickler C, Mackie GA. Action of RNase II and polynucleotide phosphorylase against RNAs containing stem-loops of defined structure. J Bacteriol 2000;182:2422–7.

[112] Wlotzka W, Kudla G, Granneman S, Tollervey D. The nuclear RNA polymerase II surveillance system targets polymerase III transcripts. EMBO J 2011;30:1790–803.

[113] Grzechnik P, Kufel J. Polyadenylation linked to transcription termination directs the processing of snoRNA precursors in yeast. Mol Cell 2008;32:247–58.

[114] Jia H, Wang X, Anderson JT, Jankowsky E. RNA unwinding by the Trf4/Air2/Mtr4 polyadenylation (TRAMP) complex. Proc Natl Acad Sci USA 2012;109:7292–7.

[115] Kadaba S, Krueger A, Trice T, Krecic AM, Hinnebusch AG, Anderson J. Nuclear surveillance and degradation of hypomodified initiator tRNAMet in S. cerevisiae. Genes Dev 2004;18:1227–40.

[116] Xu Z, et al. Bidirectional promoters generate pervasive transcription in yeast. Nature 2009;457:1033–7.

[117] Jia H, et al. The RNA helicase Mtr4p modulates polyadenylation in the TRAMP complex. Cell 2011;145:890–901.

[118] Houseley J, Tollervey D. Yeast Trf5p is a nuclear poly(A) polymerase. EMBO Rep 2006;7:205–11.

[119] Fasken MB, et al. Air1 zinc knuckles 4 and 5 and a conserved IWRXY motif are critical for the function and integrity of the Trf4/5-Air1/2-Mtr4 polyadenylation (TRAMP) RNA quality control complex. J Biol Chem 2011;286:37429–45.

[120] Mitchell P, Petfalski E, Houalla R, Podtelejnikov A, Mann M, Tollervey D. Rrp47p is an exosome-associated protein required for the 3' processing of stable RNAs. Mol Cell Biol 2003;23:6982–92.

[121] Milligan L, et al. A yeast exosome cofactor, Mpp 6, functions in RNA surveillance and in the degradation of noncoding RNA transcripts. Mol Cell Biol 2008;28:5446–57.

[122] Gavin AC, et al. Functional organization of the yeast proteome by systematic analysis of protein complexes. Nature 2002;415:141–7.

[123] Nehls P, et al. cDNA cloning, recombinant expression and characterization of polypetides with exceptional DNA affinity. Nucleic Acids Res 1998;26:1160–6.

[124] Yavuzer U, Smith GC, Bliss T, Werner D, Jackson SP. DNA end-independent activation of DNA-PK mediated via association with the DNA-binding protein C1D. Genes Dev 1998;12:2188–99.

[125] Zamir I, Dawson J, Lavinsky RM, Glass CK, Rosenfeld MG, Lazar MA. Cloning and characterization of a corepressor and potential component of the nuclear hormone receptor repression complex. Proc Natl Acad Sci USA 1997;94:14400–5.

[126] Schilders G, Raijmakers R, Raats JM, Pruijn GJ. MPP6 is an exosome-associated RNA-binding protein involved in 5.8S rRNA maturation. Nucleic Acids Res 2005;33:6795–804.

[127] Vasiljeva L, Buratowski S. Nrd1 interacts with the nuclear exosome for 3' processing of RNA polymerase II transcripts. Mol Cell 2006;21:239–48.

[128] Carroll KL, Ghirlando R, Ames JM, Corden JL. Interaction of yeast RNA-binding proteins Nrd1 and Nab3 with RNA polymerase II terminator elements. RNA 2007;13:361–73.

[129] Steinmetz EJ, Conrad NK, Brow DA, Corden JL. RNA-binding protein Nrd1 directs poly(A)-independent 3'-end formation of RNA polymerase II transcripts. Nature 2001;413:327–31.

[130] Arigo JT, Carroll KL, Ames JM, Corden JL. Regulation of yeast NRD1 expression by premature transcription termination. Mol Cell 2006;21:641–51.

[131] Houalla R, et al. Microarray detection of novel nuclear RNA substrates for the exosome. Yeast 2006;23:439–54.

[132] Thiebaut M, Kisseleva-Romanova E, Rougemaille M, Boulay J, Libri D. Transcription termination and nuclear degradation of cryptic unstable transcripts: a role for the nrd1-nab3 pathway in genome surveillance. Mol Cell 2006;23:853–64.

[133] Gudipati RK, Villa T, Boulay J, Libri D. Phosphorylation of the RNA polymerase II C-terminal domain dictates transcription termination choice. Nat Struct Mol Biol 2008;15:786–94.

[134] Vasiljeva L, Kim M, Mutschler H, Buratowski S, Meinhart A. The Nrd1–Nab3–Sen1 termination complex interacts with the Ser5-phosphorylated RNA polymerase II C-terminal domain. Nat Struct Mol Biol 2008;15:795–804.

[135] Preker P, et al. PROMoter uPstream Transcripts share characteristics with mRNAs and are produced upstream of all three major types of mammalian promoters. Nucleic Acids Res 2011;39:7179–93.

[136] Gruber JJ, et al. Ars2 links the nuclear cap-binding complex to RNA interference and cell proliferation. Cell 2009;138:328–39.

[137] Domanski M, et al. Improved methodology for the affinity isolation of human protein complexes expressed at near endogenous levels. Biotechniques 2012;0:1–6.

[138] van Hoof A, Staples RR, Baker RE, Parker R. Function of the ski4p (Csl4p) and Ski7p proteins in 3'-to-5' degradation of mRNA. Mol Cell Biol 2000;20:8230–43.

[139] Araki Y, Takahashi S, Kobayashi T, Kajiho H, Hoshino S, Katada T. Ski7p G protein interacts with the exosome and the Ski complex for 3'-to-5' mRNA decay in yeast. EMBO J 2001;20:4684–93.

[140] Wickner RB, Toh-e A. [HOK], a new yeast non-Mendelian trait, enables a replication-defective killer plasmid to be maintained. Genetics 1982;100:159–74.

[141] Ridley SP, Sommer SS, Wickner RB. Superkiller mutations in Saccharomyces cerevisiae suppress exclusion of M2 double-stranded RNA by L-A-HN and confer cold sensitivity in the presence of M and L-A-HN. Mol Cell Biol 1984;4:761–70.

[142] Brown JT, Bai X, Johnson AW. The yeast antiviral proteins Ski2p, Ski3p, and Ski8p exist as a complex in vivo. RNA 2000;6:449–57.

[143] Cheng Z, Liu Y, Wang C, Parker R, Song H. Crystal structure of Ski8p, a WD-repeat protein with dual roles in mRNA metabolism and meiotic recombination. Protein Sci 2004;13:2673–84.

[144] Orban TI, Izaurralde E. Decay of mRNAs targeted by RISC requires XRN1, the Ski complex, and the exosome. RNA 2005;11:459–69.

[145] Seago JE, Chernukhin IV, Newbury SF. The Drosophila gene twister, an orthologue of the yeast helicase SKI2, is differentially expressed during development. Mech Dev 2001;106:137–41.

[146] Zhu B, et al. The human PAF complex coordinates transcription with events downstream of RNA synthesis. Genes Dev 2005;19:1668–73.

[147] Halbach F, Rode M, Conti E. The crystal structure of S. cerevisiae Ski2, a DExH helicase associated with the cytoplasmic functions of the exosome. RNA 2012;18:124–34.

[148] Benard L, Carroll K, Valle RC, Masison DC, Wickner RB. The ski7 antiviral protein
       is an EF1-alpha homolog that blocks expression of non-Poly(A) mRNA in Saccharo-
       myces cerevisiae. J Virol 1999;73:2893–900.
[149] Tsuboi T, et al. Dom34:hbs1 plays a general role in quality-control systems by
       dissociation of a stalled ribosome at the 3' end of aberrant mRNA. Mol Cell
       2012;46:518–29.
[150] Kellis M, Birren BW, Lander ES. Proof and evolutionary analysis of ancient genome
       duplication in the yeast Saccharomyces cerevisiae. Nature 2004;428:617–24.
[151] Chen L, et al. Structure of the Dom34-Hbs1 complex and implications for no-go
       decay. Nat Struct Mol Biol 2010;17:1233–40.
[152] van den Elzen AM, et al. Dissection of Dom34-Hbs1 reveals independent functions in
       two RNA quality control pathways. Nat Struct Mol Biol 2010;17:1446–52.
[153] van Hoof A. Conserved functions of yeast genes support the duplication, degeneration
       and complementation model for gene duplication. Genetics 2005;171:1455–61.

CHAPTER TWO

Plant Exosomes and Cofactors

Heike Lange[1], Dominique Gagliardi

Institut de Biologie Moléculaire des Plantes du Centre National de la Recherche Scientifique (CNRS), Université de Strasbourg, 12 rue du général Zimmer, Strasbourg Cedex, Strasbourg, France
[1]Corresponding author: e-mail address: heike.lange@ibmp-cnrs.unistra.fr

Contents

1. Introduction   32
2. Composition of the Plant Exosome Core Complex   32
   2.1 Plant exosome subunits are often encoded by duplicated genes   32
   2.2 The exosome core complex of *A. thaliana*   33
   2.3 Arabidopsis CSL4 is dispensable for the stability of the core complex   34
   2.4 Functional specialization of exosome core subunits?   36
3. Exosome-Related Activities and Cofactors in Plants   39
   3.1 Plants may have retained a catalytically active core complex   39
   3.2 Three RRP6-like proteins in different intracellular compartments   40
   3.3 The SKI complex is conserved in plants   41
   3.4 The Arabidopsis MTR4 homologue is a nucleolar protein   42
4. Impact of Exosome-Mediated Degradation in Plants   44
   4.1 Polyadenylated substrates of the exosome have been identified   44
   4.2 A high level of redundancy among plant RNA degradation pathways   46
5. Final Remarks   48
References   48

Abstract

The exosome is a large protein complex mediating 3′–5′ RNA degradation in both nucleus and cytosol of all eukaryotic cells. It consists of nine conserved subunits forming the core complex, which associates with ribonucleolytic enzymes and other cofactors such as RNA-binding proteins or RNA helicases. Both the composition of the core exosome and its general role as a major player in RNA maturation, RNA surveillance, and RNA turnover are largely conserved between plants, human, and fungi. However, plant exosomes have some peculiar and interesting features including a catalytically active core subunit, or a certain extent of functional specialization among both core subunits and putative exosome cofactors.

The Enzymes, Volume 31
ISSN 1874-6047
http://dx.doi.org/10.1016/B978-0-12-404740-2.00002-1

1. INTRODUCTION

Regulated genome expression relies on both transcription and RNA degradation. A number of RNA degradation activities and pathways are required for the processing of functional RNAs from primary transcripts, to control the appropriate level of each individual mRNA or noncoding RNA, and to eliminate every RNA at the end of its lifetime [1]. In addition, quality control and surveillance mechanisms ensure the elimination of defective and nonfunctional transcripts. One of the main mechanisms contributing to RNA processing, RNA turnover, and RNA surveillance is 3′–5′ exoribonucleolytic RNA degradation by the exosome, a conserved protein complex found in both nucleus and cytosol of all eukaryotic cells. RNA degradation by the exosome requires accessible 3′ extremities: 3′–5′ degradation of cytosolic mRNAs needs either endonucleolytic cleavage in the body of the mRNA or shortening of the polyA tail and removal of the protecting polyA-binding proteins [2,3] (see also Chapter 9). 3′–5′ degradation of noncoding RNAs can be facilitated by the addition of short oligo-A tails, which serve as unstructured "landing pads" for the nuclear exosome [4–6] (see Chapter 4). In Arabidopsis, oligo-adenylated exosome substrates include precursors and mature forms of noncoding RNAs, maturation by-products of rRNA and miRNA processing, short transcripts mapping 5′ or 3′ of annotated mRNAs, and a number of transcripts generated from intergenic and heterochromatic regions [7,8]. The role of the exosome in RNA maturation, surveillance, and turnover; the composition of the exosome core complex; and the cooperation of the core complex with cofactors are best studied in yeast and believed to be largely conserved in animal and plant kingdoms. But growing evidence suggests that, in particular, the assembly of exosome cofactors into complexes with distinct composition and specific biological roles is more complex in multicellular eukaryotes [9–11] and maybe different in animals and plants.

2. COMPOSITION OF THE PLANT EXOSOME CORE COMPLEX

2.1. Plant exosome subunits are often encoded by duplicated genes

The eukaryotic exosome core complex consists of three heterodimers of the PH-domain proteins RRP41–RRP45, RRP42–MTR3, and RRP43–RRP46 that form a ring-like structure, to which a "cap" of three

S1/KH domain proteins, RRP4, RRP40, and CSL4, is bound (see Chapter 3). Most plant genomes encode at least one of each subunit and sometimes more than that (resource: www.phytozome.org [12]). For instance, both *MTR3* and *RRP46* genes are duplicated in *Populus trichocarpa* (poplar), *RRP43* genes are duplicated in *Zea mays* (maize), and both *RRP42* and *RRP41* are duplicated in *Sorghum bicolor*. Among the genes encoding cap proteins, *RRP4* is duplicated in *Brachypodium distachyon* (purple false brome), *Arabidopsis lyrata* (lyrate rockcress), and *Capsella rubella* (red shepherd's purse), and *RRP40* is duplicated in *Z. mays*, *Oriza sativa* (rice), and *Arabidopsis thaliana* (thale cress). Not surprisingly, the most extensive duplication of genes encoding exosome subunits is observed in the tetraploid *Glycine max* (soybean), which has two genes for each *MTR3*, *RRP42*, and *RRP40*; three genes for *RRP43*; and four genes encoding *RRP46*.

Two versions of *RRP45*, encoding the genuine RRP45b/CER7 and a slightly shorter protein, RRP45a, are conserved between *A. thaliana* and *A. lyrata* (lyrate rocke cress), but probably not in *Brassica rapa* (turnip mustard) (Fig. 2.1). In all other cases, the duplicated genes are very similar and not conserved across even closely related species, indicating independent and rather recent duplication events.

2.2. The exosome core complex of *A. thaliana*

The *Arabidopsis* exosome complex was purified using transgenic lines expressing tagged versions of AtRRP4 or AtRRP41, respectively [7]. This revealed that the plant core complex contains, as all other eukaryotic exosome complexes investigated so far, homologues of RRP4, RRP40, CSL4, RRP41, RRP42, RRP43, RRP45, RRP46, and MTR3. Under the experimental conditions used, only AtRRP40A and AtRRP45B were incorporated into the core complex. Whether alternative core complexes containing AtRRP40B and/or AtRRP45A can be assembled, for instance, in other tissues or other developmental stages, has not been determined yet. The few transcriptome data that discriminate the duplicated genes indicate that *RRP40B* might be more expressed in leaves, while *RRP40A* appears to be more expressed in inflorescences (Fig. 2.2). By contrast, *RRP45A* seems to be poorly expressed in all tissues examined (resource: www.genevestigator.com [13]). However, *Arabidopsis* single mutants downregulated for either *AtRRP45A* or *AtRRP45B* expression have no or only a mild phenotype, respectively, while simultaneous downregulation of both proteins is lethal [14]. This suggests that AtRRP45A can partially complement for loss of AtRRP45B. Moreover, expression of either AtRRP45A or AtRRP45B

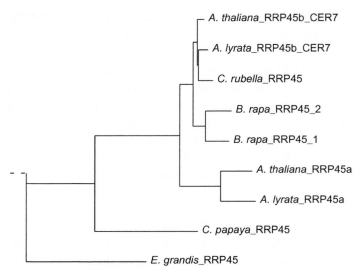

Figure 2.1 Phylogenetic tree showing that two versions of RRP45, the genuine RRP45b/ CER7 and the shorter protein RRP45a, are conserved between *Arabidopsis thaliana* and *Arabidopsis lyrata*. Most, if not all, other duplications of genes encoding exosome core subunits occurred apparently independently and recently, resulting in very similar genes such as the two versions of RRP45 found in *Brassica rapa*. Sequences were retrieved from Phytozome (www.phytozome.com). ClustalX was used for both alignment and calculation of the neighbor-joined phylogenetic tree. *A. thaliana, Arabidopsis thaliana* (thale cress); *A. lyrata, Arabidopsis lyrata* (lyrate rock cress); *C. rubella, Capsella rubella* (red shepherd's purse); *B. rapa, Brassica rapa* (turnip mustard); *C. papaya, Carica papaya* (papaya); *E. grandis, Eucalyptus grandis* (eucalyptus).

could restore the growth of a yeast *rrp45* null mutant [14]. Together, these results suggest that AtRRP45A can indeed be incorporated into a functional core complex, at least under some circumstances. This result also implied that at least one RRP45-like subunit is required for plant exosome assembly and/or function and that functional exosomes are essential for viability of *Arabidopsis*.

2.3. Arabidopsis CSL4 is dispensable for the stability of the core complex

In *Saccharomyces cerevisiae*, all nine core subunits are essential, and downregulation of individual subunits results in similar rRNA–processing phenotypes, a characteristic accumulation of a 3′-extended transcript of the 5.8S rRNA [15–17]. Moreover, the X-ray crystallographic analysis of the human exosome revealed unique interactions between all subunits

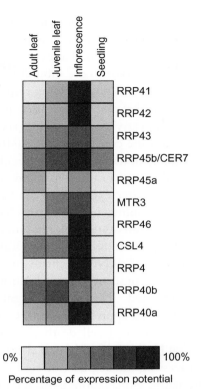

Figure 2.2 Heatmap visualizing expression levels of mRNAs encoding known or potential exosome core subunits in *Arabidopsis* seedlings, inflorescences, young leaves, and old leaves as measured by agronomics whole-genome tiling arrays (www.genevestigator.com).

of the core complex with each other [18] (see Chapter 3). It is therefore believed that all nine subunits are required to assemble a stable and functional core complex [19]. This appears to be different in *Arabidopsis* because downregulation of the AtCSL4 subunit did not result in any obvious phenotype and size fractionation from cells lacking AtCSL4 yielded nearly intact exosome complexes [7]. Interestingly, knockdown of the TbCSL4 subunit of *Trypanosome brucei* exosomes resulted neither in the degradation of other subunits nor in the disassembly of the core complex [20,21]. Therefore, in both plants and trypanosomes, the CSL4 subunit of the cap appears to be dispensable for exosome function. In fact, recent data indicate that a large part of CSL4 is also dispensable in yeast, because truncated versions of yeast CSL4, containing either only the N-terminal RRP27 domain or the C-terminal Zn-ribbon-like domain, are sufficient to promote essential exosome functions [22].

Results obtained in *Hordum vulgare* (barley) could also challenge the idea that one copy of each eukaryotic core subunit is required to assemble a functional exosome complex: While studying response reactions to the pathogen *Blumeria graminis* (abbreviated *bgh*, Barley-powdery Mildew), Xi *et al.* isolated a barley mutant carrying a deletion of six genes, among them the gene encoding the exosome core subunit RRP46 [23]. The *bcd1* (for *bgh*-induced tip cell death 1) mutation, or downregulation of HvRRP46 by virus-induced gene silencing, resulted in rapid cell death in the top 2–3 cm of developing leaves after exposure to the bacterial pathogen. In addition, both infected and noninfected *bcd1* plants had pleiotropic symptoms such as a certain proportion of sterile florets and a reduced number of tillers. On the molecular level, loss of HvRRP46 resulted in accumulation of polyadenylated, misprocessed rRNA precursors and, probably in a compensatory reaction, led to a constitutive upregulation of ribosomal proteins. The authors suggested that the misregulation of ribosome biosynthesis pathways observed upon loss of HvRRP46 renders the plants more vulnerable to stress-induced cell death [23]. However, the most striking conclusion is that HvRRP46, although clearly required for typical exosome-mediated processing and degradation reactions, is not essential in barley. Of course, alternative explanations exist: it could be possible that a second RRP46-like protein with overlapping functions is present in barley or that a copy of the related subunit MTR3 can replace HvRRP46 in the exosome core complex. In this context, it is interesting to note that, according to current genome annotations, the moss *Physcomytrella patens* has a homologue of RRP46 but apparently lacks a homologue of the closely related protein MTR3. Vice versa, the spikemoss *Selaginalla moellendorffii* possesses an MTR3 homologue but lacks a genuine RRP46 homologue.

2.4. Functional specialization of exosome core subunits?

Intriguingly, depletion of individual subunits of the plant core complex gives rise to distinct phenotypes on both the molecular level and plant development. The first plant exosome mutant characterized was the *Arabidopsis cer7* mutant isolated in a screen that intended to identify wax biosynthesis genes (*eceriferum* lines) [14]. *CER7* was mapped as an allele of *RRP45B*, encoding one of the PH-domain proteins copurifying with the exosome core complex. Both plants carrying the *cer7* point mutation and plants with a T-DNA insertion in the *RRP45B* locus have reduced levels of *CER3/WAX2* transcripts, which encode a protein of unknown function that is required for the normal production of cuticular wax in stems. This suggests that *CER7/AtRRP45B* acts as a positive regulator of *CER3* expression, maybe by degrading transcripts of a putative *CER3* repressor [14]. Notably, alterations in cuticular wax synthesis by a similar mechanism could also explain the higher sensitivity of the barley *bcd1/rrp46* mutants toward the pathogen *B. graminis* [23,24].

It is still unclear whether the function of AtRRP45B in cuticular wax production is linked to its function as an exosome subunit or due to an exosome-independent activity of the protein. However, loss of RRP45B leads to a reduced cuticular wax load and reduced seed viability, but not to major defects in growth or development, while plants lacking the RRP45A protein have normal levels of cuticular wax and normal seed viability, and are indistinguishable from wild-type plants [14]. Nevertheless, the simultaneous disruption of both *RRP45A* and *B* is lethal, which indicates that the two RRP45 proteins have both distinct and overlapping functions.

Unexpectedly, loss of AtRRP41 or AtRRP4, each of which is encoded by a single gene in *Arabidopsis*, results in growth arrest at different stages of plant development [7]. Plants carrying a wild-type and a mutant allele of the PH-ring component *AtRRP41* produce viable seeds and aborted ovules in equal rates, and the resulting progeny segregates 1:1 for wild-type and heterozygous plants. This indicates that *AtRRP41* is required for female gametogenesis. By contrast, loss of the cap component *AtRRP4* does not affect pollen or ovule development but impairs postzygotic processes: *rrp4* mutant seeds arrest at an early stage of embryo development with most seeds containing two-cell embryos and noncellularized endosperm. However, downregulation of either *AtRRP4* or *AtRRP41* by inducible RNAi leads to seedling growth arrest and plant death, demonstrating that each of the two core subunits is also essential for postembryonic growth. In a pioneering genome-wide tiling study, Belostotsky and colleagues compared the polyadenylated RNA substrates that accumulated upon downregulation of AtRRP4 or AtRRP41 with those accumulating in *csl4* mutants [7]. They found that loss of AtCSL4 affected only a minor fraction of the RNA substrates upregulated upon loss of AtRRP4 and/or AtRRP41, suggesting that exosome complexes without AtCSL4 are still mostly functional. The majority of RNA substrates were affected upon downregulation of either AtRRP4 or AtRRP41. However, some polyadenylated transcripts accumulated to distinct levels, or even in only one of the mutants, which led to the conclusion that AtRRP4 and AtRRP41 have both common and distinct substrates in plants. Hence, it appears that the two essential subunits of the core complex make different contributions to the *in vivo* activity of the plant exosome [7]. Maybe RRP41 alone, or a RRP4-less core exosome, is sufficient to promote ovule development and a few cell divisions postfertilization, while a fully assembled complex containing all nine core subunits is essential to proceed into embryo development. A possible explanation could be that AtRRP41, which in contrast to all other core subunits is catalytically active [25], can perform some basic functions as a monomer. Another, more speculative explanation would be that an alternative (sub)complex comprising RRP41 [26,27] is sufficient to promote female gametogenesis but not embryo development, but such alternative complexes have not been detected in Arabidopsis yet.

2.4.1 Take home message 1

The available data suggest that individual subunits of the *Arabidopsis* core complex make unequal contributions to exosome function, or even to complex integrity [7,14]. This challenges the traditional all-or-nothing concept of the exosome complex and opens the possibility that distinct complexes with specialized functions can be assembled in plants. Interestingly, distinct cellular distribution patterns for individual core subunits have been observed in *Drosophila melanogaster*, suggesting that complexes of variable composition may also exist in flies [28]. In fact, some of the molecular and cellular phenotypes observed upon downregulation of individual exosome core subunits in yeast and flies are rather difficult to explain with the hypothesis that the nine-subunit core complex is the only functional unit of exosome-mediated RNA degradation [26,29,30]. Therefore, Kiss and Andrulis have recently proposed an alternative concept, the "exozyme model," which suggests a flexible and dynamic assembly of functionally distinct complexes from both exosome core subunits and cofactors [26]. Interestingly, recombinant yRRP41, yRRP45, and yRRP44 (lacking the first 25 amino acids) alone can actually assemble into a stable ternary complex [27]. However, the occurrence of such alternative subcomplexes *in vivo* remains to be demonstrated.

3. EXOSOME-RELATED ACTIVITIES AND COFACTORS IN PLANTS

3.1. Plants may have retained a catalytically active core complex

Archaeal exosomes consist of three RRP41–RRP42 dimers with a phosphorolytic active site in each of the three RRP41 subunits [31]. By contrast, all PH–domain proteins of eukaryotic exosomes are catalytically inactive, at least in yeast and humans [18,22,32] (see Chapters 1 and 3). Interestingly, all residues essential for RNA binding and phosphate coordination and cleavage in archaeal RRP41 proteins are conserved in the RRP41 subunit of plants [32], and indeed, recombinant *Arabidopsis* RRP41 was shown to have a phosphorolytic activity *in vitro* [25]. Interestingly, AtRRP41 preferred polyadenylated RNA substrates [25], which would fit well to the fact that many plant exosome substrates are tagged by oligo–adenylation [7,8,33]. Hence, it is possible that plant exosomes have retained a catalytic activity similar to archaeal exosomes. However, the activity of purified or reconstituted plant exosome complexes has not been determined yet. In addition, it is unknown whether and to which degree the activity of AtRRP41 contributes to exosome-mediated RNA degradation processes *in vivo*.

The enzymatic activity of yeast and human exosomes relies exclusively on the assembly of the core complex with catalytically active cofactors [9,10,18,32,34]. In yeast, a single essential RNase R–like protein, Rrp44p/Dis3p, binds tightly to both nuclear and cytoplamic exosomes and has both a hydrolytic 3′ to 5′ exoribonuclease and an endonuclease activity conferred by an N-terminal PIN domain [22,35,36]. The PIN domain is also essential for binding of Rrp44p to the core complex [27,36]. Humans have two distinct RRP44 homologues, named hDIS3 and hDIS3L, which interact with either nucleoplasmic or cytoplasmic exosomes, respectively [10]. Plants also possess two homologues of RRP44, but because none of the proteins copurified with the *Arabidopsis* exosome core complex [7], their function as exosome cofactors remains to be demonstrated. However, one of the proteins, the nuclear isoform AtRRP44 [37], possesses all domains necessary for both endo- and exoribonucleolytic activity and also for the association with the exosome core complex. Moreover, AtRRP44 is essential for female gametogenesis, similar to the core subunit RRP41 [37]. Together, these data strongly suggest that AtRRP44 is a functional homologue of hDIS3, but this needs to be confirmed by experimental data. The second RRP44-like protein of plants, termed SOV (for suppressor of varicose) was recently shown to play a role in cytoplasmic 3′–5′ mRNA degradation [37]. Importantly, it lacks the N-terminal PIN domain, which was shown to be essential for binding of RRP44-like proteins to the exosome core complex [10,27,36]. Moreover, SOV is located in cytoplasmic granules, although it is unclear in which type of granules [37]. P-bodies, one of the best-characterized type of cytoplasmic granules, contain enzymes involved in deadenylation, decapping, and 5′–3′ degradation [38,39]. By contrast, exosome core components are absent from P-bodies and were also not observed in other types of cytoplasmic granules. Hence, both domain structure and intracellular localization suggest that the function of SOV in plant mRNA degradation is independent of the core exosome. In other words, the cytoplasmic exosome of plants may have retained a phosphorolytic activity residing in the RRP41 core subunit but possibly lacks the hydrolytic, yRRP44/hDIS3L-dependent activity of cytoplasmic exosomes in yeast and humans.

3.2. Three RRP6-like proteins in different intracellular compartments

Besides RRP44, exosomes of yeast and animals bind another active exoribonuclease, the RNase D homologue RRP6 [16]. Yeast Rrp6p is a strictly nuclear protein and plays a central role in poly(A)-assisted RNA degradation because a variety of polyadenylated transcripts accumulate upon Rrp6p downregulation [6]. Human and fly RRP6 homologues are also predominantly nuclear, although a minor fraction of both Drosophila and human RRP6 is detected in the cytoplasm [16]. In contrast to fungi, insects, and mammals, which all have only one RRP6 protein, three RRP6-like proteins are found in plants: RRP6L1, RRP6L2, and RRP6L3 [33]. While RRP6L1 and RRP6L2 are related to yeast and human proteins, RRP6L3 proteins form an evolutionary distant, plant-specific subgroup. All plant species encode the RRP6L3 protein (resource: www.phytozome.com), which appears to be cytoplasmic in *A. thaliana* [33]. Whether RRP6L3 can associate with cytoplasmic exosomes and the role it plays in cytoplasmic RNA degradation have not yet been determined. Next to RRP6L3, all plants encode either RRP6L1 or RRP6L2, and many species among both monocotyledons and dicotyledons encode both isoforms. *Arabidopsis* RRP6L1 is located in the nucleoplasm but excluded from nucleoli, while

RRP6L2 is enriched in nucleoli with only a minor fraction in the nucleoplasm. Hence, each of the *Arabidopsis* RRP6L proteins occupies a distinct
intracellular compartment. Whether the partitioning of RRP6 proteins is
conserved in other plant species is not known.

   Arabidopsis mutants downregulated for the expression of either RRP6L1
or RRP6L2 have no obvious growth defects. However, downregulation of
RRP6L2 results in a mild accumulation of 5.8S rRNA precursors, and in the
accumulation of a polyadenylated fragment of the 5′ external transcribed
spacer, a by-product of rRNA maturation which is a known substrate of
both yeast and human RRP6 proteins. Therefore, RRP6L2 functions in
polyadenylation-assisted RNA decay in *Arabidopsis* as does Rrp6p in yeast,
despite the failure of RRP6L2 to complement the yeast mutant. Interestingly, RRP6L1 rescued the growth defect of a temperature-sensitive yeast
rrp6 mutant, suggesting that it also shares some functions with the yeast protein. However, no accumulation of the 5′ ETS was observed in *Arabidopsis*
rrp6l1 mutants, and the 5′ ETS levels observed in *rrp6l1–rrp6l2* double mutants resembled those observed in *rrp6l2* single mutants. Hence, although
some functional overlap cannot be excluded, the data available so far suggest
that RRP6L2 and RRP6L1 have distinct functions in plants [33].

3.3. The SKI complex is conserved in plants

In addition to active ribonucleases, many other cofactors such as RNA-
binding proteins or RNA helicases are important for exosome-mediated
RNA degradation pathways [19,40]. Some proteins physically interact
with either cytoplasmic or nuclear exosomes, thereby modulating both
substrate specificity and activity of the complex. Other cofactors, such as
the nuclear TRAMP complex (see Chapter 4), assist exosome-mediated
RNA degradation by modifying the RNA substrates [4–6].

   All functions of the yeast cytoplasmic exosome require interaction of the
core with the SKI complex consisting of Ski2p, Ski3p, and two copies of Ski8p
[3,41–44]. While Ski2p is an RNA helicase, Ski3p and Ski8p have no catalytic
activity and are thought to promote protein–protein interactions via their
tetratricopeptide and WD40 domains, respectively [45]. In addition, both
mRNA turnover and quality control functions of the cytoplasmic exosome
require Ski7p, which shares some homology to eukaryotic translation factor
EF1α and physically interacts with both the exosome core complex and the
SKI complex [46,47]. Due to a poor level of sequence conservation, it is
difficult to identify SKI3 and SKI8 homologues in plant genomes.

However, a SKI2 homologue is clearly present in all plant species (resource: www.phytozome.com). A recent study showed that an *Arabidopsis* SKI8 homologue, named VIP3 (for Vernalization IndePendence 3), coprecipitates with both AtSKI2 and a SKI3 homologue [48]. Moreover, downregulation of VIP3/SKI8 resulted in stabilization of mRNAs, among them known substrates of the *Arabidopsis* core exosome. Hence, both the composition and the function of the SKI complex in mRNA turnover appear to be conserved in *Arabidopsis*. Potential plant homologues of SKI7 or other, plant-specific partners of the cytoplasmic exosome may exist but have not been characterized yet.

Although loss of yeast Ski2p, Ski3p, or Ski8p slightly affects the decay rates of mRNA and NMD reporter substrates, null mutants grow normal under standard laboratory conditions [2,3,41]. By contrast, downregulation of either SKI2 or SKI8/VIP3 in *Arabidopsis* results in dwarf growth [48]. In addition, *Arabidopsis ski8* mutants have a flower development and early flowering phenotype which is likely due to a second, exosome-independent function of the protein as a component of the conserved nuclear Paf1c complex [48]. A recent genetic study is now suggesting that mutations in both human SKI2 and human SKI3 genes are associated with a rare congenital bowel disorder [49]. Hence, a functional SKI complex is required for normal development of plants and mammals.

Interestingly, a second SKI2-like protein, ISE2, is present in all plants but absent in animals or fungi. According to sequence analysis, ISE2 is a distant member of the SKI2/MTR4 RNA helicase family. However, the characteristic arch domain of SKI and MTR4 proteins [50,51], which, next to other functions, may also provide an interface for binding to the exosome complex, is not well conserved in ISE2. *Arabidopsis ISE2* is essential for embryogenesis and required for the formation of plasmodesmata [52]. Originally, reporter studies implicated ISE2 in cytoplasmic RNA degradation or, more specifically, in posttranscriptional gene silencing [52]. However, recent data demonstrate that ISE2 functions in chloroplasts [53] suggesting that the protein has, despite its similarity to SKI2, acquired a different function in plants.

3.4. The Arabidopsis MTR4 homologue is a nucleolar protein

The nuclear exosome also interacts with various partners, some of which are organized in protein complexes themselves. In yeast and human, ribonuclease RRP6 recruits two RNA-binding proteins, RRP47/LRP1/C1D and

MPP6, to the core complex [54–58]. RRP47 has the capacity of binding both double-stranded DNA and RNA, while MPP6 was characterized as a single-strand RNA-binding protein with a preference for poly-U [55,56]. A BLAST search with yeast or human MPP6 protein fails to identify a sequence homologue in *Arabidopsis*. By contrast, RRP47 appears to be conserved in plants. The interaction of RRP47 with RRP6 depends on an N-terminal PMC2NT domain that is present in *Arabidopsis* RRP6L2 but not conserved in RRP6L1 or RRP6L3 [33]. Indeed, only RRP6L2 interacted with the *Arabidopsis* homologue of Rrp47 (AtRRP47) (L. Philippe, J. Canaday, and D. Gagliardi, unpublished). However, studies addressing intracellular distribution and function of AtRRP47 in plants are lacking to date.

The most prominent cofactor of the yeast nuclear exosome is the TRAMP complex, responsible for the oligo-adenylation of exosome substrates prior to degradation by RRP6 and the nuclear exosome. Yeast TRAMP consists of one of the polyA polymerases Trf4p or Trf5p, one of the two RNA-binding proteins Air1p or Air2p, and the RNA helicase Mtr4p [4,5,59] (see Chapter 4). Interestingly, the single MTR4 homologue encoded in humans is part of a TRAMP-like complex in the nucleolus and also part of a nucleoplasmic complex, named NEXT (for Nuclear Exosome Targeting), with distinct composition and function [11]. In addition, it has been detected in a trimeric complex with RRP6 and MPP6 [58].

Among potential TRAMP or NEXT components, only a MTR4 homologue, AtMTR4, was characterized in plants. AtMTR4 is a mostly nucleolar protein with only a small portion detected in the nucleoplasm. Loss of AtMTR4 results in the overaccumulation of 3′ extended 5.8S rRNA precursors, misprocessed species of the 18S rRNA, and the 5′ ETS, an unfunctional maturation by-product of rRNA biogenesis [60]. This indicated that AtMTR4 functions predominantly in rRNA surveillance. Indeed, mature rRNAs accumulate to wild-type levels, which argues against an essential role of AtMTR4 in rRNA trimming. However, loss of AtMTR4 diminished the production rate of mature rRNAs [60]. Moreover, mutants lacking MTR4 show general growth retardation and pleiotropic developmental defects such as abnormal vein patterns in cotyledons, pointy first leafs, or short roots. Actually, this combination of developmental defects is characteristic for ribosomal protein mutants and mutants lacking rRNA-processing or ribosome assembly factors such as APUM23 or Nucleolin [61–63]. Hence, both molecular data and growth phenotype indicate that AtMTR4 is a cofactor of the nucleolar exosome and is required for efficient rRNA biogenesis in *Arabidopsis*. Whether the requirement for normal ribosome biogenesis rates is linked to its role in rRNA surveillance or linked to a possible function in 3' processing of small nucleolar RNAs (snoRNAs) is unknown for the moment. Unlike yeast and human MTR4, AtMTR4 is not essential, maybe due to the presence of another MTR4 homologue in *Arabidopsis*, HEN2. Downregulation of HEN2 resulted in distinct developmental defects and did not affect the accumulation of rRNA-processing intermediates [60,64]. However, the simultaneous disruption of both HEN2 and AtMTR4 expression is apparently lethal [60]. The data available so far suggest that AtMTR4 and HEN2 may have both common and specific roles.

3.4.1 Take home message 2

The current data suggest that activities and organization of plant exosome complexes differ in various aspects from their counterparts in yeast and human. First, the plant core complex may have retained a phosphorolytic activity that is unique among eukaryotic exosomes. Second, only nuclear exosomes are likely associated with an RRP44 homologue, while the cytoplasmic RRP44 homologue SOV has probably lost the capacity to bind the core complex. Third, a plant-specific RRP6 homologue, RRP6L3, is present in all plant genomes and appears to encode a cytoplasmic protein, at least in *Arabidopsis*. Finally, two other RRP6-like proteins with distinct intracellular distribution and function are present in *Arabidopsis* and in many other plant species. Other exosome cofactors have been poorly studied. A cytoplasmic SKI complex seems conserved in plants, and AtMTR4 shares at least some functions with its counterparts in yeast and human. But whether nuclear cofactors like AtMTR4 assemble into protein complexes like TRAMP or NEXT remains to be determined.

4. IMPACT OF EXOSOME-MEDIATED DEGRADATION IN PLANTS

4.1. Polyadenylated substrates of the exosome have been identified

Based on their accumulation as polyadenylated transcripts upon downregulation of the core subunits RRP41 and RRP4, approximately 1100 substrates of the *Arabidopsis* exosome have been identified by genome-wide tiling arrays [7]. A major group of polyadenylated exosome targets are rRNA precursor transcripts and maturation by-products removed during rRNA processing such as external and internal transcribed spacer regions. Importantly, these substrates accumulate also upon downregulation of the RNA helicase AtMTR4 and, albeit to lower levels, upon downregulation of the ribonuclease RRP6L2, both of which are predominantly nucleolar proteins [33,60]. This indicates that, in *Arabidopsis*, the removal of incompletely processed or misprocessed rRNA precursors and rRNA maturation by-products is ensured by a nucleolar exosome machinery comprising a dedicated RRP6-like protein and only one of two MTR4 homologues.

Other stable RNAs identified as plant exosome substrates are several small nuclear RNAs (snRNA), some snoRNAs, MRP/7-2 RNA, and 7SL RNA [7]. In many cases, both correctly processed and 3′-extended precursor transcripts can accumulate as polyadenylated transcripts, suggesting that the plant exosome is involved in both turnover of structural RNAs and removal of misprocessed species. Alternatively, 3′-extended species could be intermediates of polyadenylation-assisted 3′ trimming by the exosome. Intermediates and by-products of miRNA processing were also identified as exosome targets [7], but whether the plant exosome participates in the biogenesis and turnover of small RNAs, as recently suggested for the *Neurospora* exosome [65], remains to be investigated. Interestingly, no tRNAs were observed among the polyadenylated exosome substrates, with the exception of tRNA-Tyr [7]. This particular tRNA undergoes multiple base modification steps during its maturation. It was therefore suggested that the plant exosome may participate in quality control of hypomodified tRNAs [7] as it was observed in yeast [5,66–69].

Another important function of the plant exosome is the rapid removal of a large variety of heterochromatic transcripts from centromeric or pericentromeric regions, novel transcripts with no protein-coding potential and no predicted function derived from nonannotated regions, and short-lived transcripts of 100–600 nt corresponding to 5′ regions of known mRNAs [7]. Similar unstable noncoding RNAs, termed CUTs (cryptic unstable transcripts) or PROMPTs (promoter upstream transcripts), have also been observed in yeast and mammals, respectively [6,70]. Some yeast CUTs have been proposed to be regulators of gene expression because their synthesis might regulate transcription of neighboring genes [71,72]. It remains to be explored to what extent the production of these novel transcripts has a regulatory function in plants. One of the noncoding exosome substrates is stress-induced in wild-type plants, alongside several other transcripts derived from unannotated regions [73,74]. However, the fact that the vast majority of this "dark matter of the transcriptome" is observed only upon downregulation of the exosome demonstrates that the efficient elimination of these spurios transcripts is one of the prominent tasks of exosome-mediated RNA degradation in plants [7,75].

Finally, downregulation of plant exosome subunits affected also a number of mRNA loci, some of which were deregulated in either only the RRP4-depleted or the RRP41-depleted sample [7]. These could be true mRNA substrates of the exosome or mRNAs up- or downregulated due to secondary effects caused by the depletion of functional exosomes. A significant fraction of upregulated mRNA regions corresponded to sense and antisense transcripts derived from intronless pseudogenes or appeared to be irregular 3' read-through transcripts from protein-coding genes. Other upregulated regions corresponded to introns suggesting that the plant exosome participates in the removal of excised introns or in the degradation of incompletely spliced mRNAs, or both.

4.2. A high level of redundancy among plant RNA degradation pathways

As mentioned at the beginning, mRNA degradation by the cytoplasmic exosome requires either deadenylation or endonucleolytic cleavage in the body of the mRNA. Because such degradation intermediates would mostly escape the detection by tiling arrays, not much is known about the function of the cytoplasmic exosome in plants. A recent study used a novel random-primed RNA seq method to show that loss of AtSKI8, a component of the cytoplasmic SKI complex, slightly affects mRNA turnover in *Arabidopsis* [48]. The rather mild effect is not surprising because it is believed that the default pathway for eukaryotic mRNA degradation is decapping and degradation by 5'–3' exoribonucleases of the XRN family (see Chapter 5). In Arabidopsis, the known cytoplasmic 5'–3' exoribonuclease is XRN4 [1,76]. However, despite its putative role as the main player in cytoplasmic RNA degradation, only a small number of mRNAs overaccumulate upon downregulation of XRN4 [76–78]. It was therefore suggested that 3'–5' degradation by the plant exosome can largely compensate for a defective 5'–3' pathway [76]. As in other eukaryotes, 3'–5' degradation by the plant exosome could also be important for the rapid elimination of defective mRNAs following their detection by specific quality control pathways such as nonsense-mediated decay, nonstop decay, or ARE-mediated decay (reviewed in Ref. [79]), but this has not been directly investigated yet. Furthermore, the exosome is believed to degrade 5' mRNA fragments generated by RISC-induced cleavage, while XRN4 participates in the removal of both 5′ and 3′ fragments [76,78,80–82].

Redundancy or cooperation of both 3′–5′ and 5′–3′ degradation pathways is definitely observed in plant nuclei. In addition to the nuclear exosome, two nuclear 5′–3′ exoribonucleolases, XRN2 and XRN3, are involved in both rRNA processing and degradation of rRNA maturation by-products [83] (see also Chapter 7). A clear example of cooperation between both 3′–5′ and 5′–3 degradation pathways is the degradation of the 5′ ETS: Degradation of the 1.8-kb-long 5′ external transcribed spacer of the *Arabidopsis* 35S rRNA precursor is initiated by two endonucleolytic cleavages at P and P′ sites, respectively [33,60,83]. One of the resulting fragments accumulates only in *xrn2* single and *xrn2/3* double mutants and is therefore a substrate of the 5′–3′ pathway. The second fragment accumulates only upon downregulation of AtMTR4, RRP6L2, and exosome core subunits and appears to be an exclusive substrate of the 3′-5′ pathway [7,60]. These data demonstrate that both pathways are necessary for a complete degradation of the *Arabidopsis* 5′ ETS. Furthermore, some of the aberrant rRNA-processing intermediates detected in *xrn2* single and *xrn2/3* double mutants become polyadenylated [83], indicating that they are tagged for 3′–5′ degradation by the exosome. This indicates that although each of the main degradation pathways may have its preferred set of substrates, the majority of rRNA-processing intermediates and maturation by-products are probably degraded from both 5′ and 3′ ends.

Cooperation and redundancy between the two RNA degradation pathways might be particularly important for the elimination of rRNA maturation by-products, since rRNAs are transcribed at high rates and the rapid removal of rRNA maturation by-products is a major task of RNA degradation in the nucleus. However, it appears that both miRNA processing intermediates and mRNA 3′-read-through transcripts are also targets of both XRN2/3 and the exosome [7,84,85]. Hence, a high level of redundancy between both 5′–3′ and 3′–5′ pathways may generally account for the, relative to the size of the Arabidopsis genome, low numbers of transcripts that were observed as exclusive substrates of individual XRN proteins or the exosome, respectively. This cooperation/redundancy of the two main degradation pathways ensures the rapid and efficient elimination of nonfunctional RNAs and also prevents the uncontrolled production of small RNAs from accumulating degradation intermediates [60,78,85]. Therefore, the current data probably underestimate the impact of the exosome on RNA turnover, RNA quality control, and RNA surveillance.

5. FINAL REMARKS

The present data suggest that the degradation of the large amounts of by-products produced during rRNA maturation and processing of non-coding RNAs is a main task of the plant exosome. A plethora of other exosome substrates have been identified and are believed to be mostly nuclear. However, cofactors involved in recognition and degradation of other substrates than rRNA-related transcripts have not been identified yet. Moreover, almost nothing is known about the impact of the exosome and its cofactors in mRNA metabolism. The recent identification of SKI components has now opened the possibility to study the role of the cytoplasmic exosome, for example, in mRNA quality control pathways such as NMD. It will be also very interesting to investigate the relationship between the cytoplasmic exosome and SOV, the PIN-less RRP44 homologue detected in cytoplasmic granules, and the role of the plant-specific RRP6L3 protein. Another fascinating topic of future research is to unravel the contribution of the core exosome activity in both cytoplasmic and nuclear RNA metabolism.

REFERENCES

[1] Houseley J, Tollervey D. The many pathways of RNA degradation. Cell 2009;136:763–76.

[2] Masison DC, Blanc A, Ribas JC, Carroll K, Sonenberg N, Wickner RB. Decoying the cap- mRNA degradation system by a double-stranded RNA virus and poly(A)-mRNA surveillance by a yeast antiviral system. Mol Cell Biol 1995;15:2763–71.

[3] Anderson JS, Parker RP. The 3′ to 5′ degradation of yeast mRNAs is a general mechanism for mRNA turnover that requires the SKI2 DEVH box protein and 3′ to 5′ exonucleases of the exosome complex. EMBO J 1998;17:1497–506.

[4] LaCava J, Houseley J, Saveanu C, Petfalski E, Thompson E, Jacquier A, et al. RNA degradation by the exosome is promoted by a nuclear polyadenylation complex. Cell 2005;121:713–24.

[5] Vanacova S, Wolf J, Martin G, Blank D, Dettwiler S, Friedlein A, et al. A new yeast poly (A) polymerase complex involved in RNA quality control. PLoS Biol 2005;3:e189.

[6] Wyers F, Rougemaille M, Badis G, Rousselle J-C, Dufour M-E, Boulay J, et al. Cryptic pol II transcripts are degraded by a nuclear quality control pathway involving a new poly (A) polymerase. Cell 2005;121:725–37.

[7] Chekanova JA, Gregory BD, Reverdatto SV, Chen H, Kumar R, Hooker T, et al. Genome-wide high-resolution mapping of exosome substrates reveals hidden features in the Arabidopsis transcriptome. Cell 2007;131:1340–53.

[8] Lange H, Sement FM, Canaday J, Gagliardi D. Polyadenylation-assisted RNA degradation processes in plants. Trends Plant Sci 2009;14:497–504.

[9] Chlebowski A, Tomecki R, López MEG, Séraphin B, Dziembowski A. Catalytic properties of the eukaryotic exosome. Adv Exp Med Biol 2011;702:63–78.

[10] Tomecki R, Kristiansen MS, Lykke-Andersen S, Chlebowski A, Larsen KM, Szczesny RJ, et al. The human core exosome interacts with differentially localized processive RNases: hDIS3 and hDIS3L. EMBO J 2010;29:2342–57.

[11] Lubas M, Christensen MS, Kristiansen MS, Domanski M, Falkenby LG, Lykke-Andersen S, et al. Interaction profiling identifies the human nuclear exosome targeting complex. Mol Cell 2011;43:624–37.

[12] Goodstein DM, Shu S, Howson R, Neupane R, Hayes RD, Fazo J, et al. Phytozome: a comparative platform for green plant genomics. Nucleic Acids Res 2012;40: D1178–86.

[13] Zimmermann P, Hirsch-Hoffmann M, Hennig L, Gruissem W. GEN-EVESTIGATOR. Arabidopsis microarray database and analysis toolbox. Plant Physiol 2004;136:2621–32.

[14] Hooker TS, Lam P, Zheng H, Kunst L. A core subunit of the RNA-processing/degrading exosome specifically influences cuticular wax biosynthesis in Arabidopsis. Plant Cell 2007;19:904–13.

[15] Mitchell P, Petfalski E, Shevchenko A, Mann M, Tollervey D. The exosome: a conserved eukaryotic RNA processing complex containing multiple 3' → 5' exoribonucleases. Cell 1997;91:457–66.

[16] Allmang C, Petfalski E, Podtelejnikov A, Mann M, Tollervey D, Mitchell P. The yeast exosome and human PM-Scl are related complexes of 3' → 5' exonucleases. Genes Dev 1999;13:2148–58.

[17] van Hoof A, Parker R. The exosome: a proteasome for RNA? Cell 1999;99:347–50.

[18] Liu Q, Greimann JC, Lima CD. Reconstitution, activities, and structure of the eukaryotic RNA exosome. Cell 2006;127:1223–37.

[19] Lykke-Andersen S, Brodersen DE, Jensen TH. Origins and activities of the eukaryotic exosome. J Cell Sci 2009;122:1487–94.

[20] Estévez AM, Kempf T, Clayton C. The exosome of Trypanosoma brucei. EMBO J 2001;20:3831–9.

[21] Estévez AM, Lehner B, Sanderson CM, Ruppert T, Clayton C. The roles of intersubunit interactions in exosome stability. J Biol Chem 2003;278:34943–51.

[22] Schaeffer D, Tsanova B, Barbas A, Reis FP, Dastidar EG, Sanchez-Rotunno M, et al. The exosome contains domains with specific endoribonuclease, exoribonuclease and cytoplasmic mRNA decay activities. Nat Struct Mol Biol 2009;16:56–62.

[23] Xi L, Moscou MJ, Meng Y, Xu W, Caldo RA, Shaver M, et al. Transcript-based cloning of RRP46, a regulator of rRNA processing and R gene-independent cell death in barley-powdery mildew interactions. Plant Cell 2009;21:3280–95.

[24] Zabka V, Stangl M, Bringmann G, Vogg G, Riederer M, Hildebrandt U. Host surface properties affect prepenetration processes in the barley powdery mildew fungus. New Phytol 2008;177:251–63.

[25] Chekanova JA, Shaw RJ, Wills MA, Belostotsky DA. Poly(A) tail-dependent exonuclease AtRrp41p from Arabidopsis thaliana rescues 5.8 S rRNA processing and mRNA decay defects of the yeast ski6 mutant and is found in an exosome-sized complex in plant and yeast cells. J Biol Chem 2000;275:33158–66.

[26] Kiss DL, Andrulis ED. The exozyme model: a continuum of functionally distinct complexes. RNA 2011;17:1–13.

[27] Bonneau F, Basquin J, Ebert J, Lorentzen E, Conti E. The yeast exosome functions as a macromolecular cage to channel RNA substrates for degradation. Cell 2009;139:547–59.

[28] Graham AC, Kiss DL, Andrulis ED. Differential distribution of exosome subunits at the nuclear lamina and in cytoplasmic foci. Mol Biol Cell 2006;17:1399–409.

[29] Houalla R, Devaux F, Fatica A, Kufel J, Barrass D, Torchet C, et al. Microarray detection of novel nuclear RNA substrates for the exosome. Yeast 2006;23:439–54.

[30] Kiss DL, Andrulis ED. Genome-wide analysis reveals distinct substrate specificities of Rrp6, Dis3, and core exosome subunits. RNA 2010;16:781–91. http://dx.doi.org/10.1261/rna.1906710.

[31] Lorentzen E, Walter P, Fribourg S, Evguenieva-Hackenberg E, Klug G, Conti E. The archaeal exosome core is a hexameric ring structure with three catalytic subunits. Nat Struct Mol Biol 2005;12:575–81.

[32] Dziembowski A, Lorentzen E, Conti E, Seraphin B. A single subunit, Dis3, is essentially responsible for yeast exosome core activity. Nat Struct Mol Biol 2007;14:15–22.

[33] Lange H, Holec S, Cognat V, Pieuchot L, Le Ret M, Canaday J, et al. Degradation of a polyadenylated rRNA maturation by-product involves one of the three RRP6-like proteins in Arabidopsis thaliana. Mol Cell Biol 2008;28:3038–44.

[34] Tomecki R, Drazkowska K, Dziembowski A. Mechanisms of RNA degradation by the eukaryotic exosome. Chembiochem 2010;11:938–45.

[35] Lebreton A, Tomecki R, Dziembowski A, Séraphin B. Endonucleolytic RNA cleavage by a eukaryotic exosome. Nature 2008;456:993–6.

[36] Schneider C, Leung E, Brown J, Tollervey D. The N-terminal PIN domain of the exosome subunit Rrp44 harbors endonuclease activity and tethers Rrp44 to the yeast core exosome. Nucleic Acids Res 2009;37:1127–40.

[37] Zhang W, Murphy C, Sieburth LE. Conserved RNaseII domain protein functions in cytoplasmic mRNA decay and suppresses Arabidopsis decapping mutant phenotypes. Proc Natl Acad Sci USA 2010;107:15981–5.

[38] Kulkarni M, Ozgur S, Stoecklin G. On track with P-bodies. Biochem Soc Trans 2010;38:242–51.

[39] Xu J, Chua N-H. Processing bodies and plant development. Curr Opin Plant Biol 2011;14:88–93.

[40] Lebreton A, Séraphin B. Exosome-mediated quality control: substrate recruitment and molecular activity. Biochim Biophys Acta 2008;1779:558–65.

[41] van Hoof A, Frischmeyer PA, Dietz HC, Parker R. Exosome-mediated recognition and degradation of mRNAs lacking a termination codon. Science 2002;295:2262–4.

[42] Mitchell P, Tollervey D. An NMD pathway in yeast involving accelerated deadenylation and exosome-mediated 3′ → 5′ degradation. Mol Cell 2003;11:1405–13.

[43] Synowsky SA, Heck AJR. The yeast Ski complex is a hetero-tetramer. Protein Sci 2008;17:119–25.

[44] Brown JT, Bai X, Johnson AW. The yeast antiviral proteins Ski2p, Ski3p, and Ski8p exist as a complex in vivo. RNA 2000;6:449–57.

[45] Wang L, Lewis MS, Johnson AW. Domain interactions within the Ski2/3/8 complex and between the Ski complex and Ski7p. RNA 2005;11:1291–302.

[46] Araki Y, Takahashi S, Kobayashi T, Kajiho H, Hoshino S, Katada T. Ski7p G protein interacts with the exosome and the Ski complex for 3′-to-5′ mRNA decay in yeast. EMBO J 2001;20:4684–93.

[47] Takahashi S, Araki Y, Sakuno T, Katada T. Interaction between Ski7p and Upf1p is required for nonsense-mediated 3′-to-5′ mRNA decay in yeast. EMBO J 2003;22:3951–9.

[48] Dorcey E, Rodriguez-Villalon A, Salinas P, Santuari L, Pradervand S, Harshman K, et al. Context-dependent dual role of SKI8 homologs in mRNA synthesis and turnover. PLoS Genet 2012;8:e1002652.

[49] Fabre A, Charroux B, Martinez-Vinson C, Roquelaure B, Odul E, Sayar E, et al. SKIV2L mutations cause syndromic diarrhea, or trichohepatoenteric syndrome. Am J Hum Genet 2012;90:689–92.

[50] Weir JR, Bonneau F, Hentschel J, Conti E. Structural analysis reveals the characteristic features of Mtr4, a DExH helicase involved in nuclear RNA processing and surveillance. Proc Natl Acad Sci USA 2010;107:12139–44.

[51] Jackson RN, Klauer AA, Hintze BJ, Robinson H, van Hoof A, Johnson SJ. The crystal structure of Mtr4 reveals a novel arch domain required for rRNA processing. EMBO J 2010;29:2205–16.

[52] Kobayashi K, Otegui MS, Krishnakumar S, Mindrinos M, Zambryski P. INCREASED SIZE EXCLUSION LIMIT 2 encodes a putative DEVH box RNA helicase involved in plasmodesmata function during Arabidopsis embryogenesis. Plant Cell 2007;19:1885–97.

[53] Burch-Smith TM, Brunkard JO, Choi YG, Zambryski PC. Organelle-nucleus cross-talk regulates plant intercellular communication via plasmodesmata. Proc Natl Acad Sci USA 2011;108:E1451–60.

[54] Butler JS, Mitchell P. Rrp6, rrp47 and cofactors of the nuclear exosome. Adv Exp Med Biol 2011;702:91–104.

[55] Mitchell P, Petfalski E, Houalla R, Podtelejnikov A, Mann M, Tollervey D. Rrp47p is an exosome-associated protein required for the 3′ processing of stable RNAs. Mol Cell Biol 2003;23:6982–92.

[56] Milligan L, Decourty L, Saveanu C, Rappsilber J, Ceulemans H, Jacquier A, et al. A yeast exosome cofactor, Mpp 6, functions in RNA surveillance and in the degradation of noncoding RNA transcripts. Mol Cell Biol 2008;28:5446–57.

[57] Schilders G, Raijmakers R, Raats JMH, Pruijn GJM. MPP6 is an exosome-associated RNA-binding protein involved in 5.8S rRNA maturation. Nucleic Acids Res 2005;33:6795–804.

[58] Schilders G, van Dijk E, Pruijn GJM. C1D and hMtr4p associate with the human exosome subunit PM/Scl-100 and are involved in pre-rRNA processing. Nucleic Acids Res 2007;35:2564–72.

[59] Houseley J, Tollervey D. Yeast Trf5p is a nuclear poly(A) polymerase. EMBO Rep 2006;7:205–11.

[60] Lange H, Sement FM, Gagliardi D. MTR4, a putative RNA helicase and exosome co-factor, is required for proper rRNA biogenesis and development in Arabidopsis thaliana. Plant J 2011;68:51–63.

[61] Petricka JJ, Nelson TM. Arabidopsis nucleolin affects plant development and patterning. Plant Physiol 2007;144:173–86.

[62] Abbasi N, Kim HB, Park N-I, Kim H-S, Kim Y-K, Park Y-I, et al. APUM23, a nucleolar Puf domain protein, is involved in pre-ribosomal RNA processing and normal growth patterning in Arabidopsis. Plant J 2010;64:960–76.

[63] Byrne ME. A role for the ribosome in development. Trends Plant Sci 2009;14:512–9.

[64] Western TL, Cheng Y, Liu J, Chen X. HUA ENHANCER2, a putative DExH-box RNA helicase, maintains homeotic B and C gene expression in Arabidopsis. Development 2002;129:1569–81.

[65] Xue Z, Yuan H, Guo J, Liu Y. Reconstitution of an Argonaute-dependent small RNA biogenesis pathway reveals a handover mechanism involving the RNA exosome and the exonuclease QIP. Mol Cell 2012;46:299–310.

[66] Alexandrov A, Chernyakov I, Gu W, Hiley SL, Hughes TR, Grayhack EJ, et al. Rapid tRNA decay can result from lack of nonessential modifications. Mol Cell 2006;21:87–96.

[67] Kadaba S, Wang X, Anderson JT. Nuclear RNA surveillance in Saccharomyces cerevisiae: Trf4p-dependent polyadenylation of nascent hypomethylated tRNA and an aberrant form of 5S rRNA. RNA 2006;12:508–21.

[68] Schneider C, Anderson JT, Tollervey D. The exosome subunit Rrp44 plays a direct role in RNA substrate recognition. Mol Cell 2007;27:324–31.

[69] Wang X, Jia H, Jankowsky E, Anderson JT. Degradation of hypomodified tRNAiMet in vivo involves RNA-dependent ATPase activity of the DExH helicase Mtr4p. RNA 2008;14:107–16.

[70] Preker P, Nielsen J, Kammler S, Lykke-Andersen S, Christensen MS, Mapendano CK, et al. RNA exosome depletion reveals transcription upstream of active human promoters. Science 2008;322:1851–4.

[71] Neil H, Malabat C, d' Aubenton-Carafa Y, Xu Z, Steinmetz LM, Jacquier A. Widespread bidirectional promoters are the major source of cryptic transcripts in yeast. Nature 2009;457:1038–42.

[72] Xu Z, Wei W, Gagneur J, Perocchi F, Clauder-Münster S, Camblong J, et al. Bidirectional promoters generate pervasive transcription in yeast. Nature 2009;457:1033–7.

[73] Matsui A, Ishida J, Morosawa T, Okamoto M, Kim J-M, Kurihara Y, et al. Arabidopsis tiling array analysis to identify the stress-responsive genes. Methods Mol Biol 2010;639:141–55.

[74] Zeller G, Henz SR, Widmer CK, Sachsenberg T, Rätsch G, Weigel D, et al. Stress-induced changes in the Arabidopsis thaliana transcriptome analyzed using whole-genome tiling arrays. Plant J 2009;58:1068–82.

[75] Belostotsky D. Exosome complex and pervasive transcription in eukaryotic genomes. Curr Opin Cell Biol 2009;21:352–8.

[76] Souret FF, Kastenmayer JP, Green PJ. AtXRN4 degrades mRNA in Arabidopsis and its substrates include selected miRNA targets. Mol Cell 2004;15:173–83.

[77] Olmedo G, Guo H, Gregory BD, Nourizadeh SD, Aguilar-Henonin L, Li H, et al. ETHYLENE-INSENSITIVE5 encodes a $5' \rightarrow 3'$ exoribonuclease required for regulation of the EIN3-targeting F-box proteins EBF1/2. Proc Natl Acad Sci USA 2006;103:13286–93.

[78] Gregory BD, O'Malley RC, Lister R, Urich MA, Tonti-Filippini J, Chen H, et al. A link between RNA metabolism and silencing affecting Arabidopsis development. Dev Cell 2008;14:854–66.

[79] Schaeffer D, Clark A, Klauer AA, Tsanova B, van Hoof A. Functions of the cytoplasmic exosome. Adv Exp Med Biol 2011;702:79–90.

[80] Orban TI, Izaurralde E. Decay of mRNAs targeted by RISC requires XRN1, the Ski complex, and the exosome. RNA 2005;11:459–69.

[81] Ibrahim F, Rohr J, Jeong W-J, Hesson J, Cerutti H. Untemplated oligoadenylation promotes degradation of RISC-cleaved transcripts. Science 2006;314:1893.

[82] Shen B, Goodman HM. Uridine addition after microRNA-directed cleavage. Science 2004;306:997.

[83] Zakrzewska-Placzek M, Souret FF, Sobczyk GJ, Green PJ, Kufel J. Arabidopsis thaliana XRN2 is required for primary cleavage in the pre-ribosomal RNA. Nucleic Acids Res 2010;38:4487–502.

[84] Gy I, Gasciolli V, Lauressergues D, Morel J-B, Gombert J, Proux F, et al. Arabidopsis FIERY1, XRN2, and XRN3 are endogenous RNA silencing suppressors. Plant Cell 2007;19:3451–61.

[85] Kurihara Y, Schmitz RJ, Nery JR, Schultz MD, Okubo-Kurihara E, Morosawa T, et al. Surveillance of 3′ noncoding transcripts requires FIERY1 and XRN3 in Arabidopsis. G3 (Bethesda) 2012;2:487–98.

CHAPTER THREE

Structure and Activities of the Eukaryotic RNA Exosome

Elizabeth V. Wasmuth[*,†], Christopher D. Lima[*,1]

*Structural Biology Program, Sloan-Kettering Institute, New York, USA
†Louis V. Gerstner Jr. Graduate School of Biomedical Sciences, Memorial Sloan-Kettering Cancer Center, 1275 York Avenue, New York, USA
1Corresponding author: e-mail address: limac@mskcc.org

Contents

1. Introduction   53
2. Global Architecture of the Eukaryotic Exosome Core   56
3. RNase PH-Like Domains Comprise a PH-Like Ring in Eukaryotic Exosomes   56
4. S1 and KH Domains Cap the PH-Like Ring   59
5. Rrp44, a Hydrolytic Endoribonuclease and Processive Exoribonuclease   61
6. Rrp44 and the 10-Component Exosome   65
7. Rrp6, a Eukaryotic Exosome Subunit with Distributive Hydrolytic Activities   67
8. Rrp44, Rrp6, and the 11-Component Nuclear Exosome   70
9. Conclusions   71
Acknowledgments   71
References   71

Abstract

The composition of the multisubunit eukaryotic RNA exosome was described more than a decade ago, and structural studies conducted since that time have contributed to our mechanistic understanding of factors that are required for 3′-to-5′ RNA processing and decay. This chapter describes the organization of the eukaryotic RNA exosome with a focus on presenting results related to the noncatalytic nine-subunit exosome core as well as the hydrolytic exo- and endoribonuclease Rrp44 (Dis3) and the exoribonuclease Rrp6. This is achieved in large part by describing crystal structures of Rrp44, Rrp6, and the nine-subunit exosome core with an emphasis on how these molecules interact to endow the RNA exosome with its catalytic activities.

> ## 1. INTRODUCTION

3′-to-5′ RNA decay is an evolutionarily conserved process in all known kingdoms of life, and the family of enzymes that catalyze RNA decay share mechanistic and structural relationships. In eukaryotes, nuclear and

53

cytoplasmic 3′ to 5′ decay is catalyzed by an essential multisubunit complex termed the RNA exosome [1,2] (Fig. 3.1). The RNA exosome includes a noncatalytic core formed by six subunits (Rrp41, Rrp45, Rrp42, Rrp43, Mtr3, and Rrp46) that share similarity to RNase PH and three subunits (Csl4, Rrp4, and Rrp40) that share similarity to proteins containing S1/KH RNA-binding domains. The exosome core associates with two hydrolytic endo- and exoribonucleases (Rrp44 and Rrp6) that catalyze processive and distributive 3′-to-5′ exoribonuclease activities as well as endoribonuclease activities.

The RNA exosome is essential in budding yeast and its subunit composition is largely conserved from yeast to human [3–5]. In yeast,

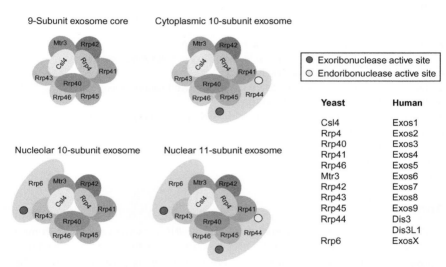

Figure 3.1 Schematics of eukaryotic exosomes. Cartoon schematics depicting subunit compositions and general architecture of the 9-subunit exosome core (upper left), the cytoplasmic 10-subunit exosome (upper right), the nucleolar 10-subunit exosome (lower left), and nuclear 11-subunit exosome (lower right). Subunits are labeled and color coded and include the PH-like ring subunits Mtr3 (orange), Rrp42 (red), Rrp41 (purple), Rrp45 (blue), Rrp46 (green), and Rrp43 (yellow); the S1/KH-domain proteins Csl4 (light blue), Rrp4 (green), and Rrp40 (pink); the catalytic subunits Rrp44 (gray) and Rrp6 (gray-blue). The S1/KH protein ring is shown on the top of the PH-like ring with Rrp44 shown below the PH-like ring to reflect structural models of the complex. Rrp6 is shown on the other side of the complex below the PH-like ring, although there is no definitive structural data for this complex. The exoribonuclease active sites are depicted by red circles in Rrp44 and Rrp6, and the endoribonuclease active site of Rrp44 is depicted with a yellow circle. The names of subunits are indicated on the right under headings for yeast and human although human proteins are often referred to by the corresponding yeast nomenclature. (See color plate section in the back of the book.)

nuclear RNA exosomes include the nine-subunit exosome core, Rrp44, and Rrp6, while cytoplasmic exosomes appear to include only the exosome core and Rrp44 [3,6]. Interestingly, human encodes two Rrp44 homologs as well as Rrp6, and while each can associate with the RNA exosome core, they exhibit distinct subcellular localizations [7]. This observation suggests that RNA exosome subunit composition may be dynamically regulated or that distinct exosomes exist for distinct functions. For instance, subcellular localization patterns suggest the existence of a nucleolar human exosome that includes the RNA exosome core and Rrp6.

Components of the eukaryotic RNA exosome share evolutionary relationships to bacterial and archaeal factors that catalyze 3′–5′ RNA decay. In bacteria, 3′-to-5′ RNA decay is catalyzed by RNase II and RNase R, two processive hydrolytic enzymes that share similarity to Rrp44; RNase D, a distributive hydrolytic enzyme that shares similarity to Rrp6; and PNPase, a processive phosphorolytic exoribonuclease that shares similarities to the noncatalytic human RNA exosome core [8,9]. PNPase is a multidomain protein that homooligomerizes as a trimer to form a two-ring structure that features a prominent central channel. The top ring is formed by S1/KH domains, while the bottom ring is formed by PH domains that harbor the 3′-to-5′ processive phosphorolytic active sites. RNA must pass through the central channel to enter the phosphorolytic chamber [10].

Archaeal exosomes are also processive phosphorolytic enzymes [11] but are composed of up to four individually encoded proteins that oligomerize to form a two-ring structure analogous to PNPase. In this case, six PH-domain subunits form the bottom ring while three S1/KH-domain proteins form the top ring [12,13]. No RNase II or RNase D family members have yet been identified in archaea [14,15]. Analogous to PNPase, archaeal exosome rings possess a central channel that guides RNA substrates into the phosphorolytic chamber and active sites [16].

The eukaryotic RNA exosome core is structurally related to bacterial PNPase and archaeal exosomes, although it is composed of nine distinct subunits. Furthermore, unlike PNPase and archaeal exosomes, the nine-subunit eukaryotic exosome core is devoid of catalytic or phosphorolytic activities [17,18]. Therefore, it seems that the RNA exosome core has diverged mechanistically from its bacterial and archaeal cousins, dropping phosphorolytic catalytic capacity in its core in favor of interactions with the hydrolytic endo- and exoribonuclease Rrp44 and the hydrolytic exoribonuclease Rrp6 as well as protein cofactors such as the TRAMP and SKI complexes [2], which presumably add additional layers of regulation to 3′-to-5′ decay pathways in eukaryotes. Structural comparisons between the eukaryotic RNA exosome core and enzymes from bacteria and archaea have been extensively discussed elsewhere [19], so this chapter focuses on the structure and functions of the eukaryotic exosome core; its associated ribonucleases, Rrp44 and Rrp6; and their role in forming catalytically competent cytoplasmic, nuclear, and nucleolar exosomes.

2. GLOBAL ARCHITECTURE OF THE EUKARYOTIC EXOSOME CORE

The structure of the human nine-subunit exosome core (Exo9) revealed a pseudohexameric six-component ring composed of the RNase PH-like proteins Rrp41, Rrp45, Rrp42, Rrp43, Mtr3, and Rrp46 that is capped by a three-component ring formed by the S1/KH-domain proteins Csl4, Rrp4, and Rrp40 (Fig. 3.2; [17]). This structure revealed overall architectural similarities to bacterial PNPase [21–23] and archaeal exosomes [12,16] including a prominent central channel. In the human exosome structure, Rrp4 bridges Rrp41 and Rrp42, Rrp40 bridges Rrp45 and Rrp46, and Csl4 contacts Mtr3 and, to a lesser extent, Rrp43. While archaeal exosomes form stable and catalytically active six-subunit RNase PH-subunit rings, eukaryotic exosomes require at least one cap protein to form stable complexes *in vitro*. The general architecture and subunit composition of the human Exo9 core is likely conserved across eukaryotic phylogeny based on sequence analysis and conservation of individual subunits in organisms ranging from budding yeast to man [17]. Furthermore, some human exosome core subunits can complement deletion of the corresponding yeast genes [24,25].

3. RNase PH-LIKE DOMAINS COMPRISE A PH-LIKE RING IN EUKARYOTIC EXOSOMES

RNase PH domains comprise a βαβα-fold and are conserved in RNase PH, and PNPase in prokaryotes, archaeal exosomes, and eukaryotic exosomes (Fig. 3.3). Exosome subunits with structural homology to RNase PH are thus referred to as "PH-like" proteins and include Rrp41, Rrp42, Rrp43, Rrp45, Rrp46, and Mtr3. The six-component PH-like ring in eukaryotes consists of three distinct heterodimer pairs that are arranged in a head-to-tail configuration. Although eukaryotic organisms have diverged

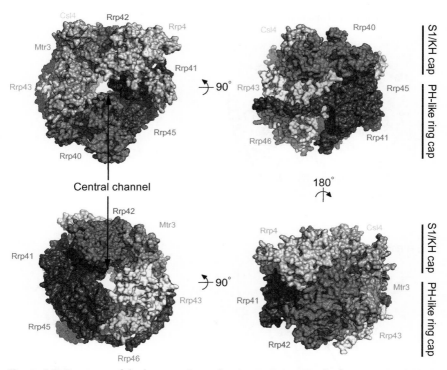

Figure 3.2 Structure of the human nine-subunit exosome core. Surface representations of the human exosome core (PDB 2NN6) from the top (upper left) and bottom (lower left) in addition to two orthogonal views depicting the structure from side views to enable visualization of each exosome subunit in the complex. The subunits are labeled and color coded as in Fig. 3.1 showing Mtr3 (orange), Rrp42 (red), Rrp41 (purple), Rrp45 (blue), Rrp46 (green), and Rrp43 (yellow), and the S1/KH-domain proteins Csl4 (light blue), Rrp4 (green), and Rrp40 (pink). The central channel is apparent in the left panels and indicated by a label and arrows. The positions of the S1/KH cap and PH-like ring are indicated in the right panels with lines and labels. All structure depictions generated with the program PyMol [20]. (See color plate section in the back of the book.)

to encode these subunits in six distinct genes, Rrp41, Mtr3, and Rrp46 share higher sequence and structural similarity to archaeal Rrp41 or PNPase RNase PH 2-like proteins while Rrp42, Rrp43, and Rrp45 share greater similarity to archaeal Rrp42 or PNPase RNase PH 1-like proteins. In eukaryotes, Rrp41 pairs with Rrp45, Rrp43 pairs with Rrp46, and Mtr3 pairs with Rrp42 (Fig. 3.3).

The majority of amino acid side chains that form the phosphorolytic active site and RNA-binding surfaces in bacterial PNPase and the archaeal exosome are not conserved in eukaryotic exosomes; however, amino acid

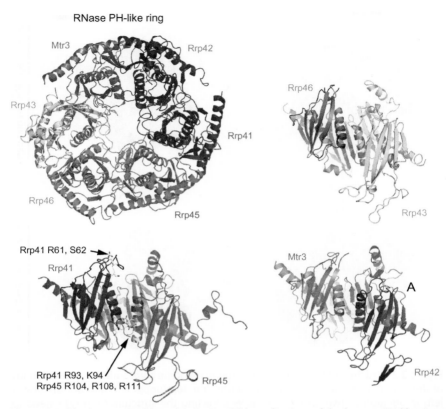

Figure 3.3 The human PH-like subunit ring. Ribbon diagram of the human PH-like ring (PDB 2NN6) depicting β strands as arrows, α helices as coiled ribbons, and connecting elements as thin tubes (upper left). The view of the intact PH-like ring is from the "top" as presented in Figs. 3.1 and 3.2 with subunits labeled and color coded with Mtr3 (orange), Rrp42 (red), Rrp41 (purple), Rrp45 (blue), Rrp46 (green), and Rrp43 (yellow). The Rrp41/Rrp45 heterodimer is shown lower left in a side view as if the viewer were inside the central channel looking outward. The subunits are labeled and color coded as before with amino acid side chains implicated in RNA-binding interactions labeled and side chains colored yellow for Rrp41 Arg61, Ser62, Arg93, and Lys94 and Rrp45 Arg104, Arg108, and Arg111. The Rrp46/Rrp43 and Mtr3/Rrp42 heterodimers are shown in a similar orientation with subunits labeled and color coded as before. (See color plate section in the back of the book.)

side chains that comprise one potential RNA-binding surface are conserved in the Rrp41–Rrp45 heterodimer interface (Fig. 3.2). The location of this putative RNA-binding site was evident in sequence alignments and by location of a tungstate ion in crystals of the human exosome core [17]. Furthermore, point mutations that disrupt this anion-binding site (Rrp41 R95E/R96E and Rrp41 K62E/S63D) attenuate RNA decay activity of cytoplasmic exosomes (Exo9 core plus Rrp44) *in vitro* [26]. These residues are not required *in vivo* as these and other point mutations in the central channel and RNA-binding surface do not lead to a growth defect in *Saccharomyces cerevisiae* [18,27]. Given the structural similarities between bacterial and archaeal exosomes, it remains possible that additional residues within the PH-like ring contribute to RNA binding as several basic residues not involved in intersubunit contacts face the central channel and are conserved across eukaryotes [17].

4. S1 AND KH DOMAINS CAP THE PH-LIKE RING

Rrp4, Rrp40, and Csl4 form a ring and cap the PH-like ring of the eukaryotic exosome core. Each of these subunits contain an N-terminal domain (NTD) that makes extensive contacts to their respective RNase PH-like 2 binding partner, tethering the cap proteins onto the PH-like ring (Fig. 3.4; [17]). The Rrp4 NTD interacts with Rrp41, the Rrp40 NTD binds to Rrp46, and the Csl4 NTD contacts Mtr3. Rrp4 and Rrp40 include two putative RNA-binding domains, KH type I and S1, while Csl4 contains an S1 domain and C-terminal zinc ribbon fold. KH type I domains [29] feature a $\beta1-\alpha1-\alpha2-\beta2-\beta3-\alpha3$ secondary structure topology and tertiary structure that consists of three β strands that form a sheet and pack against three α helices. Single-stranded RNA (ssRNA) typically binds KH type I domains via surfaces formed by residues within helix $\alpha1$, a conserved GXXG motif between helices $\alpha1$ and $\alpha2$; helix $\alpha2$, the variable loop between strands $\beta2$ and $\beta3$; and residues within strand $\beta2$ [30]. However, this motif is not conserved in eukaryotic Rrp4 and Rrp40; instead, a unique and conserved GXNG motif is found between strands $\beta7$ and $\beta8$. The GXNG motif is buried at the interface between the S1 and KH domains, and unless RNA binding induces a dramatic conformational change to expose the GXNG motif, this surface likely contributes to the structural stability of the protein rather than to RNA binding.

S1 domains [31] contain an OB (oligonucleotide/oligosaccharide binding) fold with a five-stranded β-sheet coiled to form a closed β-barrel [32]. OB domains generally bind nucleic acid through surfaces composed of positively charged and hydrophobic residues on the solvent exposed β-sheet [33]. Rrp40 exhibits the typical β-barrel, with long loops between $\beta3$ and $\beta4$ that project into the central channel [17,34]. A comparably long loop between $\beta3$ and $\beta4$ is also present in Rrp4; however, both of these regions are disordered in the structure of the human exosome. These loops project toward the central channel and contain several basic

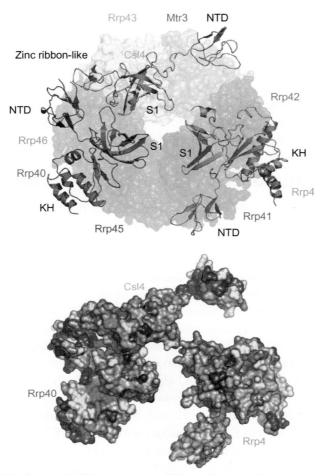

Figure 3.4 The human S1/KH protein ring. Ribbon diagram of the human S1/KH proteins on top of a transparent surface representation of the PH-like ring shown from the "top" with subunits labeled and color coded as in previous figures with Csl4 (light blue), Rrp4 (green), and Rrp40 (pink) (top). The N-terminal domain (NTD), S1, KH, and zinc ribbon-like domains are labeled. Note that the S1 domains from each S1/KH protein face the central channel. The bottom panel depicts just the S1/KH proteins in surface representation in the same orientation as in the top panel indicating sequence conservation colored from red (variable) to blue (conserved) as calculated by ConSurf [28] from manually assembled sequence alignments [17]. Note that the most conserved surfaces are located on the S1 domains that face the central channel. (See color plate section in the back of the book.)

residues that are highly conserved throughout eukaryotes and thus may be responsible for binding or guiding RNA into the central channel.

The zinc ribbon domain of Csl4 contains a three-stranded β-sheet and is structurally related to archaeal Csl4; however, the four cysteine residues that coordinate Zn^{2+} in archaea are not conserved in eukaryotes, nor is Zn^{2+} coordinated in human Csl4. While this domain in not essential *in vivo*, its deletion results in conditional defects in nonsense-mediated decay in budding yeast [35]. Furthermore, while Csl4 is essential *in vivo*, budding yeast strains expressing Csl4 truncations that lack either the NTD or S1 domain are still viable. These data are perhaps consistent with the observation that Csl4 is not stably associated with exosomes reconstituted *in vitro* [36]. In contrast, each of the Rrp4 and Rrp40 NTD, S1, and KH domains is essential in budding yeast [35].

The arrangement of the cap subunits on the PH-like six-subunit ring in the human exosome positions conserved and positively charged putative RNA-binding S1-domain surfaces toward the central channel, while the NTDs and KH domains are located on the periphery of the complex (Fig. 3.4). While RNA binding cannot be detected *in vitro* for Rrp40 [34], reconstituted exosomes lacking one cap subunit exhibit weaker RNase protection patterns compared to a complete Exo9 core [36]. Alignment of multiple crystal structures of the archaeal exosome from *Sulfolobus solfataricus* indicates that the cap proteins, specifically Rrp4, sometimes deviates from the threefold symmetry observed in other crystal structures, suggesting limited conformational flexibility [13,37]. Furthermore, Csl4 and Rrp40, and not the PH-like subunits, display increased flexibility when the crystal structure of the human exosome core is compared to a 14 Å cryo-electron microscopy (EM) reconstruction of reconstituted Exo10 from budding yeast [17,36]. It remains unclear if these differences are biologically relevant or if they simply reflect subtle differences between human and yeast RNA exosomes.

5. RRP44, A HYDROLYTIC ENDORIBONUCLEASE AND PROCESSIVE EXORIBONUCLEASE

Rrp44 includes five domains, an N-terminal Pilus-forming N-terminus (PIN) domain that contains an active site capable of endoribonuclease activity [38], two cold-shock domains (CSD1 and CSD2), a central ribonuclease domain (RNB) that catalyzes processive 3′-to-5′ exoribonuclease activity, and a C-terminal S1 domain (Fig. 3.5). Humans and other higher eukaryotes

A

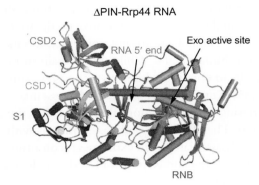

B

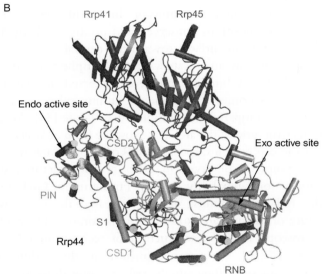

Figure 3.5 The endo- and exoribonuclease Rrp44. (A) Schematic representation of the Rrp44 polypeptide indicating the PIN, CSD1, CSD2, RNB, and S1 domains with labels and color coded pink, green, orange, blue, and purple, respectively. Amino acid numbering is for *Saccharomyces cerevisiae* Rrp44. Below the schematic is the structure of yeast Rrp44 bound to RNA (PDB 2VNU) in cartoon ribbon representation with β strands as arrows, loops as ribbons, and helices as solid tubes. The domains are labeled and color coded as in the schematic. The exoribonuclease active site is indicated by a label and arrow with key residues colored green in stick and surface representation. RNA is shown as a red ribbon with bases indicated as sticks with the 5′ end labeled and indicated by an arrow; the 3′ end is buried in the exoribonuclease active site. (B) Structure of full-length yeast Rrp44 in complex with yeast Rrp41 and Rrp45 (PDB 2WP8) in cartoon ribbon representation color coded as in the top panel with the PIN domain in pink. The endoribonuclease active site is labeled and indicated by an arrow with key residues colored yellow in stick and surface representation. (See color plate section in the back of the book.)

encode three homologs of budding yeast Rrp44, including Dis3 and Dis3L, which are localized to the nucleus and cytoplasm, respectively [7,39]. The third paralog, Dis3L2, lacks a PIN domain and has not been extensively characterized to date. Human Dis3 is most similar to yeast Rrp44, and budding yeast strains depleted of Rrp44 can be partially rescued by human Dis3, but not Dis3L [7]. Dis3 and Dis3L associate with exosomes *in vivo* as evidenced by their ability to pull down other exosome core subunits as well as Mtr4 in lysate from human cells [7,39].

Rrp44 is structurally and mechanistically related to bacterial RNase II and RNase R [40], and a crystal structure of budding yeast Rrp44 lacking the PIN domain determined in complex with RNA [41] revealed many overall similarities to RNase II in apo- and RNA-bound states (Fig. 3.5; [42]) including conservation of CSD1, CSD2, RNB, and S1 domains. Comparison of the structures reveals that Rrp44 and RNase II RNB active sites make similar contacts to the four ribonucleotides upstream of the 3′ OH. Based on structural similarities within their active sites, it is likely that Rrp44 employs the same two–metal–ion catalytic mechanism as RNase II to hydrolyze RNA 3′-to-5′, resulting in the release of 5′ nucleotide monophosphates [19]. One of two catalytic magnesium ions is visible in the Rrp44 structure and is coordinated by Asp543 and Asp552. Although the second magnesium is not detected in the Rrp44–RNA structure, it is likely to be coordinated by Asp551 and Asp549. In fact, mutation of Asp551 to Asn retains RNA binding but leads to loss of exoribonuclease activity *in vitro* and a slow growth phenotype *in vivo* [18].

Several basic residues within the RNB domain line the active site and contact RNA via contacts to the RNA backbone. An exception is Tyr595, which participates in base stacking interactions with the 3′ most nucleotide, while bases 1–5 stack with each other [41]. The prominence of protein contacts to the RNA backbone and absence of base-specific interactions are consistent with the enzyme's lack of sequence specificity; however, several direct or water-mediated hydrogen bonds between protein residues and the 2′ hydroxyl groups of nucleotides 2, 3, 4, 6, and 8 impart specificity in recognition of ssRNA rather than ssDNA as Rrp44 binds ssRNA 50-fold better than ssDNA [41]. Rrp44 can degrade duplex RNAs with 3′ single-stranded overhangs as short as four nucleotides but is most efficient with overhangs of at least 10 nucleotides [41], similar to bacterial RNase R [43]. The requirement of a 3′ single-stranded overhang is consistent with the structures insofar as the channel to the RNB active site is only large enough to accommodate ssRNA.

Many similarities are shared between Rrp44 and RNase II; nevertheless, differences are readily apparent with respect to how they engage RNA via the CSDs and S1 domain. In RNase II, RNA threads through a channel formed by the CSDs and S1 domain into the RNB active site. In addition to contacts to the 3′ end via the RNB domain, CSD2 and the S1 domain make additional contacts to ssRNA that contribute to the enzyme's processive activity [42]. In contrast, the Rrp44–RNA complex shows the 5′ end of RNA in a path perpendicular to that observed in RNase II in a channel formed at the interface between the RNB domain and CSD1 which is 15 Å closer to the RNB domain compared to RNase II. The mode of RNA binding by RNase II explains its processive activities, while the binding mode observed for Rrp44 may facilitate alternative routes for RNA ingress including substrates exiting the central channel of the eukaryotic exosome core. Although it is unclear if the S1/KH-domain ring of the eukaryotic Exo9 core functions in a manner analogous to the CSDs and S1 domain in RNase II with respect to ssRNA binding, it is interesting to note that the S1/KH ring and CSDs/S1 ring share overall architectural similarities, particularly with respect to their positions over an extended channel in the Rrp44 RNB domain and the exosome PH-like ring.

The structure of the yeast Rrp44 PIN domain was revealed in a crystal structure of full-length Rrp44 bound to Rrp41 and Rrp45 (Fig. 3.5; [26]). The PIN domain consists of a central twisted five-stranded β-sheet flanked by α helices and is similar to the overall fold observed in other RNase H family members [44,45]. PIN domains that catalyze nuclease activity include four conserved acidic residues that coordinate two divalent cations to cleave nucleic acid via a similar mechanism involving two-metal-ion catalysis [46–49]. The endoribonucleolytic active site of the Rrp44 PIN domain from budding yeast consists of Asp91, Glu120, Asp171, and Asp198 (Fig. 3.5).

Rrp44 endoribonuclease activity is dependent on the integrity of the PIN active site and can be observed *in vitro* in the presence of millimolar manganese concentrations; however, an active site point mutation (D171N) that abolishes endonuclease activity *in vitro* does not display a growth defect *in vivo* [38]. These observations make the biological function of the endoribonuclease activity unclear, but it is important to note that a combination of mutations (D551N/D171N) that simultaneously disrupt exoribonuclease and endoribonuclease activities results in synthetic lethality [35,38]. The four acidic residues are conserved in human Dis3 but not in Dis3L, as it only retains two of the four acidic residues. Consistent with this observation, only Dis3 appears catalytically competent for endoribonucleolytic function [7]. In addition to its catalytic activities, the PIN domain interacts directly with Rrp41/Rrp45 heterodimer as exemplified in the structure of Rrp44 in complex with Rrp41 and Rrp45 (Fig. 3.5). While endoribonuclease activity is not essential, deletion of the PIN domain is lethal in budding yeast suggesting that contacts to Rrp41/Rrp45 and presumably the exosome core are essential for growth [35].

6. RRP44 AND THE 10-COMPONENT EXOSOME

Models of the eukaryotic 10-component cytoplasmic exosome Exo10^{44} have been proposed based on the X-ray structure of the human nine-component exosome core [17], the crystal structure of the budding yeast Rrp41-Rrp45-Rrp44 trimer (Fig. 3.5; [26]), and negative-stain and cryo-EM structures of apo budding yeast Rrp44 bound to the core exosome (Exo10^{44}) [36,50]. These models reveal that Rrp44 is anchored to the bottom of exosome core through extensive interactions between the PIN domain of Rrp44 and Rrp41/Rrp45 in addition to contacts between the Rrp44 CSD1 and Rrp43 (Fig. 3.6).

How does RNA engage the activities of Rrp44 when Rrp44 is associated with the exosome core? Biochemical studies revealed that charge-swap mutations in Rrp41 and the central channel diminished Rrp44 activity in the presence of the exosome core [26]. Furthermore, a 12-Å resolution cryo-EM structure of gold-labeled RNA bound to a catalytically inactive Exo10 from budding yeast showed the gold label to be located in the center of the exosome coincident with the location of the conserved channel and additional density not present in apo reconstructions was observed within the channel of RNA-bound structures [36]. These data are consistent with the hypothesis that RNA transits through the exosome core and central channel to engage the Rrp44 exoribonuclease active site.

The path of RNA to the Rrp44 exoribonuclease active site can be further explored through alignment of crystal structures of human Exo9 [17], yeast Rrp44 in complex with RNA [41], and yeast Rrp44–Rrp41–Rrp45 [26]. In this model, RNA could transit through and exit the central channel of the exosome to reach Rrp44 yet would require a ~45° turn around CSD1 before entering the RNB channel and active site (Fig. 3.6). While this path is consistent with RNase protection assays that indicate RNA substrates require at least 31–34 single-stranded nucleotides at the 3′ end to be

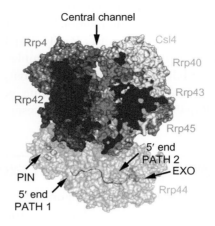

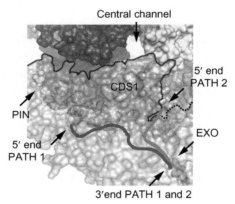

Figure 3.6 Model of the 10-subunit cytoplasmic exosome. The top panel depicts a side view of the 10-subunit exosome in surface representation that was constructed by aligning Rrp41/Rrp45 from the yeast Rrp41/Rrp45/Rrp44 and human nine-subunit exosome core structures. To enable visualization of the exosome core central channel, the Mtr3 subunit was removed from the complex and parts of Rrp42 and Rrp43 were removed from view (dark areas). The subunits are colored as before except Rrp4 is now depicted in gray. The central channel is labeled and indicated by an arrow at the top of the channel. Rrp44 is depicted in transparent gray with endoribonuclease (yellow) and exoribonuclease (green) sites labeled with side chains in surface representation. Two RNA paths are depicted in Rrp44, one derived from the structure of RNase II in complex with RNA (blue; PDB 2IX1) which passes through the CSDs and S1 domain and the other derived from the dPIN-Rrp44/RNA complex (yellow; PDB 2NVU) which passes by CSD1 and the RNB domain. Bottom panel shows an orthogonal view looking up into the exosome central channel indicating the position of CDS1 which appears to block a direct path from the exosome central channel to the RNB active site (green). RNA paths, PIN and EXO active sites are labeled as in the top panel. (See color plate section in the back of the book.)

engaged by a 10-component exosome [26], it is also conceivable that RNA binding and/or additional protein–protein contacts between Rrp44 and Exo9 might induce conformational changes to facilitate a more direct RNA path through the exosome central channel to the Rrp44 RNB active site.

Although this model is attractive, the importance of the central channel for exosome function *in vivo* remains unclear as channel lining mutations exhibit no apparent phenotype [18,27] and because models of Rrp44 in complex with Exo9 indicate that both endoribonuclease and exoribonuclease active sites remain exposed to solvent (Fig. 3.6). These observations suggest that the channel is not essential *in vivo* or that alternative routes exist for RNA to access the Rrp44 active sites. In fact, recent biochemical studies revealed that the exosome core modulates both Rrp44 endo- and exoribonuclease activities because both activities are attenuated in Rrp44-associated 10-subunit exosomes [27]. Furthermore, Rrp44 endo- and exoribonuclease activities are dependent on the integrity of the central channel as channel-occluding mutations severely diminish both RNase activities and binding to RNA. Existing models for Exo10^{44} cannot fully explain these observations, and further structural work will be required to understand how the exosome core regulates access to both Rrp44 exo- and endoribonuclease active sites.

7. RRP6, A EUKARYOTIC EXOSOME SUBUNIT WITH DISTRIBUTIVE HYDROLYTIC ACTIVITIES

Rrp6 is associated with the nuclear exosome (Exo11$^{44/6}$) in budding yeast and humans, although recent evidence suggests the existence of a nucleolar exosome in human cells consisting of Rrp6 and the Exo9 core which we denote Exo10^{6} [7]. Rrp6 is the only nonessential subunit of the exosome; however, budding and fission yeast strains lacking Rrp6 exhibit a temperature-sensitive growth phenotype and accumulate many nuclear RNA precursors [51,52]. Rrp6 is involved in 3' end processing of snRNAs, snoRNAs [52,53], pre-rRNAs [54], destruction of aberrant nuclear RNAs, and the degradation of cryptic unstable transcripts that result from bidirectional transcription [55,56]. In fission yeast, Rrp6 and the nuclear exosome cooperate with the RITS complex to induce constitutive heterochromatin spreading at centromeres [57,58] and to silence meiotic genes in vegetative cells [59,60]. Evidence for exosome-independent functions of Rrp6 have been reported in budding yeast, but these findings are restricted to nuclear RNAs [61,62]. In other species, including trypanosomes [63], Drosophila [5], and humans [7,64], Rrp6 has been detected in the cytoplasm, although the biological implications of these findings have not been fully characterized.

Rrp6 is composed of a PMC2NT domain, a NTD, a DEDD-Y exoribonuclease domain (EXO), a helicase and RNase D carboxy terminal (HRDC) domain, and a C-terminal domain (CTD). The PMC2NT domain is required for interaction with its nuclear cofactor, Rrp47 [65]. The EXO of Rrp6 is related to RNase D and members of the DEDD-Y nuclease family that are so named for four conserved acidic residues, DEDD, and a conserved tyrosine that coordinate two metals to catalyze distributive, hydrolytic $3'$-to-$5'$ exoribonuclease activity via a two–metal–ion mechanism [66]. HRDC domains are posited to bind nucleic acid. While it remains unclear if the Rrp6 HRDC binds RNA, deletion of this domain in human Rrp6 results in diminished catalytic activity [67]. The CTD contains a nuclear localization sequence and is not necessary for interaction with the exosome [61]. At present, no structural information exists for the PMC2NT and CTD domains and both domains are predicted to be unstructured.

Crystal structures of Rrp6 fragments that include the NTD, EXO, and HRDC domains were determined from budding yeast and human (Fig. 3.7; [66,67]). The NTD wraps around the EXO domain and forms a platform with a linker that connects the EXO and HRDC domains. The EXO core shares the α/β-fold observed in the Klenow fragment of DNA polymerase I from *Escherichia coli* [68] as well as conserved DEDD residues in the EXO domain that are required for coordination of the divalent metal ions for two–metal–ion catalysis. In Rrp6, the conserved tyrosine side chain activates a nucleophilic water molecule for cleavage of the phosphodiester. Mutating any of the conserved active site residues either abolishes or severely attenuates exoribonuclease activity *in vitro* [69]. The human and yeast Rrp6 structures show the EXO domain and active site to be solvent exposed, thus potentially explaining Rrp6 distributive activity as these structures lack channels that could bind or guide ssRNA into the active site. The HRDC domain resembles the first HRDC domain of *E. coli* RNase D [70] and consists of five α helices. The HRDC confers substrate specificity, as disruption of the EXO/HRDC interface in budding yeast Rrp6 (D457A) results in deficiencies in $3'$ end processing of nuclear RNAs, such as snRNAs, but not in the clearance of the $5'$ ETS fragment of pre-rRNA and maturation of 5.8S rRNA [71]. Although

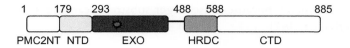

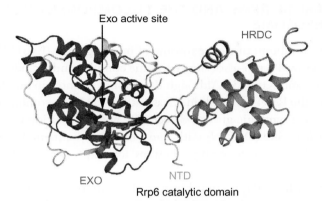

Rrp6 catalytic domain

Figure 3.7 The exoribonuclease Rrp6. Schematic of the Rrp6 polypeptide indicating the PMC2NT (white), NTD (light blue), EXO (dark blue), HRDC (purple), and C-terminal (CTD; white) domains with the exoribonuclease active site colored red. Amino acid numbering is for human Rrp6. Lower panel depicts a cartoon ribbon representation of the human Rrp6 catalytic domain structure with α helices in cartoon ribbon, loops as thin ribbons, and β strands as arrows (PDB 3SAF). Domains are colored and labeled as in the schematic with the linker between the EXO and HRDC domains in gray. The EXO active site is labeled and indicated with an arrow with key side chains colored red and shown in stick representation; the magnesium ion is shown as a small green sphere. (See color plate section in the back of the book.)

Rrp6 fragments containing the NTD/EXO/HRDC domains exhibit similar catalytic activities when compared to full-length Rrp6, it is worth noting that Rrp6 fragments that include the NTD, EXO, and HRDC do not complement growth defects observed in budding yeast strains lacking *RRP6* [66].

Although the structures of the catalytic domains from budding yeast and human share many similarities throughout the EXO and HRDC domains, the human Rrp6 active site appears more solvent exposed when compared to the active site in yeast Rrp6 [67]. This structural difference is attributed to a difference in linker length between the EXO and HRDC domains—in yeast and lower eukaryotes, the linker is 26 residues, but in humans, the linker is only 10 residues long [66,67]. The net effect of a shorter linker is a more solvent exposed human Rrp6 active site that is able to degrade structured RNAs more efficiently than yeast Rrp6 *in vitro*, presumably because these larger substrates can access the human Rrp6 active site [67].

8. RRP44, RRP6, AND THE 11-COMPONENT NUCLEAR EXOSOME

The 11-subunit nuclear exosome of budding yeast $(Exo11^{44/6})$ is the best characterized Rrp6-associated exosome complex and includes the Exo9 core, Rrp44, and Rrp6 [3]. Although Rrp6 is not essential in budding yeast, mutation of the Rrp44 endoribonuclease site (D171N) in combination with *rrp6Δ* leads to a synthetic growth defect [38], and inactivation of the Rrp44 exoribonuclease site (D551N) with *rrp6Δ* results in synthetic lethality [18]. These data suggest overlapping functions for Rrp6 and Rrp44 activities *in vivo*.

Insights to how Rrp6 contributes to the activities of the exosome are beginning to emerge. As described above, association of Rrp44 with the Exo9 core attenuates the RNA binding and exoribonuclease activities of Rrp44 in $Exo10^{44}$ *in vitro* in a manner dependent on the exosome core and central channel [27]. Interestingly, addition of Rrp6 in $Exo11^{44/6}$ stimulates the endoribonuclease and endoribonuclease activities of Rrp44, independent of Rrp6 catalytic activity, and addition of Rrp44 that contains an inactivating mutation in the EXO severely inhibits Rrp6 in $Exo11^{44exo-/6}$ [27]. These observations suggest that the exosome core and central channel mediate a dynamic interplay between Rrp44 and Rrp6 activities.

How does Rrp6 interact with the exosome? Structural details regarding this issue remain unknown; however, yeast two-hybrid studies suggest that Rrp6 interacts with the PH-like ring proteins Rrp41, Rrp43, Rrp46, and Mtr3 [72], and a 35-Å resolution negative-stain EM structure of the *Leishmania tarentolae* exosome purified from native sources positions Rrp6 toward the top of exosome core interacting with the cap proteins [73]. It is unknown if the Rrp6-binding surface is conserved from trypanosomes to humans. Also in question is whether Exo11 is the sole nuclear exosome in humans, as the endogenous stoichiometry of human Dis3 to Rrp6 was roughly estimated to be 1:10 in human cells [7]. Thus, it remains unclear if Rrp6-associated exosome cores represent a predominant species or if free Rrp6 is abundant in human cells. Reconstitution of Rrp6 associated 10- or 11-subunit nuclear exosomes from human or other higher eukaryote has not been reported to date. Further studies are required to address how the exosome core modulates Rrp6 activity and how Rrp6 stimulates Rrp44 ribonuclease activities. Structural studies addressing the aforementioned points could provide key insights to the biological functions of nuclear and nucleolar exosomes.

9. CONCLUSIONS

Structures and models for the eukaryotic exosome core illustrate striking architectural similarities to RNA-degrading enzymes in bacteria and archaea with respect to PH-like and S1/KH-domain rings that stack to form a prominent central channel wide enough to accommodate ssRNA substrates. In bacterial PNPase and archaeal exosomes, this channel harbors both RNA-binding surfaces and phosphorolytic active sites that confer processivity to these complexes by providing at least two RNA-binding surfaces that prevent RNA substrates from diffusing away from the complex between successive rounds of cleavage.

Eukaryotic exosomes appear to use the same strategy to engage RNA substrates by utilizing a noncatalytic core to bind and guide ssRNA substrates through the central channel to engage the RNase activities of Rrp44 and Rrp6, at least for the yeast exosome [27]. It remains unknown if this feature is conserved among all eukaryotic exosomes or if all RNA substrates engage the exosome core in a similar manner. Although it is clear that the exosome core can modulate the activities of Rrp44 and Rrp6, very little is known regarding how factors such as the TRAMP and SKI complexes change or affect substrate specificity or the activities of the RNA exosome. While much has been learned since discovery of the eukaryotic exosome, it will remain a significant challenge to determine the structural basis for eukaryotic exosome RNase activities as well as its association with cofactors that modulate its nuclear and cytoplasmic functions in RNA processing and decay.

ACKNOWLEDGMENTS

Research reported in this publication was supported by the National Institute of General Medical Sciences of the National Institutes of Health under award numbers F31GM097910 (E.V.W) and R01GM079196 (C.D.L). The content is solely the responsibility of the authors and does not necessarily represent the official views of the National Institutes of Health.

REFERENCES

[1] Mitchell P, Petfalski E, Tollervey D. The 3′ end of yeast 5.8S rRNA is generated by an exonuclease processing mechanism. Genes Dev 1996;10:502–13.

[2] Houseley J, LaCava J, Tollervey D. RNA-quality control by the exosome. Nat Rev Mol Cell Biol 2006;7:529–39.

[3] Allmang C, Petfalski E, Podtelejnikov A, Mann M, Tollervey D, Mitchell P. The yeast exosome and human PM-Scl are related complexes of 3′ → 5′ exonucleases. Genes Dev 1999;13:2148–58.

[4] Schneider C, Anderson JT, Tollervey D. The exosome subunit Rrp44 plays a direct role in RNA substrate recognition. Mol Cell 2007;27:324–31.

[5] Graham AC, Kiss DL, Andrulis ED. Differential distribution of exosome subunits at the nuclear lamina and in cytoplasmic foci. Mol Biol Cell 2006;17:1399–409.

[6] Mitchell P, Petfalski E, Shevchenko A, Mann M, Tollervey D. The exosome: a conserved eukaryotic RNA processing complex containing multiple 3′ → 5′ exoribonucleases. Cell 1997;91:457–66.

[7] Tomecki R, Kristiansen MS, Lykke-Andersen S, Chlebowski A, Larsen KM, Szczesny RJ, et al. The human core exosome interacts with differentially localized processive RNases: hDIS3 and hDIS3L. EMBO J 2010;29:2342–57.

[8] Carpousis AJ. The Escherichia coli RNA degradosome: structure, function and relationship in other ribonucleolytic multienzyme complexes. Biochem Soc Trans 2002;30:150–5.

[9] Symmons MF, Jones GH, Luisi BF. A duplicated fold is the structural basis for polynucleotide phosphorylase catalytic activity, processivity, and regulation. Structure 2000;8:1215–26.

[10] Nurmohamed S, Vaidialingam B, Callaghan AJ, Luisi BF. Crystal structure of Escherichia coli polynucleotide phosphorylase core bound to RNase E, RNA and manganese: implications for catalytic mechanism and RNA degradosome assembly. J Mol Biol 2009;389:17–33.

[11] Evguenieva-Hackenberg E, Walter P, Hochleitner E, Lottspeich F, Klug G. An exosome-like complex in Sulfolobus solfataricus. EMBO Rep 2003;4:889–93.

[12] Lorentzen E, Walter P, Fribourg S, Evguenieva-Hackenberg E, Klug G, Conti E. The archaeal exosome core is a hexameric ring structure with three catalytic subunits. Nat Struct Mol Biol 2005;12:575–81.

[13] Büttner K, Wenig K, Hopfner KP. Structural framework for the mechanism of archaeal exosomes in RNA processing. Mol Cell 2005;20:461–71.

[14] Mian IS. Comparative sequence analysis of ribonucleases HII, III, II PH and D. Nucleic Acids Res 1997;25:3187–95.

[15] Zuo Y, Deutscher P. Exoribonuclease superfamilies: structural analysis and phylogenetic distribution. Nucleic Acids Res 2001;29:1017–26.

[16] Lorentzen E, Dziembowski A, Lindner D, Seraphin B, Conti E. RNA channelling by the archaeal exosome. EMBO Rep 2007;8:470–6.

[17] Liu Q, Greimann JC, Lima CD. Reconstitution, activities, and structure of the eukaryotic RNA exosome. Cell 2006;127:1223–37.

[18] Dziembowski A, Lorentzen E, Conti E, Séraphin B. A single subunit, Dis3, is essentially responsible for yeast exosome core activity. Nat Struct Mol Biol 2007;14:15–22.

[19] Januszyk K, Lima CD. Jensen TH, editor. Structural Components and architectures of RNA exosomes, vol. 702. New York: Landes Bioscience and Springer Science; 2010. p. 9–28.

[20] Delano WL. The PyMOL molecular graphics system, San Carlos, CA: DeLano Scientific; 2002. http://www.pymol.org.

[21] Ishii R, Nureki O, Yokoyama S. Crystal structure of the tRNA processing enzyme RNase PH from Aquifex aeolicus. J Biol Chem 2003;278:32397–404.

[22] Harlow LS, Kadziola A, Jensen KF, Larsen S. Crystal structure of the phosphorolytic exoribonuclease RNase PH from Bacillus subtilis and implications for its quaternary structure and tRNA binding. Protein Sci 2004;13:668–77.

[23] Shi Z, Yang WZ, Lin-Chao S, Chak KF, Yuan HS. Crystal structure of Escherichia coli PNPase: central channel residues are involved in processive RNA degradation. RNA 2008;14:2361–71.

[24] Brouwer R, Allmang C, Raijmakers R, van Aarssen Y, Egberts WV, Petfalski E, et al. Three novel components of the human exosome. J Biol Chem 2001;276: 6177–6184.

[25] Mitchell P, Tollervey D. Musing on the structural organization of the exosome complex. Nat Struct Mol Biol 2000;7:843–6.

[26] Bonneau F, Basquin J, Ebert J, Lorentzen E, Conti E. The yeast exosome functions as a macromolecular cage to channel RNA substrates for degradation. Cell 2009;139:547–59.

[27] Wasmuth EV, Lima CD. Exo- and Endoribonucleolytic Activities of Yeast Cytoplasmic and Nuclear RNA Exosomes Are Dependent on the Noncatalytic Core and Central Channel, Molecular Cell 2012;http://dx.doi.org/10.1016/j.molcel.2012.07.012.

[28] Landau M, Mayrose I, Rosenberg Y, Glaser F, Martz E, Pupko T, et al. ConSurf 2005: the projection of evolutionary conservation scores of residues on protein structures. Nucleic Acids Res 2005;33:W299–302.

[29] Siomi H, Matunis MJ, Michael WM, Dreyfuss G. The pre-mRNA binding K protein contains a novel evolutionarily conserved motif. Nucleic Acids Res 1993;21:1193–8.

[30] Valverde R, Edwards L, Regan L. Structure and function of KH domains. FEBS J 2008;275:2712–26.

[31] Subramanian AR. Structure and functions of ribosomal protein S1. Prog Nucleic Acid Res Mol Biol 1983;28:101–42.

[32] Worbs M, Bourenkov GP, Bartunik HD, Huber R, Wahl MC. An extended RNA binding surface through arrayed S1 and KH domains in transcription factor NusA. Mol Cell 2001;7:1177–89.

[33] Schubert M, Edge RE, Lario P, Cook MA, Strynadka NC, Mackie GA, et al. Structural characterization of the RNase E S1 domain and identification of its oligonucleotide-binding and dimerization interfaces. J Mol Biol 2004;341:37–54.

[34] Oddone A, Lorentzen E, Basquin J, Gasch A, Rybin V, Conti E, et al. Structural and biochemical characterization of the yeast exosome component Rrp40. EMBO Rep 2007;8:63–9.

[35] Schaeffer D, Tsanova B, Barbas A, Reis FP, Dastidar EG, Sanchez-Rotunno M, et al. The exosome contains domains with specific endoribonuclease, exoribonuclease and cytoplasmic mRNA decay activities. Nat Struct Mol Biol 2009;16:56–62.

[36] Malet H, Topf M, Clare DK, Ebert J, Bonneau F, Basquin J, et al. RNA channelling by the eukaryotic exosome. EMBO Rep 2010;11:936–42.

[37] Lu C, Ding F, Ke A. Crystal structure of the S. solfataricus archaeal exosome reveals conformational flexibility in the RNA-binding ring. PLoS One 2010;5:e8739.

[38] Lebreton A, Tomecki R, Dziembowski A, Séraphin B. Endonucleolytic RNA cleavage by a eukaryotic exosome. Nature 2008;456:993–6.

[39] Staals RH, Bronkhorst AW, Schilders G, Slomovic S, Schuster G, Heck AJ, et al. Dis3-like 1: a novel exoribonuclease associated with the human exosome. EMBO J 2010;29:2358–67.

[40] Cheng ZF, Deutscher MP. Purification and characterization of the Escherichia coli exoribonuclease RNase R. Comparison with RNase II. J Biol Chem 2002;277:21624–9.

[41] Lorentzen E, Basquin J, Tomecki R, Dziembowski A, Conti E. Structure of the active subunit of the yeast exosome core, Rrp44: diverse modes of substrate recruitment in the RNase II nuclease family. Mol Cell 2008;29:717–28.

[42] Frazão C, McVey CE, Amblar M, Barbas A, Vonrhein C, Arraiano CM, et al. Unravelling the dynamics of RNA degradation by ribonuclease II and its RNA-bound complex. Nature 2006;443:110–4.

[43] Vincent HA, Deutscher MP. Insights into how RNase R degrades structured RNA: analysis of the nuclease domain. J Mol Biol 2009;387:570–83.

[44] Arcus VL, Bäckbro K, Roos A, Daniel EL, Baker EN. Distant structural homology leads to the functional characterization of an archaeal PIN domain as an exonuclease. J Biol Chem 2004;279:16471–8.

[45] Nowotny M, Gaidamakov SA, Crouch RJ, Yang W. Crystal structures of RNase H bound to an RNA/DNA hybrid: substrate specificity and metal-dependent catalysis. Cell 2005;121:1005–16.

[46] Steitz TA, Steitz JA. A general two-metal-ion mechanism for catalytic RNA. Proc Natl Acad Sci USA 1993;90:6498–502.

[47] De Vivo M, Dal Peraro M, Klein ML. Phosphodiester cleavage in ribonuclease H occurs via an associative two-metal-aided catalytic mechanism. J Am Chem Soc 2008;130:10955–62.

[48] Huntzinger E, Kashima I, Fauser M, Saulière J, Izaurralde E. SMG6 is the catalytic endonuclease that cleaves mRNAs containing nonsense codons in metazoan. RNA 2008;14:2609–17.

[49] Eberle AB, Lykke-Andersen S, Mühlemann O, Jensen TH. SMG6 promotes endonucleolytic cleavage of nonsense mRNA in human cells. Nat Struct Mol Biol 2009;16:49–55.

[50] Wang HW, Wang J, Ding F, Callahan K, Bratkowski MA, Butler JS, et al. Architecture of the yeast Rrp44 exosome complex suggests routes of RNA recruitment for 3′ end processing. Proc Natl Acad Sci USA 2007;104:16844–9.

[51] Kim DU, Hayles J, Kim D, Wood V, Park HO, Won M, et al. Analysis of a genome-wide set of gene deletions in the fission yeast Schizosaccharomyces pombe. Nat Biotechnol 2010;28:617–23.

[52] Allmang C, Kufel J, Chanfreau G, Mitchell P, Petfalski E, Tollervey D. Functions of the exosome in rRNA, snoRNA and snRNA synthesis. EMBO J 1999;18:5399–410.

[53] van Hoof A, Lennertz P, Parker R. Yeast exosome mutants accumulate 3′-extended polyadenylated forms of U4 small nuclear RNA and small nucleolar RNAs. Mol Cell Biol 2000;20:441–52.

[54] Allmang C, Mitchell P, Petfalski E, Tollervey D. Degradation of ribosomal RNA precursors by the exosome. Nucleic Acids Res 2000;28:1684–91.

[55] Neil H, Malabat C, d'Aubenton-Carafa Y, Xu Z, Steinmetz LM, Jacquier A. Widespread bidirectional promoters are the major source of cryptic transcripts in yeast. Nature 2009;457:1038–42.

[56] Wyers F, Rougemaille M, Badis G, Rousselle JC, Dufour ME, Boulay J, et al. Cryptic pol II transcripts are degraded by a nuclear quality control pathway involving a new poly (A)polymerase. Cell 2005;121:725–37.

[57] Bühler M, Haas W, Gygi SP, Moazed D. RNAi-dependent and -independent RNA turnover mechanisms contribute to heterochromatic gene silencing. Cell 2007;129:707–21.

[58] Reyes-Turcu FE, Zhang K, Zofall M, Chen E, Grewal SI. Defects in RNA quality control factors reveal RNAi-independent nucleation of heterochromatin. Nat Struct Mol Biol 2011;18:1132–8.

[59] Harigaya Y, Tanaka H, Yamanaka S, Tanaka K, Watanabe Y, Tsutsumi C, et al. Selective elimination of messenger RNA prevents an incidence of untimely meiosis. Nature 2006;442:45–50.

[60] Zofall M, Yamanaka S, Reyes-Turcu FE, Zhang K, Rubin C, Grewal SI. RNA elimination machinery targeting meiotic mRNAs promotes facultative heterochromatin formation. Science 2012;335:96–100.

[61] Callahan KP, Butler JS. Evidence for core exosome independent function of the nuclear exoribonuclease Rrp6p. Nucleic Acids Res 2008;36:6645–55.

[62] Callahan KP, Butler JS. TRAMP complex enhances RNA degradation by the nuclear exosome component Rrp6. J Biol Chem 2010;285:3540–7.

[63] Haile S, Cristodero M, Clayton C, Estévez AM. The subcellular localisation of trypanosome RRP6 and its association with the exosome. Mol Biochem Parasitol 2007;151:52–8.

[64] Lejeune F, Li X, Maquat LE. Nonsense-mediated mRNA decay in mammalian cells involves decapping, deadenylating, and exonucleolytic activities. Mol Cell 2003;12:675–87.

[65] Stead JA, Costello JL, Livingstone MJ, Mitchell P. The PMC2NT domain of the catalytic exosome subunit Rrp6p provides the interface for binding with its cofactor Rrp47p, a nucleic acid-binding protein. Nucleic Acids Res 2007;35:5556–67.

[66] Midtgaard SF, Assenholt J, Jonstrup AT, Van LB, Jensen TH, Brodersen DE. Structure of the nuclear exosome component Rrp6p reveals an interplay between the active site and the HRDC domain. Proc Natl Acad Sci USA 2006;103:11898–903.

[67] Januszyk K, Liu Q, Lima CD. Activities of human RRP6 and structure of the human RRP6 catalytic domain. RNA 2011;17:1566–77.

[68] Ollis DL, Brick P, Hamlin R, Xuong NG, Steitz TA. Structure of large fragment of Escherichia coli DNA polymerase I complexed with dTMP. Nature 1985;313:762–6.

[69] Assenholt J, Mouaikel J, Andersen KR, Brodersen DE, Libri D, Jensen TH. Exonucleolysis is required for nuclear mRNA quality control in yeast THO mutants. RNA 2008;14:2305–13.

[70] Zuo Y, Wang Y, Malhotra A. Crystal structure of Escherichia coli RNase D, an exoribonuclease involved in structured RNA processing. Structure 2005;13:973–84.

[71] Phillips S, Butler JS. Contribution of domain structure to the RNA 3′ end processing and degradation functions of the nuclear exosome subunit Rrp6p. RNA 2003;9:1098–107.

[72] Lehner B, Sanderson CM. A protein interaction framework for human mRNA degradation. Genome Res 2004;14:1315–23.

[73] Cristodero M, Böttcher B, Diepholz M, Scheffzek K, Clayton C. The Leishmania tarentolae exosome: purification and structural analysis by electron microscopy. Mol Biochem Parasitol 2008;159:24–9.

CHAPTER FOUR

TRAMP Stimulation of Exosome

Peter Holub, Stepanka Vanacova

CEITEC—Central European Institute of Technology, Masaryk University, Brno, 625 00, Czech Republic

Contents

1.	TRAMP Is a Cofactor of Nuclear Exosome	78
2.	RNA Substrate Repertoire of TRAMP	79
	2.1 Quality control of ribosomes and the role of TRAMP on rDNA loci	82
	2.2 Surveillance of stable ncRNAs	82
	2.3 Turnover of unstable noncoding RNAs	83
3.	TRAMP Biochemistry and Structure	84
	3.1 Structure of the TRAMP complex	84
	3.2 RNA-binding properties of TRAMP subunits	86
	3.3 Enzymatic activities of the TRAMP complex	87
4.	TRAMP Complex in Activation of the Exosome	88
5.	TRAMP Complexes in Different Organisms	89
	Acknowledgments	90
	References	90

Abstract

In order to control and/or enhance the specificity and activity of nuclear surveillance and degradation, exosomes cooperate with the polyadenylation complex called TRAMP. Two forms of TRAMP operate in budding yeast, TRAMP4 and TRAMP5. They oligoadenylate defective or precursor forms of RNAs and promote trimming or complete degradation by exosomes. TRAMPs target a wide variety of nuclear transcripts. The known substrates include the noncoding RNAs originating from pervasive transcription from diverse parts of the yeast genome. Although TRAMP and exosomes can be triggered to a subset of their targets via the RNA-binding complex Nrd1, it is still not completely understood how TRAMP recognizes other aberrant RNAs. The existence of TRAMP-like complexes in other organisms indicates the importance of nuclear surveillance for general cell biology. In this chapter, we review the current understanding of TRAMP function and substrate repertoire. We discuss the advances in TRAMP biochemistry with respect to its catalytic activities and RNA recognition. Finally, we speculate about the possible mechanisms by which TRAMP activates exosomes.

1. TRAMP IS A COFACTOR OF NUCLEAR EXOSOME

Exosomes act on extremely diverse spectrum of RNAs in nucleus and cytoplasm. Their activity has to be tightly regulated in order to prevent unspecific RNA trimming or degradation. Eukaryotes have evolved specialized cofactors that modulate exosome activity and specificity. In budding yeast, the activity of nuclear exosome is promoted and regulated by at least two complexes, the *TRf–Air–Mtr4* Polyadenylation complex (TRAMP) and the Nrd1–Nab3–Sen1 RNA-binding complex [1–4]. The TRAMP complex modifies target RNAs with short 3'-end poly(A) tails and activates the exosome *in vivo* and *in vitro* [1,2,5,6], Fig. 4.1. This is an opposite function to the role of canonical nuclear polyadenylation of mRNAs that promotes mRNA stability, nucleo-cytoplasmic export, and translation. Polyadenylation is a universal modification involved in degradation and quality control of bacterial RNAs implying that the role in mRNA stabilization is evolutionarily younger. For instance, in *Escherichia coli*, poly(A) polymerase (PAP) or PNPase add poly(A) tails to stimulate RNA degradation by RNaseR or PNPase itself [7–9]. It has been speculated that the mRNA stabilization function evolved together with the compartmentalization of a eukaryotic cell [2,10].

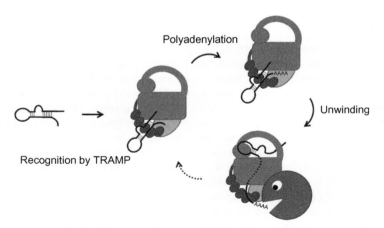

Figure 4.1 Schematic view on the TRAMP composition and mechanism of exosome activation. TRAMP recognizes and binds aberrant RNAs via Air1/2p zinc knuckle domains (in blue), Trf4/5p (in orange) add short poly(A) tails, secondary structures are resolved by Mtr4p helicase (in green), and the tailed RNAs are targeted to exosome (in purple) for degradation. RNAs undergo several cycles of oligoadenylation and partial trimming before they are completely digested. (See color plate section in the back of the book.)

The tails added by TRAMP could range from a couple of AMPs to a stretch of couple of hundred (A)s *in vitro* and *in vivo* [2,11,12]. However, the optimal and most represented length *in vivo* is between 3 and 5 nt [11,12] (see Section 3.3). Such tails are too short to bind the canonical poly(A)-binding protein Pab1p, which typically stabilizes polyadenylated mRNAs and requires a minimal length of 12 (As) to bind [13]. The TRAMP-added tails thus remain unprotected and provide a landing platform for exosomal degradation.

At least two distinct TRAMP complexes exist in budding yeast, TRAMP4, made up of three proteins: a poly(A) polymerase Trf4p, a zinc knuckle (ZnK) protein Air2p, and the DExH-box RNA helicase Mtr4p, and TRAMP5, which contains Trf5p, Air1p, and Mtr4p [1–3,14]. Although Trf4/5p possess the conserved nucleotidyl transferase motif, the PAP activity is observed only in complex with Air1/2 proteins [2]. In contrast to the canonical poly(A) polymerase Pap1p, Trf4/5p lack any recognizable RNA-binding domain. The RNA binding is instead mediated by Air1/2p that were shown to bind RNA *in vitro* [15] (Section 3.2). Airs possess dual roles within TRAMP; in addition to their RNA binding, these ZnK proteins mediate interaction between Mtr4p and Trf4p [15,16].

Mtr4p is a DExH-box RNA helicase [17,18] with multiple functions. It modulates PAP activity of TRAMP [11] and is needed for TRAMP-mediated exosome activation *in vitro* [1,2,19]. Although Mtr4p is an integral component of TRAMP, it can also be copurified with exosomes [1] and it apparently interacts also with the deadenylation complex Ccr4-Not [20]. Due to its higher expression levels compared to the rest of the TRAMP, it has been speculated that Mtr4p may have additional TRAMP-independent functions [21]. In contrast to the rest of the TRAMP, *MTR4* is essential for viability [22] and homologues in other species reveal high degree of sequence conservation [23]. In addition, it is a part of two independent exosome-activating complexes in humans [24], suggesting that Mtr4p is the key subunit in the exosome activation pathway (see Section 5).

2. RNA SUBSTRATE REPERTOIRE OF TRAMP

The TRAMP–exosome surveillance pathway targets a wide range of RNA species produced by all three RNA polymerases. Whereas only few mRNA targets have been studied [25–27], most TRAMP substrates represent non–protein coding RNAs such as precursors of rRNAs, tRNAs, snRNAs, snoRNAs, as well as stable and unstable long noncoding

RNAs, antisense RNAs, and RNAs originating from heterochromatic regions [1,3,5,6,12,21,27–31], see Table 4.1 Transcriptome-wide analyses of strains lacking *TRF4* or *TRF5* indicate their partially overlapping functions [29]. Whereas TRAMP4 acts primarily in the nucleoplasm, TRAMP5 operates in the nucleolus [14,30]. It is still not fully understood as to what extent TRAMP4 and TRAMP5 substitute each other. Overexpression of Trf5p can suppress the phenotype of *trf4Δ* strains on some TRAMP4 substrates [21,30]. Double deletion of TRF4 and TRF5 is lethal [41], however expression of catalytically inactive *TRF4* mutant can partially rescue this lethality [3] and suppress degradation and processing phenotypes of a subset of TRAMP targets [3,27,29]. Interestingly, deletion of both RNA-binding subunits *air1Δ* and *air2Δ* is viable at 30 °C but exhibits mRNA export defects [42] which may correlate with the mRNA export defects observed in strains lacking Rrp6p.

In spite of the extensive knowledge on the diversity of RNAs targeted by TRAMPs, it is still unclear how TRAMP identifies its substrates. Bioinformatics analyses of RNA targets identifed either by microarrays of RNAs stabilized upon *TRF4* or *TRF5* deletion or by sequencing analysis of RNAs crosslinked to Trf4p failed to identify any specific binding motifs [3,12,29]. In some cases, the complex preferentially targets misfolded RNAs such as tRNAs [2,12,29]; however, this likely represents only a small portion of TRAMP targets. As most of the TRAMP RNA substrates form stable or transient small ribonucleoprotein particle (RNP) complexes or larger RNPs such as pre-ribosomes [1,3,30,31,43], it is likely that TRAMP–exosome pathway monitors the dynamics of RNP assembly competing with proper protein binders. If faulty or unstable RNPs are encountered, TRAMP could compete out bound proteins, label RNAs by oligo(A) tail, and recruit exosomes for their destruction.

The Nrd1p–Nab3p–Sen1p complex recruits TRAMP and exosome to a subset of their targets via specific sequence motifs on substrate RNAs [4,12,35,36,44–47]. The Nrd1 complex is involved in transcription termination and processing of snRNAs, snoRNAs, and certain mRNAs [26,34–36,48,49]. TRAMP and Nrd1 cooperate on a wide spectrum of RNAs such as cryptic unstable transcripts (CUTs), pre-tRNAs, snoRNAs, antisense RNAs (asRNAs), and a panel of mRNAs [26,27,32,35,36]. The Nrd1 complex is also able to activate TRAMP–exosome activity *in vitro* on RNA substrates containing specific Nrd1p–Nab3p sequences [4]. Although, the Nrd1 complex is able to copurify TRAMP and exosome from yeast extracts [4,21], the nature of interaction between the three complexes is currently unknown.

Table 4.1 Yeast TRAMP RNA substrates

RNA polymerase	RNA type	RNA substrate	Reference
Pol I	Pre–rRNA		[1,6,12,14,21,22,29,30]
Pol II	mRNA	*HSP104, CTH2, NRD1, NAB2,* histone mRNAs	[12,25–27,29,32,33]
	Pre-snoRNA	*U14, U24, sn*R40, etc.	[3,4,12,14,28–31,34]
	Pre-snRNA	*U3, U4,*	[12,29,31]
	mRNA introns		[29]
	CUTs	Best characterized *NEL025C*	[3,12,28,35,36]
	asRNAs	*GAL10as, CAF17as, DBP2as, MAL32as, PCH2as HPF1as,* etc.	[12,21,29,37,38]
	Telomeric ncRNA		[12,21,29,37]
	Centromeric ncRNA		[12,21,29]
	rDNA intergenic ncRNA	*IGS1-R*	[12,21]
	RNase P RNA component	*RPR1*	[12]
	Ty1 mRNA		[29]
	Telomerase RNA	*TLC1*	[12,29,39]
Pol III	tRNA, pre-tRNAs		[2,5,6,12,29,40]
	5S rRNA		[6,14]

TRAMP plays a role in the surveillance of RNAs produced by all three yeast nuclear RNA polymerases. Summary of TRAMP RNA targets in *Saccharomyces cerevisiae* with references to the original works.

2.1. Quality control of ribosomes and the role of TRAMP on rDNA loci

The ribosome biogenesis involves a number of processing steps with diverse pre-rRNA intermediates. Exosome together with TRAMP plays a crucial role in several trimming and degradation events, preventing the assembly of faulty pre-ribosomes [50]. Strains lacking Rrp6p display accumulation of rRNA precursors that are polyadenylated by TRAMP [1,51] and Mtr4p depletion leads to hyperadenylation of large subunit (LSU) rRNA precursors [30] which had been polyadenylated by Trf4p and/or Trf5p, but inefficiently degraded because of the absence of exosome recruitment provided by Mtr4p [30]. Because not all rRNA processing defects of Mtr4p depletion are observed in *trf4* and *trf5* deletion strains, Mtr4p can probably also act independently of the rest of the TRAMP [1,22,30]. In strains defective in nuclear export of ribosomal subunits, incorrectly processed forms LSU rRNA assembled in 60S pre-ribosomal subunits concentrate to subnucleolar structures called No-body, where aberrant particles are identified and destroyed [30]. This represents the first experimental example of TRAMP-mediated surveillance of whole RNP complexes.

TRF4 and *TRF5* were originally identified in a synthetic lethal screen with *top1* mutations, where double mutants showed rDNA condensation phenotypes and mitotic segregation defects [41,52–54]. Although the synthetic lethality with *top1* was not reproduced in the more recent work of Houseley et al. [21], mutants in Trf4p and exosome subunits displayed instability at rDNA loci [21,55]. TRAMP appears to be recruited to rDNA via ncRNA IGS-1R [21] where it has yet unidentified role. Because TRAMP and exosome cofactor mutants are hypersensitive to DNA damage [56,57], it was speculated that these two factors also help to promote DNA repair [21,58].

2.2. Surveillance of stable ncRNAs

The best-characterized example of a TRAMP–exosome pathway target is the hypomodifed $tRNA_i^{Met}$. Strains lacking specific tRNA 1-methyladenosine58 methyltransferase activity produce $tRNA_i^{Met}$ lacking methylation at A58 residue, which is highly unstable at high temperatures [5]. The hypomethylated $tRNA_i^{Met}$ is polyadenylated by TRAMP and subsequently degraded by the exosome *in vivo* and *in vitro* [2,5,6,59]. The $tRNA_i^{Met}$ displays unique structural features among other tRNAs where

A58 methylation is key to establish the correct tertiary contacts [60]. It was predicted that the lacking modification destabilizes the tRNA tertiary contacts, making it susceptible to action of both TRAMP and the key exosome nuclease Rrp44p [2,59,60]. Similar effect was observed for tRNA$^{Arg(ACG)}$ that fails to fold properly even in wild type cells and thus becomes a regular target for Nrd1–TRAMP–exosome pathway [12]. In addition to misfolded or hypomodified mature tRNAs, The TRAMP–exosome degradation pathway is responsible for the elimination of a spectrum of abundant or misprocessed pre-tRNAs [12,29,40]. tRNAs are typically transcribed with 5′ and 3′ trailers; some contain introns and a subset of tRNAs undergoes substitutional editing at the anticodon loop. All these forms can be bound by the Nrd1 complex, which in turn recruits the TRAMP–exosome machinery for their destruction [12].

Nrd1p, TRAMP, and the exosome also cooperate in the 3′-end processing of several snRNAs and snoRNA precursors [4,31,32,34–36,44]. Observed removal of unprocessed snRNAs, degradation of spliced introns [29], and genetic and physical interaction with spliceosomal factors [61–63] offer an attractive hypothesis that the preassembled or recycled spliceosomal RNPs undergo quality control by TRAMP and the exosome before they enter the pre-mRNA splicing process.

Another interesting phenotype observed in TRAMP and exosome mutants is the reversible telomere shortening upon deletion of *TRF4* or *RRP6* [29]. It is not currently known whether this is a consequence of the reported accumulation and processing defects of the telomerase RNA component TLC1, or stabilization of cryptic telomeric transcripts [12,29,39].

Eukaryotic genomes are populated with retroviral elements whose expression and transposition have to be strictly regulated in order to prevent their uncontrolled expansion. In yeast, the most abundant type is the Ty1 family that sporadically undergoes retrotransposition in wild type laboratory strains [64]. *TRF4* and *TRF5* deletion causes an accumulation of Ty1 transcripts, however it remains unclear whether the nuclear RNA degradation pathway is directly involved in the control of Ty1 retrotransposition [29].

2.3. Turnover of unstable noncoding RNAs

Nrd1–TRAMP–exosome pathway seems to promote nuclear turnover of a large number of mRNAs and unstable ncRNAs, which has been interpreted as an alternative way by which budding yeast may tone gene expression [12].

Transcriptome-wide studies in wild type as well as exosome mutant cells uncovered the existence of pervasive or hidden transcription in yeast as well as in mammals [65–71]. It was initially believed that pervasive transcription represents a by-product of transcriptionally active chromatin. However, cryptic transcription also frequently occurs at heterochromatic loci such as telomeres and centromeres [12,21,29]. It appears that at least some of these ncRNAs can modulate gene expression of adjacent protein-coding genes; thus, the production and stability of such "sterile" transcripts need to be tightly controlled.

The largest proportion of hidden transcription in yeast is represented by the so-called CUTs [3]. They originate from almost as many transcription units as there are mRNA coding genes [3,28,66,68,72]. These RNAs typically initiate from bidirectional promoters associated with protein-coding genes or from intergenic regions, and in some cases they overlap mRNA coding genes, either in the sense or antisense orientation [36,66,68]. CUTs contain a number of specific sequences that are recognized by Nrd1p—Nab3p heterodimer [35], which recruits TRAMP to add poly(A) tails and induces CUT degradation by exosome.

In addition to CUTs, TRAMP–exosome controls stability of yet another type of ncRNAs which are oriented antisense (as) to mRNAs [12]. Whereas the function of CUTs in gene regulation still remains speculative, several examples show that asRNAs regulate gene expression of adjacent sense-oriented protein-coding genes [12,37,38,73]. It has been proposed that asRNAs have the potential to recruit chromatin modification factors such as histone deacetylases which then negatively regulate transcription on specific gene loci [38]. Moreover, there is growing evidence that links the production of cryptic or antisense transcripts to the role in chromatin organization. It was proposed that unstable ncRNAs recruit the surveillance machinery to DNA to stimulate chromatin modifications and remodeling. However, mechanistic details of TRAMP or exosome role in this process are currently uknown.

3. TRAMP BIOCHEMISTRY AND STRUCTURE

3.1. Structure of the TRAMP complex

The structure of the assembled TRAMP complex has not been solved yet. The main obstacle is the poor behavior of recombinant subunits. Neverthe-less, progress has been made in solving structures of individual subunits or their fragments.

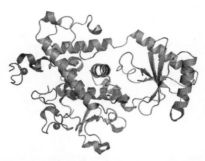

Figure 4.2 Structure of the Trf4p fragment (in green), comprising catalytic and central domains, forming a heterodimer with zinc knuckles four and five of Air2p (in blue). The Trf4p catalytic aspartates coordinating magnesium ions are shown in red, and zinc ions are in gray. The structure shown is of PDB 3NYB [74]. (See color plate section in the back of the book.)

The structure of Trf4p catalytic and central domain fragment together with two out of five ZnKs of Air2p (Fig. 4.2) was solved by Hamill and colleagues [74]. The fold of Trf4p is similar to other noncanonical RNA polymerases from the Polβ superfamily of nucleotidyl transferases. The ZnKs 4 and 5 of Air2p interact with the central domain of Trf4p. The importance of the ZnK 5 and the linker between ZnK 4 and 5 for interaction between Air and Trf4 proteins has been further demonstrated by coimmunoprecipitations from *Saccharomyces cerevisiae* lysates [15,75]. NMR solution structure analysis of Air2p fragment containing all five ZnKs revealed independently folded ZnKs separated by unstructured flexible linkers [15]. It is possible that individual ZnKs have specific roles in either RNA or protein binding.

The most interesting so far has been the elucidation of Mtr4p structure, which has been solved alone and in complex with ADP and RNA [19,76]. Whereas the central ATPase core shows fold typical for DExH helicases, there is a unique structure formed by insertion of 265 amino acids that protrudes from the globular core forming the Arch domain with an arm-like structure and a fist (also termed the KOW domain) at the end (Fig. 4.3). The Arch is dispensable for ATPase and helicase activities of Mtr4p as well as for the whole TRAMP assembly [19,76]. Instead, it binds RNA and is required for exosome activation *in vitro* [15,19] (Section 4).

Despite the lack of knowledge of the complete TRAMP structure, additional information about individual subunits interactions is available from pull-down experiments of various truncation and point mutants [15]. Such analysis revealed the importance of Air2 protein for bridging

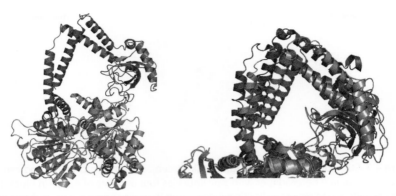

Figure 4.3 Structure of the Mtr4p helicase. Left: The canonical helicase core is shown in green. In red is the unusual insertion forming an Arch-like structure. The globular domain at the tip of the Arch is termed the KOW domain. Right: Alignment of three available crystal structures of Mtr4p suggests flexibility of the arch domain. The structures shown are of PDB 2XGJ [76] and 3L9O [19]. (See color plate section in the back of the book.)

Trf4p and Mtr4p [15,16]. In particular, the N-terminus of Air2p appears critical for Mtr4p binding [15], while the C-terminus including ZnK 5 and adjacent linker between ZnK 4 and 5 are important for interaction with Trf4p [15,74,75].

3.2. RNA-binding properties of TRAMP subunits

The central question is how TRAMP recognizes and binds its extremely diverse spectrum of RNAs. The key factors responsible for RNA binding and specificity are the Air proteins that can bind RNA *in vitro* [15]. NMR titration studies with tRNA$_i^{Met}$ revealed that the ZnKs 2, 3, and 4 are responsible for tRNA$_i^{Met}$ binding [15], whereas ZnK 1 was proposed to bind different substrates by Hamill and colleagues [74]. ZnKs are known RNA-binding domains of viral nucleocapsid proteins [77]; however, structural and sequence alignment analyses indicate that the RNA-binding ZnKs of Air2p use a different binding mode [15]. They lack the hydrophobic pocket, which accommodates guanine base in the viral proteins. Instead, they have large electropositive patches that could be responsible for the recognition of a broad spectrum of RNA substrates based on interactions with the sugar-phosphate backbone [15]. Nevertheless, the exact mechanism of distinguishing between correct and aberrant RNAs remains to be revealed by additional structural and binding studies.

The unique Arch domain of Mtr4p, particularly its middle part, the KOW domain, represents the second identified RNA-binding domain within the TRAMP complex [19,76]. KOW domain is dispensable for yeast viability, however strains expressing Arch-less or KOW-less Mtr4p have severe growth defects [15,19]. The second RNA-binding region of Mtr4p resides within the helicase core [76], which displays a marked preference for binding to poly(A) RNA relative to an oligoribonucleotide of the same length or a random sequence [17,18].

Unlike the canonical poly(A) polymerase Pap1p, Trf4p has no distinct RNA-binding domain. However, the human homologues PAPD5 and PAPD7 display PAP activity alone [78,79], with the C-terminal region of PAPD5 being responsible for RNA binding [78]. Thus, we cannot exclude the possibility that yet unidentified regions also support RNA binding in yeast Trf4p or Trf5p proteins.

3.3. Enzymatic activities of the TRAMP complex

The TRAMP complex exhibits two key enzymatic activities, poly(A)-polymerase and RNA helicase. Trf4p was proposed to have also exonuclease and dRP-lyase activities [58,80], however these have not been convincingly demonstrated. Trf4p was originally considered to be a DNA polymerase [81], however this was later disproven [82].

The minimal complex required for polyadenylation activity *in vitro* consists of the heterodimer Trf4p–Air2p [2,3,15]. While Trf4p is the poly (A)-polymerase, Air2p provides the RNA binding (Section 3.2). Although highly selective for adenosine triphosphate (ATP), the complex can incorporate low amounts of guanosine monophosphate (GMP) or uridine monophosphate (UMP) *in vivo* and *in vitro* [1,12]. Unlike the canonical poly(A)-polymerase Pap1p, the activity of Trf4p is distributive with approximately two nucleotides per binding event added [2,11].

TRAMP is able to distinguish between native and hypomodified tRNA$_i$[Met] and tRNA[Ala] *in vitro* and *in vivo*, accepting only the latter as a substrate [2,5,6]. Although Air2p ZnKs 2, 3, and 4 were identified as responsible for hypomodified tRNA$_i$[Met] binding, mutations disrupting the fold of individual ZnK 2 or 3 had no effect on the activity of purified mutant TRAMPs [15]. On the other hand, mutation of ZnK 4 and ZnK 5 fully abolishes polyadenylation [15,75]. Whereas ZnK 4 mutation probably disrupts RNA binding, mutants in ZnK 5 destroy interaction with Trf4p revealing the specialization of individual ZnKs for particular roles within the complex [15].

Kinetic experiments on the *in vitro* reconstructed recombinant TRAMP complex show that the polyadenylation activity of Trf4p is, apart from Air2p, also modulated by the interaction with the helicase Mtr4p [11].

While the heterodimer Trf4p/Air2p activity is more or less uniform over 10 nucleotides added, Mtr4p enhances the polyadenylation rate and ATP binding and reduces the dissociation from substrate until approximately 4th A nucleotide added [11]. This is in agreement with the average length of tails observed on Trf4p RNA targets *in vivo* [12].

Similarly, the helicase activity of Mtr4p is modulated by the Trf4p/Air2p heterodimer. The helicase alone shows preference for oligo(A) binding and displays ATP (or dATP) dependent 3′–5′ helicase activity [17]. The helicase activity is stimulated by the Trf4p/Air2p heterodimer even in the absence of polyadenylation activity (i.e., with catalytically dead Trf4p). The activation is achieved via increasing the affinity for ATP and the strand–separation rate constant k_{unw}^{max}, while the affinity for RNA is not changed [83]. When the assembled TRAMP complex encounters the substrate with short overhang (e.g., one nucleotide), it is first polyadenylated by Trf4p until it reaches overhang of five nucleotides (four As added). Such modified substrate becomes accessible for Mtr4p for unwinding by its helicase activity. These results revealed how poly(A)-polymerase and helicase are intrinsically fine-tuned for maximum cooperation [83] and to control the length of poly(A) tails *in vivo* possibly to prevent binding of the stabilizing Pab1p.

4. TRAMP COMPLEX IN ACTIVATION OF THE EXOSOME

The exosome alone is unable to efficiently progress through structured substrates. This leads to low activity observed *in vitro* [1,2] and accumulation of aberrant polyadenylated RNAs after deletion of exosome cofactors *in vivo*. In the nucleus, TRAMP activity can influence two exonuclease components of the exosome, Rrp44p and/or Rrp6p. Several lines of evidence suggested that TRAMP acts mostly in processes linked to Rrp6p function. Accordingly, recombinant Rrp6p shows preference for polyadenylated RNAs *in vitro* [84] and its activity is enhanced in the presence of TRAMP [85]. However, the work of Schneider et al. showed that recombinant Rrp44p could also be activated by TRAMP *in vitro* [59].

The TRAMP-mediated activation is not limited to RNA oligoadenylation but requires the physical presence of the TRAMP complex for processive and efficient RNA degradation *in vitro* [1,2]. Furthermore, TRAMP polyadenylation activity is dispensable for degradation of certain RNAs *in vitro* and *in vivo* [29,85]. It was proposed that TRAMP-mediated A-tailing is mostly needed for degradation of highly structured substrates to provide an unstructured landing platform for exosome channeling [29].

The direct TRAMP–exosome activation is mediated by Mtr4p. The main role of Mtr4p was initially seen in unwinding secondary RNA structures on complicated substrates [1,2] and possibly releasing protein–RNA complexes. However, Mtr4p can act independently of its helicase activity, as ATP is not required for exosome activation on some RNAs *in vitro* [85]. In this respect, the Arch domain of Mtr4p, in particular the RNA-binding KOW domain, was shown to be essential for exosome activation on hypomodified $tRNA_i^{Met}$ substrate *in vitro* [15]. Yeast strains expressing Arch-less Mtr4p show processing defects of both Rrp6p- and Rrp44p-specific substrates. In particular, they accumulate 5.8S + 30 pre-rRNA and 5'-ETS [19]. It remains speculative whether Arch makes direct protein–protein contacts with Rrp6p or other exosome components or whether its role mainly relies on the RNA-binding capacity. We can envisage that, thanks to the arm flexibility of the Arch stem (Fig. 4.3), polyadenylated RNAs emerging from Trf4p/Air2p are bound by KOW that transports them directly to Rrp6p or the exosome.

Weir and colleagues [76] identified a surface near the Mtr4p RNA exit channel, which is conserved not only among different Mtr4 proteins but also in Ski2p helicases [86]. Both Mtr4p and Ski2p cooperate and copurify with exosome core; thus, it is tempting to picture the RNA exit surface of Mtr4p positioned on top of the exosome cap structure RNA entrance.

5. TRAMP COMPLEXES IN DIFFERENT ORGANISMS

The homologues of *S. cerevisiae* TRAMP components can be found across species from yeast to human. Whereas a single highly conserved *MTR4* (*SKIV2L2* in mammals) gene is found in genomes across the eukaryotic kingdom, multiple homologues of noncanonical PAPs and diverse ZnK protein genes exist in different species (reviewed elsewhere [87,88]). Almost identical complex of Mtr4-Air1 and noncanonical PAP Cid14 or Cid12 functions in rRNA processing, turnover of unstable transcripts, chromosome segregation, and heterochromatic gene silencing in *Schizosaccharomyces pombe* [16,62,89–94]. Two *Drosophila melanogaster* homologues of Trf4p (DmTRF4-1 and DmTRF4-2) are localized in the nucleolus and involved in snRNA degradation [95]. It is not yet known whether DmTRF4s form a TRAMP-like complex. Plant AtMtr4 in *Arabidopsis thaliana* is also primarily nucleolar and together with AtRRP6L2 cooperate in several steps of rRNA maturation and surveillance [96].

It appears that increasing complexity of an organism correlates with diversification of RNA degradation/surveillance machines. At least two exosome-activating complexes operate in the mammalian nucleus, the hTRAMP and the Nuclear EXosome Targeting (NEXT) complex [24]. Whereas hMtr4/SKIV2L2 is an integral component of both complexes, only hTRAMP possesses polyadenylation activity [24,78]. Similarly to yeast, the human TRAMP complex contains hMtr4, the noncanonical poly(A) polymerase PAPD5/Trf4-2/TUTase3, and a ZnK protein ZCCHC7 [24,75]. The human TRAMP is localized in the nucleolus and like its yeast counterpart it was implicated in the processing and surveillance of rRNA, snoRNA, and snRNA precursors [63,97–99], histone mRNA degradation [100], and modification of miRNAs [97]. Whereas ZCCHC7 is strictly nucleolar, PAPD5 and hMTR4 are distributed throughout the nucleus [75]. In nucleoplasm, hMtr4/SKIV2L2 forms the NEXT complex with two additional RNA-binding factors ZCCHC8 and RBM7 [24,101]. NEXT cooperates with human exosome on the turnover of pervasive transcripts, such as the PROMoter uPstream Transcripts (PROMPTs) [24,71,102]. In addition, several transcriptome-wide screens uncovered that many types of stable and unstable noncoding RNAs are modified by nontemplated stretch of UMPs or AMPs [12,103]. It is likely that similarly to yeast, human TRAMP or other complexes containing noncanonical PAPs mediate processing and/or degradation of cryptic RNAs. It is tempting to speculate that hTRAMP or NEXT interacts with small RNA pathways to act in chromatin modifications and remodeling. If this was the case, RNA surveillance pathways could be involved in higher order functions such as development, immunity, antiviral defense, and others.

ACKNOWLEDGMENTS

This work was supported by the Wellcome Trust grant (084316/Z/07/Z) to S.V., Czech Science Foundation grant (P305/11/1095), and CEITEC—Central European Institute of Technology grant (CZ.1.05/1.1.00/02.0068) from European Regional Development Fund. PH is in receipt of the Brno City Municipality Scholarship for Talented Ph.D. Students.

REFERENCES

[1] LaCava J, et al. RNA degradation by the exosome is promoted by a nuclear polyadenylation complex. Cell 2005;121:713–24.
[2] Vanacova S, et al. A new yeast poly(A) polymerase complex involved in RNA quality control. PLoS Biol 2005;3:e189.

[3] Wyers F, et al. Cryptic pol II transcripts are degraded by a nuclear quality control pathway involving a new poly(A) polymerase. Cell 2005;121:725–37.

[4] Vasiljeva L, Buratowski S. Nrd1 interacts with the nuclear exosome for 3′ processing of RNA polymerase II transcripts. Mol Cell 2006;21:239–48.

[5] Kadaba S, Krueger A, Trice T, Krecic AM, Hinnebusch AG, Anderson J. Nuclear surveillance and degradation of hypomodified initiator tRNAMet in S. cerevisiae. Genes Dev 2004;18:1227–40.

[6] Kadaba S, Wang X, Anderson JT. Nuclear RNA surveillance in Saccharomyces cerevisiae: Trf4p-dependent polyadenylation of nascent hypomethylated tRNA and an aberrant form of 5S rRNA. RNA 2006;12:508–21.

[7] Xu F, Cohen SN. RNA degradation in Escherichia coli regulated by 3′ adenylation and 5′ phosphorylation. Nature 1995;374:180–3.

[8] Cheng ZF, Deutscher MP. An important role for RNase R in mRNA decay. Mol Cell 2005;17:313–8.

[9] Regnier P, Hajnsdorf E. Poly(A)-assisted RNA decay and modulators of RNA stability. Prog Mol Biol Transl Sci 2009;85:137–85.

[10] Vanacova S, Stefl R. The exosome and RNA quality control in the nucleus. EMBO Rep 2007;8:651–7.

[11] Jia H, et al. The RNA helicase Mtr4p modulates polyadenylation in the TRAMP complex. Cell 2011;145:890–901.

[12] Wlotzka W, Kudla G, Granneman S, Tollervey D. The nuclear RNA polymerase II surveillance system targets polymerase III transcripts. EMBO J 2011;30:1790–803.

[13] Sachs AB, Davis RW, Kornberg RD. A single domain of yeast poly(A)-binding protein is necessary and sufficient for RNA binding and cell viability. Mol Cell Biol 1987;7:3268–76.

[14] Houseley J, Tollervey D. Yeast Trf5p is a nuclear poly(A) polymerase. EMBO Rep 2006;7:205–11.

[15] Holub P, et al. Air2p is critical for the assembly and RNA-binding of the TRAMP complex and the KOW domain of Mtr4p is crucial for exosome activation. Nucleic Acids Res 2012;40:5679–93.

[16] Keller C, Woolcock K, Hess D, Buhler M. Proteomic and functional analysis of the noncanonical poly(A) polymerase Cid14. RNA 2010;16:1124–9.

[17] Bernstein J, Ballin JD, Patterson DN, Wilson GM, Toth EA. Unique properties of the Mtr4p-poly(A) complex suggest a role in substrate targeting. Biochemistry 2010;49:10357–70.

[18] Bernstein J, Patterson DN, Wilson GM, Toth EA. Characterization of the essential activities of Saccharomyces cerevisiae Mtr4p, a 3′→5′ helicase partner of the nuclear exosome. J Biol Chem 2008;283:4930–42.

[19] Jackson RN, Klauer AA, Hintze BJ, Robinson H, van Hoof A, Johnson SJ. The crystal structure of Mtr4 reveals a novel arch domain required for rRNA processing. EMBO J 2010;29:2205–16.

[20] Azzouz N, Panasenko OO, Colau G, Collart MA. The CCR4-NOT complex physically and functionally interacts with TRAMP and the nuclear exosome. PLoS One 2009;4:e6760.

[21] Houseley J, Kotovic K, El Hage A, Tollervey D. Trf4 targets ncRNAs from telomeric and rDNA spacer regions and functions in rDNA copy number control. EMBO J 2007;26:4996–5006.

[22] de la Cruz J, Kressler D, Tollervey D, Linder P. Dob1p (Mtr4p) is a putative ATP-dependent RNA helicase required for the 3′ end formation of 5.8S rRNA in Saccharomyces cerevisiae. EMBO J 1998;17:1128–40.

[23] Houseley J, Tollervey D. The many pathways of RNA degradation. Cell 2009;136:763–76.

[24] Lubas M, et al. Interaction profiling identifies the human nuclear exosome targeting complex. Mol Cell 2011;43:624–37.

[25] Roth KM, Byam J, Fang F, Butler JS. Regulation of NAB2 mRNA 3'-end formation requires the core exosome and the Trf4p component of the TRAMP complex. RNA 2009;15:1045–58.

[26] Ciais D, Bohnsack MT, Tollervey D. The mRNA encoding the yeast ARE-binding protein Cth2 is generated by a novel 3' processing pathway. Nucleic Acids Res 2008;36:3075–84.

[27] Rougemaille M, et al. Dissecting mechanisms of nuclear mRNA surveillance in THO/sub2 complex mutants. EMBO J 2007;26:2317–26.

[28] Davis CA, Ares Jr. M. Accumulation of unstable promoter-associated transcripts upon loss of the nuclear exosome subunit Rrp6p in Saccharomyces cerevisiae. Proc Natl Acad Sci USA 2006;103:3262–7.

[29] San Paolo S, et al. Distinct roles of non-canonical poly(A) polymerases in RNA metabolism. PLoS Genet 2009;5:e1000555.

[30] Dez C, Houseley J, Tollervey D. Surveillance of nuclear-restricted pre-ribosomes within a subnucleolar region of Saccharomyces cerevisiae. EMBO J 2006;25:1534–46.

[31] Egecioglu DE, Henras AK, Chanfreau GF. Contributions of Trf4p- and Trf5p-dependent polyadenylation to the processing and degradative functions of the yeast nuclear exosome. RNA 2006;12:26–32.

[32] Arigo JT, Carroll KL, Ames JM, Corden JL. Regulation of yeast NRD1 expression by premature transcription termination. Mol Cell 2006;21:641–51.

[33] Reis CC, Campbell JL. Contribution of Trf4/5 and the nuclear exosome to genome stability through regulation of histone mRNA levels in Saccharomyces cerevisiae. Genetics 2007;175:993–1010.

[34] Grzechnik P, Kufel J. Polyadenylation linked to transcription termination directs the processing of snoRNA precursors in yeast. Mol Cell 2008;32:247–58.

[35] Thiebaut M, Kisseleva-Romanova E, Rougemaille M, Boulay J, Libri D. Transcription termination and nuclear degradation of cryptic unstable transcripts: a role for the nrd1-nab3 pathway in genome surveillance. Mol Cell 2006;23:853–64.

[36] Arigo JT, Eyler DE, Carroll KL, Corden JL. Termination of cryptic unstable transcripts is directed by yeast RNA-binding proteins Nrd1 and Nab3. Mol Cell 2006;23:841–51.

[37] Pinskaya M, Gourvennec S, Morillon A. H3 lysine 4 di- and tri-methylation deposited by cryptic transcription attenuates promoter activation. EMBO J 2009;28:1697–707.

[38] Camblong J, Iglesias N, Fickentscher C, Dieppois G, Stutz F. Antisense RNA stabilization induces transcriptional gene silencing via histone deacetylation in S. cerevisiae. Cell 2007;131:706–17.

[39] Noel JF, Larose S, Abou Elela S, Wellinger RJ. Budding yeast telomerase RNA transcription termination is dictated by the Nrd1/Nab3 non-coding RNA termination pathway. Nucleic Acids Res 2012;40:5625–36.

[40] Copela LA, Fernandez CF, Sherrer RL, Wolin SL. Competition between the Rex1 exonuclease and the La protein affects both Trf4p-mediated RNA quality control and pre-tRNA maturation. RNA 2008;14:1214–27.

[41] Castano IB, Heath-Pagliuso S, Sadoff BU, Fitzhugh DJ, Christman MF. A novel family of TRF (DNA topoisomerase I-related function) genes required for proper nuclear segregation. Nucleic Acids Res 1996;24:2404–10.

[42] Inoue K, Mizuno T, Wada K, Hagiwara M. Novel RING finger proteins, Air1p and Air2p, interact with Hmt1p and inhibit the arginine methylation of Npl3p. J Biol Chem 2000;275:32793–9.

[43] Dez C, Dlakic M, Tollervey D. Roles of the HEAT repeat proteins Utp10 and Utp20 in 40S ribosome maturation. RNA 2007;13:1516–27.

[44] Vasiljeva L, Kim M, Mutschler H, Buratowski S, Meinhart A. The Nrd1-Nab3-Sen1 termination complex interacts with the Ser5-phosphorylated RNA polymerase II C-terminal domain. Nat Struct Mol Biol 2008;15:795-804.

[45] Carroll KL, Pradhan DA, Granek JA, Clarke ND, Corden JL. Identification of cis elements directing termination of yeast nonpolyadenylated snoRNA transcripts. Mol Cell Biol 2004;24:6241-52.

[46] Carroll KL, Ghirlando R, Ames JM, Corden JL. Interaction of yeast RNA-binding proteins Nrd1 and Nab3 with RNA polymerase II terminator elements. RNA 2007;13:361-73.

[47] Hobor F, et al. Recognition of transcription termination signal by the nuclear polyadenylated RNA-binding (NAB) 3 protein. J Biol Chem 2011;286: 3645-57.

[48] Steinmetz EJ, Conrad NK, Brow DA, Corden JL. RNA-binding protein Nrd1 directs poly(A)-independent 3'-end formation of RNA polymerase II transcripts. Nature 2001;413:327-31.

[49] Rondon AG, Mischo HE, Kawauchi J, Proudfoot NJ. Fail-safe transcriptional termination for protein-coding genes in S. cerevisiae. Mol Cell 2009;36:88-98.

[50] Allmang C, Kufel J, Chanfreau G, Mitchell P, Petfalski E, Tollervey D. Functions of the exosome in rRNA, snoRNA and snRNA synthesis. EMBO J 1999;18: 5399-410.

[51] Kuai L, Fang F, Butler JS, Sherman F. Polyadenylation of rRNA in Saccharomyces cerevisiae. Proc Natl Acad Sci USA 2004;101:8581-6.

[52] Castano IB, Brzoska PM, Sadoff BU, Chen H, Christman MF. Mitotic chromosome condensation in the rDNA requires TRF4 and DNA topoisomerase I in Saccharomyces cerevisiae. Genes Dev 1996;10:2564-76.

[53] Wang Z, Castano IB, Adams C, Vu C, Fitzhugh D, Christman MF. Structure/function analysis of the Saccharomyces cerevisiae Trf4/Pol sigma DNA polymerase. Genetics 2002;160:381-91.

[54] Edwards S, Li CM, Levy DL, Brown J, Snow PM, Campbell JL. Saccharomyces cerevisiae DNA polymerase epsilon and polymerase sigma interact physically and functionally, suggesting a role for polymerase epsilon in sister chromatid cohesion. Mol Cell Biol 2003;23:2733-48.

[55] Wang Z, Castano IB, De Las Penas A, Adams C, Christman MF. Pol kappa: a DNA polymerase required for sister chromatid cohesion. Science 2000;289:774-9.

[56] Walowsky C, Fitzhugh DJ, Castano IB, Ju JY, Levin NA, Christman MF. The topoisomerase-related function gene TRF4 affects cellular sensitivity to the antitumor agent camptothecin. J Biol Chem 1999;274:7302-8.

[57] Hieronymus H, Yu MC, Silver PA. Genome-wide mRNA surveillance is coupled to mRNA export. Genes Dev 2004;18:2652-62.

[58] Gellon L, Carson DR, Carson JP, Demple B. Intrinsic 5'-deoxyribose-5-phosphate lyase activity in Saccharomyces cerevisiae Trf4 protein with a possible role in base excision DNA repair. DNA Repair (Amst) 2008;7:187-98.

[59] Schneider C, Anderson JT, Tollervey D. The exosome subunit Rrp44 plays a direct role in RNA substrate recognition. Mol Cell 2007;27:324-31.

[60] Basavappa R, Sigler PB. The 3 A crystal structure of yeast initiator tRNA: functional implications in initiator/elongator discrimination. EMBO J 1991;10:3105-11.

[61] Hausmann S, et al. Genetic and biochemical analysis of yeast and human cap trimethylguanosine synthase: functional overlap of 2,2,7-trimethylguanosine caps, small nuclear ribonucleoprotein components, pre-mRNA splicing factors, and RNA decay pathways. J Biol Chem 2008;283:31706-18.

[62] Bayne EH, et al. Splicing factors facilitate RNAi-directed silencing in fission yeast. Science 2008;322:602-6.

[63] Nag A, Steitz JA. Tri-snRNP-associated proteins interact with subunits of the TRAMP and nuclear exosome complexes, linking RNA decay and pre-mRNA splicing. RNA Biol 2012;9:334–42.

[64] Curcio MJ, Garfinkel DJ. Single-step selection for Ty1 element retrotransposition. Proc Natl Acad Sci USA 1991;88:936–40.

[65] David L, et al. A high-resolution map of transcription in the yeast genome. Proc Natl Acad Sci USA 2006;103:5320–5.

[66] Neil H, Malabat C, d'Aubenton-Carafa Y, Xu Z, Steinmetz LM, Jacquier A. Widespread bidirectional promoters are the major source of cryptic transcripts in yeast. Nature 2009;457:1038–42.

[67] Churchman LS, Weissman JS. Nascent transcript sequencing visualizes transcription at nucleotide resolution. Nature 2011;469:368–73.

[68] Xu Z, et al. Bidirectional promoters generate pervasive transcription in yeast. Nature 2009;457:1033–7.

[69] Core LJ, Waterfall JJ, Lis JT. Nascent RNA sequencing reveals widespread pausing and divergent initiation at human promoters. Science 2008;322:1845–8.

[70] Seila AC, et al. Divergent transcription from active promoters. Science 2008;322:1849–51.

[71] Preker P, et al. RNA exosome depletion reveals transcription upstream of active human promoters. Science 2008;322:1851–4.

[72] Houalla R, et al. Microarray detection of novel nuclear RNA substrates for the exosome. Yeast 2006;23:439–54.

[73] Houseley J, Tollervey D. The nuclear RNA surveillance machinery: the link between ncRNAs and genome structure in budding yeast? Biochim Biophys Acta 2008;1779: 239–46.

[74] Hamill S, Wolin SL, Reinisch KM. Structure and function of the polymerase core of TRAMP, a RNA surveillance complex. Proc Natl Acad Sci USA 2010;107: 15045–50.

[75] Fasken MB, et al. Air1 zinc knuckles 4 and 5 and a conserved IWRXY motif are critical for the function and integrity of the Trf4/5-Air1/2-Mtr4 polyadenylation (TRAMP) RNA quality control complex. J Biol Chem 2011;286:37429–45.

[76] Weir JR, Bonneau F, Hentschel J, Conti E. Structural analysis reveals the characteristic features of Mtr4, a DExH helicase involved in nuclear RNA processing and surveillance. Proc Natl Acad Sci USA 2010;107:12139–44.

[77] D'Souza V, Summers MF. How retroviruses select their genomes. Nat Rev Microbiol 2005;3:643–55.

[78] Rammelt C, Bilen B, Zavolan M, Keller W. PAPD5, a noncanonical poly(A) polymerase with an unusual RNA-binding motif. RNA 2011;17:1737–46.

[79] Kwak JE, Wickens M. A family of poly(U) polymerases. RNA 2007;13:860–7.

[80] Rogozin IB, Aravind L, Koonin EV. Differential action of natural selection on the N and C-terminal domains of 2'-5' oligoadenylate synthetases and the potential nuclease function of the C-terminal domain. J Mol Biol 2003;326:1449–61.

[81] Wang L, Eckmann CR, Kadyk LC, Wickens M, Kimble J. A regulatory cytoplasmic poly(A) polymerase in Caenorhabditis elegans. Nature 2002;419:312–6.

[82] Haracska L, Johnson RE, Prakash L, Prakash S. Trf4 and Trf5 proteins of Saccharomyces cerevisiae exhibit poly(A) RNA polymerase activity but no DNA polymerase activity. Mol Cell Biol 2005;25:10183–9.

[83] Jia H, Wang X, Anderson JT, Jankowsky E. RNA unwinding by the Trf4/Air2/Mtr4 polyadenylation (TRAMP) complex. Proc Natl Acad Sci USA 2012;109:7292–7.

[84] Liu Q, Greimann JC, Lima CD. Reconstitution, activities, and structure of the eukaryotic RNA exosome. Cell 2006;127:1223–37.

[85] Callahan KP, Butler JS. TRAMP complex enhances RNA degradation by the nuclear exosome component Rrp6. J Biol Chem 2010;285:3540–7.

[86] Halbach F, Rode M, Conti E. The crystal structure of S. cerevisiae Ski2, a DExH helicase associated with the cytoplasmic functions of the exosome. RNA 2012;18:124–34.

[87] Martin G, Keller W. RNA-specific ribonucleotidyl transferases. RNA 2007;13:1834–49.

[88] Houseley J, LaCava J, Tollervey D. RNA-quality control by the exosome. Nat Rev Mol Cell Biol 2006;7:529–39.

[89] Win TZ, Draper S, Read RL, Pearce J, Norbury CJ, Wang SW. Requirement of fission yeast Cid14 in polyadenylation of rRNAs. Mol Cell Biol 2006;26:1710–21.

[90] Wang SW, Stevenson AL, Kearsey SE, Watt S, Bahler J. Global role for polyadenylation-assisted nuclear RNA degradation in posttranscriptional gene silencing. Mol Cell Biol 2008;28:656–65.

[91] Buhler M, Haas W, Gygi SP, Moazed D. RNAi-dependent and -independent RNA turnover mechanisms contribute to heterochromatic gene silencing. Cell 2007;129:707–21.

[92] Bah A, Wischnewski H, Shchepachev V, Azzalin CM. The telomeric transcriptome of Schizosaccharomyces pombe. Nucleic Acids Res 2012;40:2995–3005.

[93] Moazed D, et al. Studies on the mechanism of RNAi-dependent heterochromatin assembly. Cold Spring Harb Symp Quant Biol 2006;71:461–71.

[94] Motamedi MR, Verdel A, Colmenares SU, Gerber SA, Gygi SP, Moazed D. Two RNAi complexes, RITS and RDRC, physically interact and localize to noncoding centromeric RNAs. Cell 2004;119:789–802.

[95] Nakamura R, et al. TRF4 is involved in polyadenylation of snRNAs in Drosophila melanogaster. Mol Cell Biol 2008;28:6620–31.

[96] Lange H, Sement FM, Gagliardi D. MTR4, a putative RNA helicase and exosome co-factor, is required for proper rRNA biogenesis and development in *Arabidopsis thaliana*. Plant J 2011;68:51–63.

[97] Wyman SK, et al. Post-transcriptional generation of miRNA variants by multiple nucleotidyl transferases contributes to miRNA transcriptome complexity. Genome Res 2011;21:1450–61.

[98] Berndt H, et al. Maturation of mammalian H/ACA box snoRNAs: PAPD5-dependent adenylation and PARN-dependent trimming. RNA 2012;18:958–72.

[99] Shcherbik N, Wang M, Lapik YR, Srivastava L, Pestov DG. Polyadenylation and degradation of incomplete RNA polymerase I transcripts in mammalian cells. EMBO Rep 2010;11:106–11.

[100] Mullen TE, Marzluff WF. Degradation of histone mRNA requires oligouridylation followed by decapping and simultaneous degradation of the mRNA both 5' to 3' and 3' to 5'. Genes Dev 2008;22:50–65.

[101] Gustafson MP, Welcker M, Hwang HC, Clurman BE. Zcchc8 is a glycogen synthase kinase-3 substrate that interacts with RNA-binding proteins. Biochem Biophys Res Commun 2005;338:1359–67.

[102] Preker P, et al. PROMoter uPstream Transcripts share characteristics with mRNAs and are produced upstream of all three major types of mammalian promoters. Nucleic Acids Res 2011;39:7179–93.

[103] Choi YS, Patena W, Leavitt AD, McManus MT. Widespread RNA 3'-end oligouridylation in mammals. RNA 2012;18:394–401.

CHAPTER FIVE

XRN1: A Major 5′ to 3′ Exoribonuclease in Eukaryotic Cells

Sarah Geisler, Jeff Coller[1]

Center for RNA Molecular Biology, Case Western Reserve University, Cleveland Ohio, USA
[1]Corresponding author: e-mail address: jmc71@case.edu

Contents

1.	Introduction	98
2.	XRN1 in mRNA Decay	98
3.	XRN1 in mRNA Quality Control	100
4.	XRN1 in miRNA-Mediated Decay	101
5.	XRN1 in siRNA-Mediated Decay	103
6.	Localization of XRN1 in Cells	104
7.	XRN1 in lncRNA Decay	105
8.	XRN1 in tRNA Quality Control	108
9.	XRN1 in rRNA and snoRNA Processing	108
10.	Regulation of XRN1 Activity	109
	10.1 Regulation of XRN1 activity by the scavenger decapping enzyme	109
	10.2 Regulation of XRN1 activity by lithium	109
11.	Summary	110
	Acknowledgments	110
	References	110

Abstract

The degradation of RNA is a critical aspect of gene regulation. Correspondingly, ribonucleases exist within the cell to degrade RNA in specific cellular contexts. An important and conserved ribonuclease is called XRN1. This enzyme, an exoribonuclease, degrades RNA in a processive 5′ to 3′ direction. Substrates for XRN1 include decapped mRNA, endonucleolytically cleaved mRNA, lncRNA, and some aberrant tRNAs. In addition, XRN1 serves a vital role in the processing and maturation of the 5′ ends of rRNA and snoRNAs. In this review, we discuss some of the important roles of XRN1 in the cell.

The Enzymes, Volume 31
ISSN 1874-6047
http://dx.doi.org/10.1016/B978-0-12-404740-2.00005-7

97

1. INTRODUCTION

The efficient destruction of RNA transcripts is of vital importance to eukaryotic cells for a number of reasons. Perhaps, the most obvious reason being that overall RNA transcript abundance is determined by both synthesis and destruction. Clearance of RNA transcripts by degradative activities, such as exoribonucleolytic digestion, therefore, plays an important role in dictating the overall level of a given RNA transcript, and consequently, the overall level at which a given gene is expressed [1]. Degradative activities, however, can also play a role in maintaining the fidelity of gene expression. Specifically, similar to DNA, RNA transcripts are susceptible to cellular insults, such as oxidative stress, that damage nucleotides. This damage could influence the information content and, therefore, result in errors in gene expression [2]. Unlike DNA, RNA lacks the analogous damage sensing and repair pathways which protect eukaryotic genomes from the accumulation of mutations. It makes sense then that most RNA transcripts would have short half-lives compared to the total life of the cell, so that incidental mutations that arise are cleared from the cellular pool of RNA. Moreover, the cell has evolved specific quality control pathways that rapidly degrade RNA transcripts that are perceived as aberrant. The enzymes that degrade RNA, ribonucleases (RNases), thereby help ensure the quality of cellular RNA transcripts.

Many such RNases exist in the cell, and each enzyme has specific properties [3]. Exoribonucleases, for instance, require an accessible end and can exhibit either 5' to 3' or 3' to 5' polarity. The major cytoplasmic 5' to 3' exoribonuclease is encoded by the *XRN1* gene [4]. XRN1 is highly conserved from yeast to humans in terms of sequence and function [5]. While the major function of XRN1 is to degrade mRNA, there are other important cellular functions of XRN1. In this review, we highlight some of the known substrates and roles of XRN1 in RNA metabolism (Fig. 5.1).

2. XRN1 IN mRNA DECAY

The existence of a 5' to 3' exoribonuclease activity in eukaryotic cells was first suggested by Furuichi *et al.* [6] and Shimotohno *et al.* [7]. Furuichi and Shimotohno found that 5' terminally capped mRNA are more resistant to degradation than mRNA with unblocked termini in *Xenopus oocytes* as well as in crude extracts of wheat germ and mouse cells [6,7].

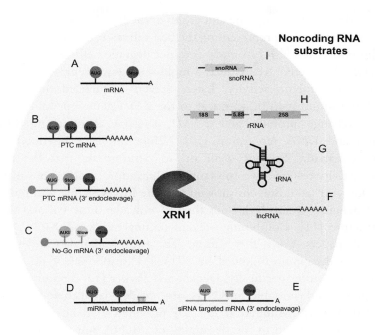

Figure 5.1 The coding and noncoding RNA substrates of XRN1. The primary substrates of XRN1 are mRNA transcripts but also a number of noncoding RNA transcripts have been demonstrated to be degraded by XRN1. (A–I) XRN1 substrates include (counter clockwise from the top): (A) deadenylated and decapped mRNAs, (B) mRNA transcripts containing a premature termination codon (PTC) that have been either (top) decapped independently of deadenylation or (bottom) the 3′ fragment generated from endonucleolytic cleavage near the PTC, (C) the 3′ fragment generated by the endonucleolytic cleavage of an mRNA targeted for No-Go decay, (D) deadenylated and decapped mRNAs targeted for miRNA-mediated repression, (E) the 3′ endonucleolytic cleavage product of siRNA-mediated repression, (F) decapped long noncoding RNAs, (G) hypomodified tRNAs, (H) 5′ ends of precursors to the 5.8S and 25S rRNAs during rRNA maturation, (I) 5′ ends of precursor to snoRNAs during snoRNA maturation. (For color version of this figure, the reader is referred to the online version of this chapter.)

Soon following these observations came the first description of this 5′ to 3′ exoribonuclease activity, ExoRNase (XRN1), by Stevens [8]. Stevens documented a 160 kDa protein that accounted for 40% of RNase activity in yeast crude extracts [8]. This activity was purified from yeast ribosomes and converted oligonucleotides into 5′ mononucleotides [9]. Importantly, the enzyme was also shown to be highly processive with no intermediate products detectable. Lastly, the end group of the RNA was deemed critical, with a monophosphate representing the best substrate; capped, tri-, di-, and nonphosphorylated 5′ RNA ends were not degraded efficiently [9]. Thus, the newly discovered XRN1 was a highly processive enzyme with 5′ to 3′ exoribonucleolytic activity. Later, Stevens sequenced the gene for *XRN1* and found it to be identical to the *SEP1* gene, encoding a DNA strand exchange protein, and *KEM1*, a gene affecting microtubule function and DNA replication [10]. While the later phenotypes still remain enigmatic, Stevens went on to demonstrate that the loss of XRN1 activity resulted in a substantial stabilization of numerous mRNAs [10]. One year later, Stevens showed that the mRNAs that accumulate in *XRN1*-deficient strains lack both their 3′ polyadenosine tail and their 5′ cap structure [11]. Combining Stevens seminal work on XRN1 with a series of papers from the Parker lab, the deadenylation-dependent pathway of decapping followed by 5′ to 3′ exoribonucleolytic digestion was laid out. More specifically, Decker and Parker [12] demonstrated that both stable and unstable mRNAs are deadenylated before they are degraded. Finally, Muhlrad *et al.* [13] provided evidence that mRNA deadenylation resulted in the decapping of mRNA and then the 5′ to 3′ digestion by XRN1. Importantly, it was later shown that the decapping enzyme cleaved the 7mGpppN cap between the alpha and beta phosphate, liberating 7mGDP and leaving behind a monophosphorylated 5′ RNA end; the ideal substrate for XRN1 [14,15].

Because mRNA lacking a cap was demonstrated to be highly unstable in other organisms, it was proposed that deadenylation-dependent decapping followed by XRN1 digestion might be a general mechanism of mRNA decay in other eukaryotes [12]. The discovery of XRN1 and decapping enzyme homologs in other yeast species, flies, mice, plants, and humans lent credence to this notion of a general pathway for mRNA decay [5]. Importantly, however, despite the high degree of XRN1 conservation, the mechanism of bulk mRNA turnover in metazoans still remains unclear. Nonetheless, these data argue that XRN1 is a conserved exoribonuclease that can play a major role in the degradation of normal mRNAs (Fig. 5.1A).

3. XRN1 IN mRNA QUALITY CONTROL

mRNA turnover pathways exist within the cell that destroy aberrant transcripts [16]. Perhaps, the best characterized "quality control" mRNA decay pathway is nonsense-mediated decay (NMD). NMD is a rapid decay pathway that responds to the presence of a nonsense mutation within the transcript [17]. Substrates for NMD arise from the introduction of point mutations that either create a nonsense mutation or a frameshift event. Other substrates include transcripts with unspliced introns, extensions in the 3'UTR, or the presence of short upstream open reading frames in the 5'UTR. In all cases, the sensing of a premature termination codon (PTC) by the ribosome elicits rapid degradation of the transcript by NMD. NMD occurs by several distinct mechanisms, but all are dependent on a group of factors referred to as UPF1–3. Targeting of an mRNA for NMD occurs either by rapid decapping of the transcript in a manner uncoupled from deadenylation or endonucleolytic cleavage of the mRNA near the PTC [18–20]. The result, in either case, is the exposure of a 5' monophosphorylated end that is recognized by XRN1 and quickly digested by this enzyme (Fig. 5.1B). In this way, XRN1 plays a vital role in ensuring the quality of mRNA transcripts.

Besides NMD, other quality control mRNA turnover pathways exist. In particular, the process of No-Go decay occurs when a dramatic stall in translational elongation is perceived [21]. Such stalling can be caused by strong secondary structures like stable hairpins within the open reading frame of an mRNA. The stalling of translational elongation results in the DOM34 and HBS1-dependent endonucleolytic cleavage of the mRNA near the site of the stall. Importantly, the 3' product of this cleavage has been demonstrated to be digested by XRN1 (Fig. 5.1C).

4. XRN1 IN miRNA-MEDIATED DECAY

RNA interference (RNAi) involves the silencing of gene expression by double-stranded RNA (dsRNA) [22]. During RNAi long precursor dsRNA molecules are processed down into 21–26 nucleotide long duplexes by cellular RNase III endonuclease enzymes. Following processing, one strand of the duplex is loaded into the multimeric RNA–induced silencing complex (RISC). The small RNAs utilized in the RNAi pathway can be either of endogenous or exogenous origin. MicroRNAs (miRNAs) are endogenous RNAi effectors which mediate the posttranscriptional repression of protein-coding genes [23]. miRNAs associate with mRNA targets through imperfect base-pairing interactions, and through this association RISC is recruited to impart repression via reduced translational efficiency and/or reduced mRNA levels. Hundreds of miRNA genes have been identified in mammalian genomes and given that a single miRNA can target multiple mRNA, the impact, therefore, of miRNA-mediated repression is widespread. Indeed with estimates suggesting that one-third of all human protein-coding genes are regulated by miRNAs, miRNAs play a very important role in the general regulation of gene expression [24].

Early observations suggested that miRNA-mediated silencing involved a reduction in protein levels with little or no change in target mRNA levels [25,26]. These findings, however, were later challenged by observations that the levels of many mRNA targets of miRNAs were indeed reduced [27]. A number of studies soon followed that suggested that miRNAs direct rapid deadenylation of target mRNAs, followed by decapping, and 5' to 3' exoribonucleolytic digestion [28–31]. Moreover, in a 2010 paper from Bartel and colleagues, it was postulated that the majority of miRNA-mediated repression in mammalian cells is through mRNA decay [32]. In their paper, Guo et al. [32] measured global protein production via ribosome profiling and compared this measure of translation to effects on mRNA levels and deduced that lowered mRNA levels accounted for most of the decrease in protein production. While the work illustrated above provided evidence that miRNA-mediated repression involved a decrease in mRNA levels through accelerated decay, there were several reports in the literature presenting evidence for miRNA-mediated repression without obvious changes in mRNA levels. Because of this conflict in the literature, researchers from the Giraldez and Green labs sought to investigate the timing of events constituting miRNA-mediated regulation. They found in zebrafish embryos as well as *Drosophila* S2 cells that early repression events involved translational repression, which was then concomitantly followed by mRNA decay [33,34]. While the relative contributions of translational repression versus mRNA degradation still remains a subject of debate, it seems that at least at some level miRNA-mediated silencing involves the degradation of targets via deadenylation, followed by decapping-dependent 5' to 3' exoribonucleolytic digestion (Fig. 5.1D).

Given the major role of XRN1 in cytoplasmic mRNA degradation, it is perhaps not surprising that XRN1 would be involved in the degradation of mRNAs targeted for miRNA-mediated repression. Direct evidence for the involvement of XRN1 in miRNA-mediated repression came from a study out of the Pasquinelli lab in 2005 [35]. Bagga et al. found that in *Caenorhabditis elegans* upon RNAi knockdown of XRN1, the reduction in mRNA levels caused by let-7 and lin-4 miRNA targeting was now attenuated. Additionally, the authors detected decay intermediates that were consistent with decapping and XRN1-mediated 5' to 3' exonucleolytic digestion up to the position of RISC on the target mRNA. These results suggest that decapping and XRN1-dependent degradation occur while RISC is still associated with the mRNA, and this association can provide a steric block to XRN1 processivity. This has led to the postulation that XRN1-dependent degradation of miRNA-targeted mRNAs might even be required for the recycling of RISC factors [36]. Evidence supporting this idea stems from a study that found that XRN1 was required for efficient RNAi in *C. elegans*, potentially because RISC complex factors were titrated out through their persistence association with the now stabilized RNAi targets [36,37].

5. XRN1 IN siRNA-MEDIATED DECAY

Small interfering RNAs (siRNAs), like miRNAs, are RNAi effectors but unlike miRNAs, siRNAs perfectly base-pair with their target [38]. Also unlike miRNAs, siRNAs elicit degradation of the target RNA through endonucleolytic cleavage of the target RNA within the base-paired region. The endonucleolytic cleavage event is mediated by RISC and more specifically by Argonaute, a core constituent of RISC. While plants contain a number of endogenous siRNA genes, in animals siRNAs generally come from exogenous sources. Due to the high efficiency of siRNA-mediated silencing, it is often used to artificially knockdown expression of a gene of interest, and therefore, has become an extremely useful laboratory tool.

Importantly, XRN1 has been implicated in mediating the downstream degradation of siRNA-targeted transcripts [39]. Orban and Izaurralde reported that siRNA-targeted mRNAs were cleared from the cell by the combined actions of XRN1, the Ski complex, and the exosome. More specifically, cleavage by Argonaute generates two RNA fragments, 5′ and 3′. In *Drosophila* S2 cells, following cleavage the 3′ cleavage product was degraded 5′ to 3′ by the XRN1 exonuclease, as evidenced by its accumulation and stabilization upon depletion of XRN1 [39]. This observation was consistent with a previous report indicating that in plants the 3′ cleavage product was degraded by the 5′ to 3′ exoribonuclease XRN4 [40]. Consistent with cleavage by Argonaute generating a substrate for XRN1 and, thus, bypassing a requirement for prior deadenylation and decapping, the 3′ fragment was demonstrated to decay independently of deadenylation and decapping [39]. XRN1s role in siRNA-mediated silencing of mRNAs, therefore, occurs independently of the bulk mRNA decay machinery, while its role in miRNA-mediated silencing occurs via acceleration of normal decay (Fig. 5.1E).

6. LOCALIZATION OF XRN1 IN CELLS

In recent years, understanding where mRNA degradation occurs within the cell has been given a great deal of investigation [41,42]. Insight into how the large repertoire of observed mRNA half-lives is achieved may come from knowing when and where degradation occurs within the cytoplasm. At present, there are two different views on where mRNA degradation takes place. The first comes from the discovery that mRNA decapping and decay factors like XRN1 aggregate into punctuate, microscopically visible structures. These structures have been given the epitaph of processing bodies or P-bodies for short. Importantly, P-bodies do not contain ribosomes nor most translational initiation factors. What they do contain is the full complement of decapping proteins, certain mRNA decay intermediates, and in higher eukaryotes, the miRNA machinery [42,43]. These data have led to a two-step model for regulating transcript stability in which ribosome dissociation first occurs, and then the mRNA is trafficked to P-bodies where mRNA decapping ensues [42]. A second model supported in the literature is that mRNA decay does not involve a fundamental transition between a polyribosome-bound state and a translationally repressed state, but rather occurs cotranslationally [44–46].

The discovery of P-bodies dates back to 1997 when Bashkirov et al. cloned the mouse homolog of XRN1. Using mouse E10 fibroblast cells, it was observed that mXRN1 localized to punctuate cytoplasmic structures [47]. At this time, the authors proposed that these structures may represent either sites for RNA turnover or sites in which the enzyme is stored until used. These theories, however, were untested. P-bodies were rediscovered in 2002 when Eystathioy et al. found that the human auto-antigen protein, GW182, was also found in cytoplasmic granules [48]. Although it was not clear at this time, GW182 is a critical factor in mediating miRNA translational control [49]. One year later, several studies simultaneously demonstrated that, in addition to XRN1, decapping factors are also found in cytoplasmic foci. Sheth and Parker [50] extended this analysis by showing that mRNA decay intermediates colocalized with these granules in a manner that is dependent on the enzymatic activity of yeast XRN1 [50]. Similar results were seen in humans [51]. Thus, the aggregation of decay factors in granules was deemed of functional significance and it was suggested that P-bodies represented the place where mRNAs were degraded.

From a quantitative stand point of view, it remains unclear to what extent mRNA decapping and 5′ to 3′ exonucleolytic decay occur in P-bodies. Furthermore, it is unclear how much of the total cellular XRN1 is localized to P-bodies. In one study, however, a quantification of AGO2 (required for the miRNA function and a component of P-bodies) in mammalian cells revealed that only less than 5% of total AGO2 is localized in P-bodies [52]. Moreover, in both yeast and high eukaryotes, it was found that P-body aggregation can be uncoupled from mRNA turnover events [53–56]. Thus, the role of P-bodies in controlling mRNA decay still remains enigmatic.

Recent evidence suggests that under normal conditions, mRNA decapping and XRN1 digestion occurs on mRNAs cotranslationally. The hypothesis that mRNA is destroyed on polysomes is not new. Early evidence was first seen in yeast. For instance, Stevens purified XRN1 from polyribosome material [9]. Moreover, Beelman and Parker investigated mRNA decay in cells treated with cycloheximide and found the mRNA accumulated as a slightly shorter species over time [45]. Although this finding was not further characterized, they proposed that the truncated mRNA was the result of decapping and digestion up to a ribosome stalled at the AUG start codon. Consistent with the hypothesis of cotranslational decapping/decay, it was also found that in humans decapping activity cosediments with polysomes [57]. Recently, we provided evidence that 5′ to 3′ exonucleolytic decay fragments are generated on polyribosomes [44]. These observations argue that under normal physiological conditions, mRNA decapping and 5′ to 3′ exonucleolytic digestion occur on polyribosomes. Interestingly, this cotranslational mRNA decay model can provide evolutionary insight on why the 5′ to 3′ exonucleolytic decay pathway became a major mRNA degradation pathway. Specifically, coupling of mRNA decay and translation ensures that mRNAs targeted for destruction are not only destroyed in a rapid and efficient manner, but also ensures that the decay machinery will not interfere with the last translocating ribosome. Thus, the final full-length peptide can still be made while the mRNA is degraded.

7. XRN1 IN lncRNA DECAY

Recent reports have expanded the repertoire of XRN1 substrates to include long noncoding RNA (lncRNA) transcripts [58–61]. With a more complete understanding of the eukaryotic transcriptome, it has become increasingly clear that eukaryotic genomes are pervasively transcribed into RNA transcripts [62–64]. The vast majority of these pervasive transcripts have been inferred to contain minimal protein-coding capacity, and thus, belong to the exploding class of lncRNAs. While these observations have been met with considerable skepticism, many lncRNAs have now been demonstrated to be *bona fide* regulators of gene expression [58,65]. For instance, lncRNAs have been implicated in regulating a variety of processes in eukaryotes including pluripotency, meiotic entry, retrotransposon silencing, imprinting, dosage compensation, and telomere length [66–71]. Further underscoring the importance of lncRNA function, altered expression of lncRNAs has been implicated in disease states such as cancer and neurological disorders [72–74]. Regulating the expression of lncRNAs is, therefore, likely to be of crucial importance to cellular physiology.

Most lncRNAs are thought to be RNA polymerase II transcripts and, similar to their mRNA cousins, have been demonstrated to be both capped and polyadenylated [58,67,75]. Because of these resemblances, lncRNAs are often described as mRNA like. It is perhaps not surprising then that lncRNAs, like mRNAs, would be substrates for decapping-dependent 5' to 3' exonucleolytic digestion by DCP2 and XRN1. One of the first studies investigating a role for XRN1 in the turnover of lncRNAs came out of the Parker lab and focused on the decay of the *SRG1* lncRNA [60]. The *SRG1* lncRNA is expressed from a region upstream of the *SER3* gene in yeast and has been implicated in modulating expression from the *SER3* promoter [76]. Because *SRG1* lncRNA was stabilized in cells lacking the decapping activity and XRN1, Thompson and Parker concluded that some lncRNAs are degraded by the same factors involved in decaying mRNAs in the cytoplasm. Another study looking at the *RTL* lncRNA, an lncRNA expressed antisense to Ty1 retrotransposons, also implicated a major role for XRN1 in its turnover. Upstream decapping of the *RTL* lncRNA, however, displayed variable dependencies on the mRNA decay factors that facilitate mRNA decapping [66]. These observations suggested that for these two lncRNAs, XRN1-dependent 5' to 3' exonucleolytic digestion contributed to their clearance from the cell, but that lncRNAs could be shunted into separate upstream decapping pathways. Along these lines, in a recent study from our lab, we demonstrate that a large number of lncRNAs are decapped by the DCP2/DCP1 holoenzyme through a pathway that occurs independently of all known decapping activators prior to XRN1-dependent clearance [59].

Additionally, reports have also implicated NMD in the turnover of lncRNAs [77,78]. For instance, a number of 5′ extended RNAs of subtelomeric genes in yeast were demonstrated to be degraded in the cytoplasm by the NMD machinery in a report from the Chanfreau lab [77]. The authors refer to these 5′ extended RNAs as cytoplasmically degraded cryptic unstable transcripts (CD-CUTs). Importantly, CD-CUTs, which appear to lack protein-coding capacity, contribute to the repression of downstream genes. Consistent with the pivotal role that XRN1 plays in the degradation of NMD substrates, CD-CUTs accumulated in the absence of XRN1, and given that CD-CUTs were demonstrated to be direct substrates for UPF1, it would be anticipated that these lncRNAs would likewise be direct substrates for XRN1. Furthermore, the Buratowski lab documented that lncRNAs referred to as stable uncharacterized transcripts in yeast accumulated in the absence of a functional NMD pathway (i.e., loss of UPF1, DCP1, or XRN1) [78].

A widespread role for XRN1 in the turnover of lncRNAs in yeast was established in a recent genome-wide study from the Morillon group [61]. In this study, van Dijk *et al.* identified nearly 1700 unannotated transcripts that accumulate in yeast in the absence of XRN1. The authors referred to these transcripts as XRN1-sensitive unstable transcripts (XUTs). Most XUTs, approximately 66%, were antisense to known genes. Interestingly, 273 genes were suggested to be negatively regulated at the transcriptional level by an antisense XUT [61]. It remains unclear, however, how stabilizing a cytoplasmic pool of antisense lncRNAs in the absence of XRN1 ultimately regulates transcription in the nucleus. It will be interesting in the future to determine how cytoplasmic lncRNA abundance can influence nuclear events.

In addition to XRN1, RAT1, the nuclear counterpart of XRN1, has likewise been implicated in the turnover of lncRNAs in yeast [59,70]; however, the relative contributions of XRN1 and RAT1 to the overall degradation of lncRNAs is currently unknown. Given that lncRNAs make up a substantial fraction of the transcribed regions of eukaryotic genomes, the addition of lncRNAs to the catalog of XRN1 and RAT1 substrates highlights the importance of 5′ to 3′ exonucleolytic activities in shaping the overall architecture of the eukaryotic transcriptome (Fig. 5.1F).

8. XRN1 IN tRNA QUALITY CONTROL

tRNA, like mRNA, is subject to quality control. Mature tRNA is normally extremely stable and extensively modified. Remarkably, there are over 100 tRNA modifications that have been described. As with any biological process, mistakes during tRNA synthesis and modification can occur. Hypomodified tRNA, therefore, are aberrant products and have been shown to be substrates for rapid degradation. In some cases, hypomodification leads to the rapid destruction of tRNA by the exosome. Importantly, however, Chernyakov et al. [79] demonstrated that certain hypomodified tRNAs are degraded rapidly by XRN1 and its nuclear homolog RAT1. Specifically, mature tRNAs lacking 7-methylguanosine and 5-methylcytidine are stabilized in cells lacking XRN1 and RAT1 activity. These data demonstrate that in addition to mRNA, XRN1 is a multifunctional monitor of RNA quality [80]. Moreover, this study was the first example of XRN1 degrading a stable mature noncoding RNA species generated by RNA polymerase III (Fig. 5.1G).

9. XRN1 IN rRNA AND snoRNA PROCESSING

The nuclear ribosomal DNA genes of all eukaryotes are arranged the same. Specifically, the small 18S rRNA and the two large rRNA genes (5.8S and 28S) are transcribed by RNA polymerase I as a single, large preribosomal RNA call the 35S rRNA. This preribosomal RNA must undergo sequential cleavage events to yield the mature rRNA. The rRNA coding regions are flanked by 5′ and 3′ spacers, and these must be excised and removed before maturation of the ribosome. Processing of the 5.8S rRNA, for example, is the result of a cleavage event between the 18S rRNA and 5.8S rRNA. This cleavage event generates a 5′ heterogenous end on the 5.8S rRNA. Henry et al. [81] demonstrated that XRN1 and RAT1 perform a trimming reaction that is required for the formation of the final 5.8S rRNA 5′ end. A number of years later, similar results were seen for maturation of the 25S rRNA 5′ end [82]. Like rRNA, snoRNA must be processed away from intervening sequences. For instance, many snoRNAs are encoded in the intronic sequences of genes or as polycistronic RNA. Following initial processing, Petfalski et al. [83] showed that both XRN1 and RAT1 are required to remove 5′ sequences on snoRNA and mature their appropriate 5′ end. Together, these studies indicate that XRN1 and RAT1 have a vital role in rRNA and snoRNA maturation that is distinct from their function in mRNA decay (Fig. 5.1H and I). Because most of these events primarily occur in the nucleus, RAT1 most likely plays the major role in rRNA and snoRNA processing. Nonetheless, the data are clear that in the absence of RAT1, XRN1 can also be involved in stable RNA maturation.

10. REGULATION OF XRN1 ACTIVITY

In general, it is not thought that XRN1 activity is regulated within the cell. Both deadenylation and decapping rates can be highly variable between mRNA species. Importantly, digestion of the transcript body by XRN1 appears to be uniform and robust. Despite this general paradigm, there are reports that XRN1 activity can be modulated under specific genetic situations. In both documented cases, the regulation of XRN1 activity results from the accumulation of an aberrant metabolic intermediate.

10.1. Regulation of XRN1 activity by the scavenger decapping enzyme

Besides the degradation of mRNA from the 5' end by XRN1, transcripts can also be degraded from the 3' end [84]. The major activity for 3' to 5' exonucleolytic digestion is the exosome; a multi-subunit complex consisting of nine core components and several accessory factors. Decay from the 3' end eventually generates an m7GpppN dinucleotide; this is eventually hydrolyzed to m7Gp and ppN by the scavenger decapping enzyme termed DCPS [85]. Surprisingly, loss of DCPS activity in yeast impedes the 5' to 3' degradation of several mRNA transcripts by XRN1 [86]. It was proposed that DCPS contributes to a feedback loop that regulates RNA stability by influencing 5' end decay. Importantly, a catalytically inactive form of DCPS has no effect on XRN1 activity, thus it was suggested that the substrate for DCPS cleavage (i.e., m7GpppN) is responsible for modulating XRN1 activity. It remains unclear how the cap dinucleotide regulates XRN1 activity.

10.2. Regulation of XRN1 activity by lithium

Lithium has long been known to have therapeutic effects in the treatment of neural diseases, but in high doses can be toxic. The molecular mechanism of lithium action, however, is unclear. Dicht *et al.* [87] demonstrated that lithium inhibits the enzyme HAL2. HAL2 is involved in sulfate assimilation and converts adenosine 3', 5' disphosphate (pAp), into 5' AMP and Pi [87]. Inactivation of HAL2 activity by lithium causes the accumulation of cellular pAp levels. Overproduction of pAp inhibits the activity of XRN1 and other similar RNA processing enzymes such as RAT1 (the nuclear homolog of XRN1). It was suggested, therefore, that the pAp-mediated inhibition of XRN1 might contribute to the effects of lithium toxicity.

11. SUMMARY

The cellular functions of XRN1 highlight two critical roles that ribonuclease play in the cell: as unrestricted digestive enzymes and as precise processing enzymes. It is interesting to observe that XRN1 can, on one hand, degrade RNA transcripts like decapped mRNA indiscriminately, while in the case of rRNA precisely removes the 5′ leader to aid in the maturation of the ribosome (Fig. 5.1). How these defined roles are achieved is unclear but most likely represents the context of the RNA being attacked by XRN1; that is, secondary structures and sequence content. New discoveries such as the regulation of lncRNA by XRN1 demonstrate that we still have much to learn about how complex the regulation of the overall transcriptome is balanced by transcription and the action of nucleases like XRN1.

ACKNOWLEDGMENTS

The authors thank and dedicate this review to the memory of Dr Audrey Stevens.

REFERENCES

[1] Ghosh S, Jacobson A. RNA decay modulates gene expression and controls its fidelity. Wiley Interdiscip Rev RNA 2010;1:351–61.
[2] Schoenberg DR, Maquat LE. Regulation of cytoplasmic mRNA decay. Nat Rev Genet 2012;13:246–59.
[3] Nicholson A, Ribonucleases (series: nucleic acids and molecular biology), Recherche. 2006; vol 26.
[4] Stevens A. 5′-Exoribonuclease 1: Xrn1. Methods Enzymol 2001;342:251–9.
[5] Jones CI, Zabolotskaya MV, Newbury SF. The 5′ → 3′ exoribonuclease XRN1/ Pacman and its functions in cellular processes and development. Wiley Interdiscip Rev RNA 2012;3:455–68.
[6] Furuichi Y, LaFiandra A, Shatkin AJ. 5′-Terminal structure and mRNA stability. Nature 1977;266:235–9.
[7] Shimotohno K, Kodama Y, Hashimoto J, Miura KI. Importance of 5′-terminal blocking structure to stabilize mRNA in eukaryotic protein synthesis. Proc Natl Acad Sci U S A 1977;74:2734–8.
[8] Stevens A. An exoribonuclease from Saccharomyces cerevisiae: effect of modifications of 5′ end groups on the hydrolysis of substrates to 5′mononucleotides. Biochem Biophys Res Commun 1978;81:656–61.
[9] Stevens A. Purification and characterization of a Saccharomyces cerevisiae exoribonuclease which yields 5′-mononucleotides by a 5′ → 3′ mode of hydrolysis. J Biol Chem 1980;255:3080–5.

[10] Larimer FW, Hsu CL, Maupin MK, Stevens A. Characterization of the XRN1 gene encoding a 5′ → 3′ exoribonuclease: sequence data and analysis of disparate protein and mRNA levels of gene-disrupted yeast cells. Gene 1992;120:51–7.

[11] Hsu CL, Stevens A. Yeast cells lacking 5′ → 3′ exoribonuclease 1 contain mRNA species that are poly(A) deficient and partially lack the 5′ cap structure. Mol Cell Biol 1993;13:4826–35.

[12] Decker CJ, Parker R. A turnover pathway for both stable and unstable mRNAs in yeast: evidence for a requirement for deadenylation. Genes Dev 1993;7:1632–43.

[13] Muhlrad D, Decker CJ, Parker R. Deadenylation of the unstable mRNA encoded by the yeast MFA2 gene leads to decapping followed by 5′ → 3′ digestion of the transcript. Genes Dev 1994;8:855–66.

[14] Beelman CA, Stevens A, Caponigro G, LaGrandeur TE, Hatfield L, Fortner DM, et al. An essential component of the decapping enzyme required for normal rates of mRNA turnover. Nature 1996;382:642–6.

[15] Dunckley T, Parker R. The DCP2 protein is required for mRNA decapping in Saccharomyces cerevisiae and contains a functional MutT motif. EMBO J 1999;18:5411–22.

[16] Isken O, Maquat LE. Quality control of eukaryotic mRNA: safeguarding cells from abnormal mRNA function. Genes Dev 2007;21:1833–56.

[17] Baker KE, Parker R. Nonsense-mediated mRNA decay: terminating erroneous gene expression. Curr Opin Cell Biol 2004;16:293–9.

[18] Muhlrad D, Parker R. Premature translational termination triggers mRNA decapping. Nature 1994;370:578–81.

[19] Huntzinger E, Kashima I, Fauser M, Saulière J, Izaurralde E. SMG6 is the catalytic endonuclease that cleaves mRNAs containing nonsense codons in metazoan. RNA 2008;14:2609–17.

[20] Eberle AB, Lykke-Andersen S, Muhlemann O, Jensen TH. SMG6 promotes endonucleolytic cleavage of nonsense mRNA in human cells. Nat Struct Mol Biol 2009;16:49–55.

[21] Doma MK, Parker R. Endonucleolytic cleavage of eukaryotic mRNAs with stalls in translation elongation. Nature 2006;440:561–4.

[22] Filipowicz W. RNAi: the nuts and bolts of the RISC machine. Cell 2005;122:17–20.

[23] Eulalio A, Huntzinger E, Izaurralde E. Getting to the root of miRNA-mediated gene silencing. Cell 2008;132:9–14.

[24] Lewis BP, Burge CB, Bartel DP. Conserved seed pairing, often flanked by adenosines, indicates that thousands of human genes are microRNA targets. Cell 2005;120:15–20.

[25] Olsen PH, Ambros V. The lin-4 regulatory RNA controls developmental timing in Caenorhabditis elegans by blocking LIN-14 protein synthesis after the initiation of translation. Dev Biol 1999;216:671–80.

[26] Wightman B, Ha I, Ruvkun G. Posttranscriptional regulation of the heterochronic gene lin-14 by lin-4 mediates temporal pattern formation in C. elegans. Cell 1993;75:855–62.

[27] Lim LP, et al. Microarray analysis shows that some microRNAs downregulate large numbers of target mRNAs. Nature 2005;433:769–73.

[28] Behm-Ansmant I, et al. mRNA degradation by miRNAs and GW182 requires both CCR4:NOT deadenylase and DCP1:DCP2 decapping complexes. Genes Dev 2006;20:1885–98.

[29] Eulalio A, et al. Deadenylation is a widespread effect of miRNA regulation. RNA 2009;15:21–32.

[30] Giraldez AJ, et al. Zebrafish MiR-430 promotes deadenylation and clearance of maternal mRNAs. Science 2006;312:75–9.

[31] Wu L, Fan J, Belasco JG. MicroRNAs direct rapid deadenylation of mRNA. Proc Natl Acad Sci U S A 2006;103:4034–9.

[32] Guo H, Ingolia NT, Weissman JS, Bartel DP. Mammalian microRNAs predominantly act to decrease target mRNA levels. Nature 2010;466:835–40.

[33] Bazzini AA, Lee MT, Giraldez AJ. Ribosome profiling shows that miR-430 reduces translation before causing mRNA decay in zebrafish. Science 2012;336:233–7.

[34] Djuranovic S, Nahvi A, Green R. miRNA-mediated gene silencing by translational repression followed by mRNA deadenylation and decay. Science 2012;336:237–40.

[35] Bagga S, et al. Regulation by let-7 and lin-4 miRNAs results in target mRNA degradation. Cell 2005;122:553–63.

[36] Valencia-Sanchez MA, et al. Control of translation and mRNA degradation by miRNAs and siRNAs. Genes Dev 2006;20:515–24.

[37] Newbury S, Woollard A. The 5′-3′ exoribonuclease xrn-1 is essential for ventral epithelial enclosure during C. elegans embryogenesis. RNA 2004;10:59–65.

[38] Dorner S, Eulalio A, Huntzinger E, Izaurralde E. Delving into the diversity of silencing pathways, In: Symposium on MicroRNAs and siRNAs: biological functions and mechanisms, EMBO Rep; 2007. p. 723–9.

[39] Orban TI, Izaurralde E. Decay of mRNAs targeted by RISC requires XRN1, the Ski complex, and the exosome. RNA 2005;11:459–69.

[40] Souret FF, Kastenmayer JP, Green PJ. AtXRN4 degrades mRNA in Arabidopsis and its substrates include selected miRNA targets. Mol Cell 2004;15:173–83.

[41] Eulalio A, Behm-Ansmant I, Izaurralde E. P bodies: at the crossroads of post-transcriptional pathways. Nat Rev Mol Cell Biol 2007;8:9–22.

[42] Franks TM, Lykke-Andersen J. The control of mRNA decapping and P-body formation. Mol Cell 2008;32:605–15.

[43] Parker R, Sheth U. P bodies and the control of mRNA translation and degradation. Mol Cell 2007;25:635–46.

[44] Hu W, Sweet TJ, Chamnongpol S, Baker KE, Coller J. Co-translational mRNA decay in Saccharomyces cerevisiae. Nature 2009;461:225–9.

[45] Beelman CA, Parker R. Differential effects of translational inhibition in cis and in trans on the decay of the unstable yeast MFA2 mRNA. J Biol Chem 1994;269:9687–92.

[46] Mangus DA, Jacobson A. Linking mRNA turnover and translation: assessing the poly-ribosomal association of mRNA decay factors and degradative intermediates. Methods 1999;17:28–37.

[47] Bashkirov VI, Scherthan H, Solinger JA, Buerstedde JM, Heyer WD. A mouse cytoplasmic exoribonuclease (mXRN1p) with preference for G4 tetraplex substrates. J Cell Biol 1997;136:761–73.

[48] Eystathioy T, Chan EKL, Tenenbaum SA, Keene JD, Griffith K, Fritzler MJ. A phosphorylated cytoplasmic autoantigen, GW182, associates with a unique population of human mRNAs within novel cytoplasmic speckles. Mol Biol Cell 2002;13:1338–51.

[49] Eulalio A, Tritschler F, Izaurralde E. The GW182 protein family in animal cells: new insights into domains required for miRNA-mediated gene silencing. RNA 2009;15:1433–42.

[50] Sheth U, Parker R. Decapping and decay of messenger RNA occur in cytoplasmic processing bodies. Science 2003;300:805–8.

[51] Cougot N, Babajko S, Seraphin B. Cytoplasmic foci are sites of mRNA decay in human cells. J Cell Biol 2004;165:31–40.

[52] Leung AK, Calabrese JM, Sharp PA. Quantitative analysis of Argonaute protein reveals microRNA-dependent localization to stress granules. Proc Natl Acad Sci U S A 2006;103:18125–30.

[53] Sweet TJ, Boyer B, Hu W, Baker KE, Coller J. Microtubule disruption stimulates P-body formation. RNA 2007;13:493–502.

[54] Decker CJ, Teixeira D, Parker R. Edc3p and a glutamine/asparagine-rich domain of Lsm4p function in processing body assembly in Saccharomyces cerevisiae. J Cell Biol 2007;179:437–49.

[55] Eulalio A, Behm-Ansmant I, Schweizer D, Izaurralde E. P-body formation is a consequence, not the cause, of RNA-mediated gene silencing. Mol Cell Biol 2007;27: 3970–81.

[56] Chu CY, Rana TM. Translation repression in human cells by microRNA-induced gene silencing requires RCK/p54. PLoS Biol 2006;4:e210.

[57] Wang Y, Liu CL, Storey JD, Tibshirani RJ, Herschlag D, Brown PO. Precision and functional specificity in mRNA decay. Proc Natl Acad Sci U S A 2002;99:5860–5.

[58] Berretta J, Morillon A. Pervasive transcription constitutes a new level of eukaryotic genome regulation. EMBO Rep 2009;10:973–82.

[59] Geisler S, Lojek L, Khalil AM, Baker KE, Coller J. Decapping of long noncoding RNAs regulates inducible genes. Mol Cell 2012;45:279–91.

[60] Thompson DM, Parker R. Cytoplasmic decay of intergenic transcripts in Saccharomyces cerevisiae. Mol Cell Biol 2007;27:92–101.

[61] van Dijk EL, Chen CL, d'Aubenton-Carafa Y, Gourvennec S, Kwapisz M, Roche V, et al. XUTs are a class of Xrn1-sensitive antisense regulatory non-coding RNA in yeast. Nature 2011;475:114–7.

[62] ENCODE Project Consortium , Birney E, Stamatoyannopoulos JA, Dutta A, Guigó R, Gingeras TR, et al. Identification and analysis of functional elements in 1% of the human genome by the ENCODE pilot project. Nature 2007;447:799–816.

[63] Nagalakshmi U, Wang Z, Waern K, Shou C, Raha D, Gerstein M, et al. The transcriptional landscape of the yeast genome defined by RNA sequencing. Science 2008;320: 1344–9.

[64] Xu Z, Wei W, Gagneur J, Perocchi F, Clauder-Münster S, Camblong J, et al. Bidirectional promoters generate pervasive transcription in yeast. Nature 2009;457: 1033–7.

[65] Ponting CP, Belgard TG. Transcribed dark matter: meaning or myth? Hum Mol Genet 2010;19:R162–8.

[66] Berretta J, Pinskaya M, Morillon A. A cryptic unstable transcript mediates transcriptional trans-silencing of the Ty1 retrotransposon in S. cerevisiae. Genes Dev 2008;22:615–26.

[67] Guttman M, Amit I, Garber M, French C, Lin MF, Feldser D, et al. Chromatin signature reveals over a thousand highly conserved large non-coding RNAs in mammals. Nature 2009;458:223–7.

[68] Hongay CF, Grisafi PL, Galitski T, Fink GR. Antisense transcription controls cell fate in Saccharomyces cerevisiae. Cell 2006;127:735–45.

[69] Loewer S, Cabili MN, Guttman M, Loh Y-H, Thomas K, Park IH, et al. Large intergenic non-coding RNA-RoR modulates reprogramming of human induced pluripotent stem cells. Nat Genet 2010;42:1113–7.

[70] Luke B, Panza A, Redon S, Iglesias N, Li Z, Lingner J. The Rat1p 5′ → 3′ exonuclease degrades telomeric repeat-containing RNA and promotes telomere elongation in Saccharomyces cerevisiae. Mol Cell 2008;32:465–77.

[71] Ponting CP, Oliver PL, Reik W. Evolution and functions of long noncoding RNAs. Cell 2009;136:629–41.

[72] Niland CN, Merry CR, Khalil AM. Emerging roles for long non-coding RNAs in cancer and neurological disorders. Front Genet 2012;3:25.

[73] Qureshi IA, Mattick JS, Mehler MF. Long non-coding RNAs in nervous system function and disease. Brain Res 2010;1338:20–35.

[74] Tsai M-C, Spitale RC, Chang HY. Long intergenic noncoding RNAs: new links in cancer progression. Cancer Res 2011;71:3–7.

[75] Khalil AM, Guttman M, Huarte M, Garber M, Raj A, Rivea Morales D, et al. Many human large intergenic noncoding RNAs associate with chromatin-modifying complexes and affect gene expression. Proc Natl Acad Sci U S A 2009;106:11667–72.

[76] Martens JA, Wu P-YJ, Winston F. Regulation of an intergenic transcript controls adjacent gene transcription in Saccharomyces cerevisiae. Genes Dev 2005;19: 2695–704.

[77] Toesca I, Nery CR, Fernandez CF, Sayani S, Chanfreau GF. Cryptic transcription mediates repression of subtelomeric metal homeostasis genes. PLoS Genet 2011;7: e1002163.

[78] Marquardt S, Hazelbaker DZ, Buratowski S. Distinct RNA degradation pathways and $3'$ extensions of yeast non-coding RNA species. Transcription 2011;2:145–54.

[79] Chernyakov I, Whipple JM, Kotelawala L, Grayhack EJ, Phizicky EM. Degradation of several hypomodified mature tRNA species in Saccharomyces cerevisiae is mediated by Met22 and the $5''$-$3''$ exonucleases Rat1 and Xrn1. Genes Dev 2008;22:1369–80.

[80] Whipple JM, Lane EA, Chernyakov I, D'Silva S, Phizicky EM. The yeast rapid tRNA decay pathway primarily monitors the structural integrity of the acceptor and T-stems of mature tRNA. Genes Dev 2011;25:1173–84.

[81] Henry Y, Wood H, Morrissey JP, Petfalski E, Kearsey S, Tollervey D. The $5'$ end of yeast 5.8S rRNA is generated by exonucleases from an upstream cleavage site. EMBO J 1994;13:2452–63.

[82] Geerlings TH, Vos JC, Raué HA. The final step in the formation of 25S rRNA in Saccharomyces cerevisiae is performed by $5' \rightarrow 3'$ exonucleases. RNA 2000;6: 1698–703.

[83] Petfalski E, Dandekar T, Henry Y, Tollervey D. Processing of the precursors to small nucleolar RNAs and rRNAs requires common components. Mol Cell Biol 1998;18: 1181–9.

[84] Liu H, Kiledjian M. Decapping the message: a beginning or an end. Biochem Soc Trans 2006;34:35–8.

[85] Wang Z, Kiledjian M. Functional link between the mammalian exosome and mRNA decapping. Cell 2001;107:751–62.

[86] Liu H, Kiledjian M. Scavenger decapping activity facilitates $5' \rightarrow 3'$ mRNA decay. Mol Cell Biol 2005;25:9764–72.

[87] Dichtl B, Stevens A, Tollervey D. Lithium toxicity in yeast is due to the inhibition of RNA processing enzymes. EMBO J 1997;16:7184–95.

CHAPTER SIX

Structures of 5′–3′ Exoribonucleases

Jeong Ho Chang*, Song Xiang*,†, Liang Tong*,1
*Department of Biological Sciences, Columbia University, New York, NY, USA
†Key Laboratory of Nutrition and Metabolism, Institute for Nutritional Sciences, Shanghai Institutes for Biological Sciences, Chinese Academy of Sciences, Shanghai, PR China
1Corresponding author: e-mail address: ltong@columbia.edu

Contents

1.	Introduction	116
2.	Crystal Structure of Rat1/Xrn2	116
3.	Crystal Structures of Xrn1	118
4.	Structural Homologs of XRNs	121
5.	Active Site of XRNs	122
6.	Structure of the Rat1–Rai1 Complex	124
7.	Structure of Rai1/Dom3Z	125
8.	Perspectives	126
Acknowledgments		127
References		127

Abstract

5′–3′ Exoribonucleases (XRNs) have important functions in RNA processing, RNA turnover and decay, RNA interference, RNA polymerase transcription, and other cellular processes. Their sequences share two highly conserved regions, CR1 and CR2. The cytoplasmic Xrn1 and the nuclear Xrn2/Rat1 are found in yeast and animals, and XRNs are found in most other eukaryotes. Crystal structures of Xrn1 and Rat1 have been reported recently, offering the first detailed information on these enzymes. The two conserved regions of XRNs form a single, large domain. CR1 has structural homology with the FEN superfamily of nucleases, while CR2 restricts access to the active site, ensuring that XRNs are exclusive exoribonucleases. The structure of Rai1, the protein partner of Rat1, revealed the presence of an active site, and further studies demonstrated that this activity is a novel mechanism for mRNA 5′-end capping quality surveillance.

The Enzymes, Volume 31
ISSN 1874-6047
http://dx.doi.org/10.1016/B978-0-12-404740-2.00006-9

115

1. INTRODUCTION

The 5'–3' Exoribonucleases (XRNs) are ubiquitously involved in RNA metabolism such as mRNA turnover, mRNA decay, RNA interference, rRNA processing, transcription termination, and maturation of other RNA species in eukaryotes [1]. XRNs prefer RNAs with 5'-end monophosphate as substrates [1–3]. Similar to most nucleases [4], XRNs require divalent metal ions in the active site for substrate hydrolysis.

Two XRNs, named Xrn1 (175 kDa) and Xrn2 (115 kDa), are present in yeast and animals. Xrn1 homologs are also found in *Drosophila* (known as Pacman) [5] and *C. elegans* [6]. Three homologs of Xrn2 are found in *Arabidopsis* (AtXrn2, AtXrn3, AtXrn4) while Xrn1 homologs may not exist in higher plants [7]. Xrn1 is primarily in the cytoplasm, and Xrn2 (more commonly known as Rat1 in yeast) is primarily in the nucleus. The amino acid sequences of the XRNs contain two highly conserved regions (CR1 and CR2) in their N-terminal segment, while sequence conservation outside these two regions is much lower (Fig. 6.1A). CR1 contains seven strictly conserved acidic residues, which coordinate the metal ions for catalysis [11–13]. The larger size of Xrn1 is due to an extended C-terminal segment that is absent in Xrn2. On the other hand, roughly half of the poorly conserved C-terminal segment in Xrn1 and Xrn2 is required for catalytic activity (Fig. 6.1A) [12,14].

In this chapter, we will describe current knowledge on the structures of the XRNs (Xrn1 and Xrn2/Rat1) and compare them to related nucleases. In addition, we will describe the structures of Rai1 (45 kDa), the activating protein partner of Rat1 [3,14,15], and the mammalian Rai1 homolog Dom3Z. The structures of Rai1/Dom3Z reveal the presence of an active site, and it has been established that Rai1/Dom3Z is a novel enzyme involved in mRNA 5'-end capping quality surveillance [8,16]. For more information on the functions of Xrn1, Xrn2/Rat1, and Rai1, please refer to this chapter, Chapters 8, and 9, respectively.

2. CRYSTAL STRUCTURE OF RAT1/XRN2

The crystal structure of *Schizosaccharomyces pombe* Rat1 in complex with its activator Rai1 was the first structure reported for XRNs [8]. CR1 and CR2 form a single, large domain, while the linker between CR1 and CR2 (~200 residues) is mostly disordered (Fig. 6.1B). CR1

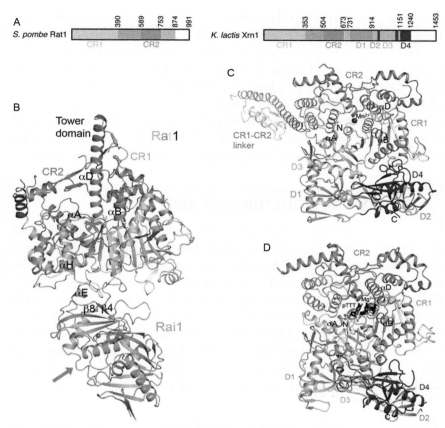

Figure 6.1 Overall structures of XRNs. (A) Schematic drawing of the domain organization of *S. pombe* Rat1 and *K. lactis* Xrn1. The domains are given different colors and labeled. (B) Schematic drawing of the structure of *S. pombe* Rat1–Rai1 complex [8]. The active site of Rat1 is indicated with the red asterisk, and the arrow points to the opening of the Rai1 active site pocket. A bound divalent metal cation in the active site of Rai1 is shown as a sphere in gray. (C) Schematic drawing of the structure of *K. lactis* Xrn1 (KlXrn1, E178Q mutant) [9]. The bound Mn^{2+} ion is shown as a sphere in black. (D) Schematic drawing of the structure of *D. melanogaster* Xrn1 (DmXrn1, D207A mutant), in complex with pTTT (black) and Mg^{2+} (gray sphere) [10]. (See color plate section in the back of the book.)

contains a seven-stranded, mostly parallel β-sheet surrounded by α helices on both faces and shares structural similarity with a few other nucleases (see Section 4) [1]. CR2, with several helices and long loops, is unlikely to be stable on its own. Instead, it wraps around the base (N-terminal end) of a long helix (αD helix) in CR1, known as the tower domain. The C-terminal end of this helix is projected ∼30 Å away from the rest of the Rat1 structure (Fig. 6.1B), while its N–terminal end contributes to the formation of the active site (see Section 5).

The poorly conserved segment (∼120 residues) following CR2 that is required for activity adds four antiparallel strands to the central β–sheet of CR1, which is followed by a long loop that traverses the entire bottom face of this central β–sheet (Fig. 6.1B). This loop also serves as the primary region of contact with Rai1 (see Section 6). Finally, an α helix at the C–terminal end of this segment interacts with helices αA and αH in CR1. The structural roles of this segment are consistent with its importance for the function of Rat1.

3. CRYSTAL STRUCTURES OF XRN1

Crystal structures of *Kluyveromyces lactis* Xrn1 (KlXrn1, E178Q mutant) free enzyme and Mn^{2+} complex (Fig. 6.1C) and *Drosophila melanogaster* Xrn1 (DmXrn1, D207A mutant) in complex with a three–nucleotide, single-stranded DNA with 5′–end monophosphate (pTTT), and Mg^{2+} (Fig. 6.1D) were reported recently [9,10]. Both structures contain ∼1200 residues, covering the region of Xrn1 that is required for exonuclease activity (Fig. 6.1A) [12]. The overall architecture of the two structures is similar, with r.m.s. distance of 2.0 Å for 509 equivalent Cα atoms of CR1 and CR2 between the two proteins (Fig. 6.2A). The structure of CR1 and CR2 of Xrn1 is similar to that of Rat1, although helix αD is

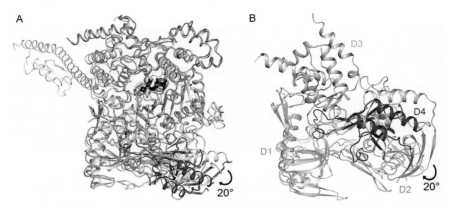

Figure 6.2 Overlay of Xrn1 structures. (A) Structural overlay of DmXrn1 in complex with pTTT and Mg^{2+} (in color) and KlXrn1 in complex with Mn^{2+} (gray). The overlay is based on equivalent residues in CR1 and CR2. (B) Detailed view of the differences in the positions of domains D2–D4 in the structures of DmXrn1 (in color) and KlXrn1 (in gray). (See color plate section in the back of the book.)

much shorter in KlXrn1 (Fig. 6.2A). Like Rat1, Xrn1 also contains a segment (∼60 residues) following CR2 that wraps around the bottom face of CR1, although it does not have the helix that interacts with helix αA of CR1 (Fig. 6.1C). The linker between CR1 and CR2 is mostly ordered in the structure of KlXrn1 (Fig. 6.1C), partly stabilized by crystal packing.

The remaining ∼510-residue segment at the C-terminal region of both structures is composed of four separate domains (named D1–D4, Fig. 6.1A). Domains D1, D2, and D4 share a common backbone fold, with a five-stranded β-barrel core, which is similar to that of Tudor, PAZ, chromo, KOW, or SH3-like domains (Fig. 6.3A–C) [17–20]. Most of these domains are involved in protein–protein, protein–peptide, or protein–nucleic acid interactions [21–23]. Typically, chromo and Tudor domains in histone-modifying enzymes recognize histone methyl-lysines [24,25]. However, the domains in Xrn1 do not contain the aromatic cage, suggesting that they are unlikely to recognize methylated proteins. Similarly, the homology to SH3-like domains is limited to the backbone fold, and the domains in Xrn1 do not bind proline-rich motifs. Domain D3 in DmXrn1 is mostly helical and contains a winged-helix (WH) motif (Fig. 6.3D) [26], and it contacts helix αD in CR1 (Fig. 6.2A). In comparison, domain D3 is mostly disordered in KlXrn1.

An overlay of the structures of KlXrn1 and DmXrn1, based on CR1 and CR2, shows that the positions of domain D1 in the two structures are similar (Fig. 6.2A). On the other hand, the positions of domains D2 and D4 are different, related by a rotation of ∼20° (Fig. 6.2B), suggesting that there may be some degree of variability in the positioning of these domains in Xrn1. The binding of a pTTT nucleotide in the DmXrn1 structure, as well as sequence differences between the two enzymes in this C-terminal segment, may contribute to the observed differences in the positions of these domains.

Domains D1 and D4 have intimate contacts with CR1, although none of the residues in these domains are located directly in the active site of Xrn1 (Fig. 6.1C and D). Internal deletion mutants of KlXrn1 lacking domains D2 or D2–D3 were inactive, and the C-terminal truncation mutant lacking domains D1–D4 was also inactive [9]. On the other hand, the C-terminal truncation mutant lacking domains D2–D4 showed weak nuclease activity. These data suggest that domain D1 is essential for activity. It is in contact with the N-terminal αA helix of CR1, which is located in the active site region. In fact, deleting the first four amino acid residues of Xrn1 essentially abolished the nuclease activity [9], indicating the functional importance of the N-terminal segment of Xrn1. Interestingly, the poorly conserved

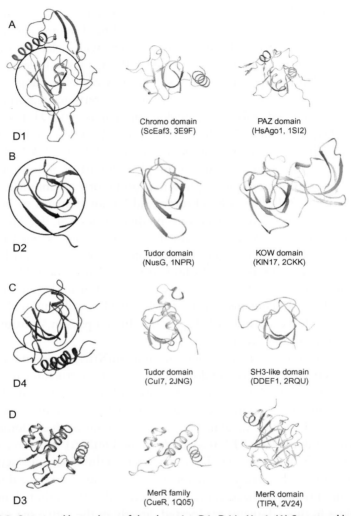

Figure 6.3 Structural homologs of the domains D1–D4 in Xrn1. (A) Structural homologs of domain D1. (B) Structural homologs of domain D2. (C) Structural homologs of domain D4. (D) Structural homologs of domain D3. The Protein Data Bank accession codes are shown for the structures. (See color plate section in the back of the book.)

C-terminal segment of Rat1 also has an α helix that contacts helix αA at the N-terminus (Fig. 6.1B), suggesting that this C-terminal segment is required for stabilization of the N-terminal segment for catalysis. In addition, domains D1–D4 of Xrn1 are also involved in binding the RNA substrate, as deletion mutants have reduced affinity for RNA [9].

4. STRUCTURAL HOMOLOGS OF XRNs

The CR1 structures of XRNs (Fig. 6.4A and B) have strong similarity to the FEN (flap endonuclease) superfamily of nucleases, including FEN-1 (Fig. 6.4C) [27,32–36], 5' Exo1 (Fig. 6.4D) [28], T5 exonuclease (Fig. 6.4E) [29], T4 RNase H (Fig. 6.4F) [30,31], 5' nuclease domain of Taq polymerase [37,38], and PIN-domain containing nucleases [39,40], despite sharing only low amino acid sequence conservation with them [1,4,41]. Most of these other enzymes are structure-specific endo- or exo(ribo) nucleases and they typically interact with substrates with branched structures such as flap, overhang, fork (Y-shape), and Holliday junction. In general, the substrates adopt a kinked downstream duplex and the cleavage occurs near the junction site on a single-stranded 5' overhang [41]. Like the XRNs, FEN superfamily nucleases contain seven conserved acidic residues, which coordinate two metal ions in the active site for catalysis (Fig. 6.4C–E).

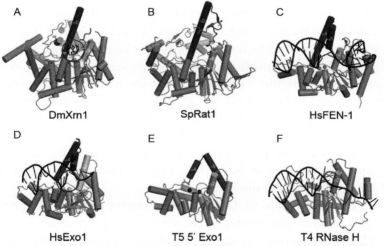

Figure 6.4 Structural homologs of XRNs. (A) Structure of DmXrn1. Only CR1 and CR2 are shown, in cyan and magenta except for helices αB (in red) and αD (in blue). The pTTT oligo is in black, and Mg^{2+} in gray. The blue circle highlights the active site region, and a small gap between helices αB and αD is blocked by CR2 in the back, in comparison to the active site region in human Exo1 (panel D). (B)Structure of *S. pombe* Rat1. (C) Structure of human FEN-1 [27]. The two Sm^{3+} ions are in magenta, and the double flap substrate DNA is in black. (D) Structure of human Exo1 [28]. The two Ba^{2+} ions are in green, and the 3' overhang DNA substrate is in black. (E) Structure of T5 phage 5' Exo1 [29]. The two Mn^{2+} ions are in black. (F) Structure of T4 phage RNase H [30,31]. The fork DNA substrate is shown in black. (See color plate section in the back of the book.)

Helix αD in the XRNs (tower domain in Rat1) is equivalent to the helical clamp or helical arch (α4 and α5 helices), and helix αB is equivalent to helix α2, in FEN-1 like proteins (Fig. 6.4A–E). These two helices provide crucial residues for substrate recognition and catalysis in all of these enzymes.

There are also important differences between XRNs and these other nucleases. First of all, the active site region of these other nucleases is much more open compared to the XRNs. Especially, a gateway formed by helices α2 and α4 (equivalent to helices αB and αD in the XRNs) interacts with the 5′ overhang of the substrate, consistent with an endonuclease activity [41]. In contrast, helices αB and αD are located closer to each other, and therefore, this gateway is much narrower in the XRNs (Fig. 6.4A and B). Moreover, this gateway is blocked by CR2 in the XRNs (e.g., compare Fig. 6.4A with 6.4D). This is the molecular basis for why the XRNs are exclusive exoribonucleases [8].

5. ACTIVE SITE OF XRNs

The active site of XRNs is located at the top of the central β-sheet of CR1, and it also involves residues from the αB and αD helices (Fig. 6.1B). The seven conserved acidic residues in CR1 are located in the center of the active site (Fig. 6.5A), mediating the coordination of two metal ions for catalysis, probably similar to the two-metal-ion mechanism proposed for other nucleases [4]. Helices αB and αD contribute positively charged and polar residues to the active site. Residues in CR2 are generally not directly involved in the active site. One exception is Trp540 in DmXrn1, whose side chain is π-stacked with the third base of the pTTT oligo (Fig. 6.5A) [10]. The equivalent residue in KlXrn1 is Trp638, but the 13-residue loop containing this residue is disordered.

The crystal structure of DmXrn1 in complex with pTTT provides molecular insights into substrate recognition and processivity of XRNs [10]. The 5′-end monophosphate of the substrate interacts extensively with Lys93, Gln97, Arg100, and Arg101 of the αD helix (Fig. 6.5B). The R100A/R101A mutant had essentially no nuclease activity. The base of the first nucleotide is π-stacked with the side chain of His41, in helix αB (Fig. 6.5B). In fact, the three bases of the pTTT oligonucleotide are in a π-stack, which is capped at the 5′-end by His41 and at the 3′-end by Trp540. No specific recognition of the bases themselves is observed, consistent with the fact that XRNs are not selective toward the sequence of the substrate. Mutation of His41 to Ala, as well as Lys93 to Ala, reduced

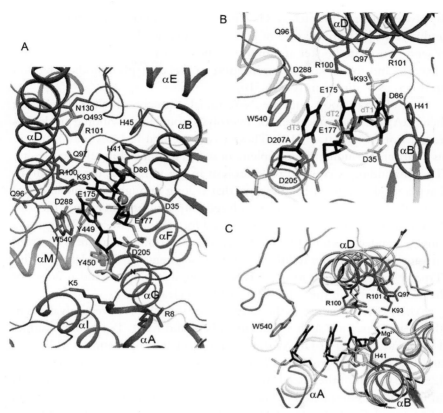

Figure 6.5 Active site of XRNs. (A) Schematic drawing of the DmXrn1 active site, in complex with pTTT and Mg²⁺ ion [10]. (B) Detailed interactions between 5'-end nucleotide of pTTT and the active site of DmXrn1. Also visible is the base stacking of the three nucleotides, capped at the 5'-end by His41 and 3'-end by Trp540. (C) Comparison of the substrate binding modes in DmXrn1 (cyan and magenta), human FEN-1 (gray), and human Exo1 (pink). (See color plate section in the back of the book)

the nuclease activity, and the major product of the H41A mutant was a dinucleotide instead of mononucleotide. More importantly, both mutants also showed intermediates in the reaction, suggesting that they are important for the processivity of the enzyme. On the other hand, the W540A mutant showed comparable activity as the wild-type enzyme, and Trp540 does not have an equivalent in FEN-1 and Exo1 as they lack CR2 (Fig. 6.5C).

Based on the structural and biochemical data, a Brownian ratchet mechanism was proposed to explain DmXrn1 processivity [10]. This mechanism has been applied to understand the physical basis of translocation or transport by processive proteins such as RNA polymerase elongation complex [42,43] and myosin [44]. In the case of DmXrn1, the side chain of His41 and the 5'-end phosphate pocket are important both for substrate recognition and cleavage and for substrate translocation to the next phosphodiester bond in the 5' to 3' direction (Fig. 6.5B).

The Trp540 loop and the N-terminal segment of Xrn1 (including helix αA) may be important for hydrolysis-coupled unwinding of duplex regions of the substrate [10]. These two structural features may function as a barrier, excluding the complement strand from reaching the active site. The unwinding is coupled to hydrolysis in the sense that the RNA must have a single-stranded 5' overhang that is long enough (≥ 3 nucleotides) to reach the active site and be hydrolyzed [10]. On the other hand, the four N-terminal residues of KlXrn1 are critical for exonuclease activity against a single-stranded substrate [9]. Therefore, the N-terminal segment is involved in more than just unwinding the duplex segments of the substrate. This Trp540 loop is unique to the XRNs, belonging to CR2, and it does not have equivalents in the other FEN-1 like nucleases.

Many of the residues in the active site of the XRNs have equivalents in the related enzymes such as human FEN-1 and Exo1. Especially, residues Lys93 and Arg100 of DmXrn1 are equivalent to those residues in FEN-1 and Exo1 that recognizes the 5'-end phosphate of the product (Fig. 6.5B) [27,28]. On the other hand, the other two residues (Gln97 and Arg101) in the 5'-end phosphate binding pocket of DmXrn1 are not conserved in FEN-1 and Exo1. His41 of DmXrn1 is equivalent to Tyr40 of FEN-1 and His36 of Exo1. Mutation of Tyr40 to Ala in FEN-1 or His36 to Ala in Exo1 reduced the nuclease activity by ~ 150- or 20-fold [27,28].

6. STRUCTURE OF THE RAT1–RAI1 COMPLEX

Rai1 is bound on the opposite face from the Rat1 active site (Fig. 6.1B), and primarily interacts with the poorly conserved C-terminal loop that traverses the bottom of CR1 [8]. Within Rai1, the α8–βE segment and strand β4 are located in the interface with Rat1 (Fig. 6.1B). Mutations in the Rat1–Rai1 interface can abolish the interaction as well as the stimulation of Rat1 by Rai1 [8]. On the other hand, Rai1 residues in the interface with Rat1 are not conserved in the mammalian Rai1 homolog, known as Dom3Z. Therefore, there are no interactions between Xrn2 and Dom3Z in mammalian cells.

The α8-βE segment of Rai1 undergoes large conformational changes upon binding to Rat1. In the structure of Rai1 alone, the αE helix is disordered, and the α8-βE loop assumes a conformation different from that of Rai1 in complex with Rat1. These conformational changes suggest an "induced fit" binding mechanism. Other than these differences the structures of Rai1 alone and in complex with Rat1 are very similar to each other, with r.m.s. distance of 0.5 Å for equivalent Cα atoms.

Rai1 does not directly contribute to the active site of Rat1 (Fig. 6.1B), and it enhances Rat1's exonuclease activity in part by increasing the enzyme's stability [15]. At the same time, the Rat1 enzyme may be inherently less active [45], and Rai1 may also indirectly help to organize the active site of Rat1, possibly through indirectly stabilizing the N-terminal segment (including helix αA).

7. STRUCTURE OF RAI1/DOM3Z

Rai1 homologs are found in most eukaryotes. The structure of *S. pombe* Rai1 is composed of two highly twisted, mixed β-sheets, and several α helices that cover up some of the exposed surfaces of the β-sheets (Fig. 6.6A) [8]. The overall structure of Dom3Z is similar to that of Rai1, consistent with their sequence conservation [8]. The r.m.s. distance between equivalent Cα atoms in Rai1 and Dom3Z is 1.4 Å.

A striking feature of the Rai1 structure is a large pocket in its surface. The pocket is about 15 Å in diameter and 10 Å deep, with a divalent cation (Mg^{2+} or Mn^{2+}) bound at the bottom (Fig. 6.6B). The side chains of highly conserved Glu150, Asp201, and Glu239; the main-chain carbonyl of Leu240; and two water molecules form the octahedral coordination sphere of the cation (Fig. 6.6C). Other residues located in this pocket are also conserved among Rai1 homologs (Fig. 6.6B), despite the fact that the overall sequence conservation among these proteins is rather weak. In fact, the few residues that are conserved between Rai1 and Dom3Z homologs are all located in this pocket. Therefore, the structural observations and sequence analyses suggest that the pocket is the active site of Rai1. Rai1 and its homologs (including Dom3Z) possess enzymatic activity.

Further, biochemical studies demonstrate that Rai1 has RNA 5′-end pyrophosphohydrolase (PPH) activity toward RNA with 5′-end triphosphate group (pppRNA) as well as decapping activity toward capped but unmethylated RNA (GpppRNA), releasing GpppN as the product [8,16]. Rai1/Dom3Z is involved in a novel quality surveillance

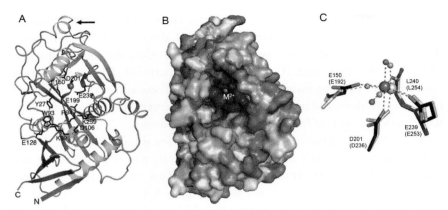

Figure 6.6 Structural and sequence analyses suggest Rai1 has an active site. (A) Schematic drawing of the structure of *S. pombe* Rai1 [8]. The two β-sheets are in cyan and green. A bound divalent cation is shown as a sphere in gray. The black arrow indicates the interface with Rat1. (B) Molecular surface of Rai1, colored based on sequence conservation (blue: most conserved, red: least conserved, other colors: intermediate conservation) [46]. Conserved residues are located in a pocket in the structure. The metal ion is at the bottom of the pocket (gray sphere). (C) Overlay of the metal ion coordination sphere in the structure of Rai1 (in black) and mouse Dom3Z (in gray, with residue numbers in parentheses). Glu192 in Dom3Z coordinates the metal ion through a water molecule. (See color plate section in the back of the book.)

mechanism for mRNA 5′-end capping, mediating the removal of mRNAs with aberrant 5′-end caps. Additional information on the function of Rai1 can be found in Chapter 9.

8. PERSPECTIVES

The XRN structures present several key components that are important for their 5′–3′ exoribonuclease activity. First, a conserved pocket captures the 5′-end phosphate of the substrate. This pocket is only large enough for a phosphate group and cannot accommodate triphosphate or cap group, thereby ensuring substrate specificity. Second, the side chain of His41 π-stacks with the base of the first nucleotide. The 5′-end phosphate binding pocket and His41 are important for substrate translocation and processivity. Third, the N-terminal segment of the enzymes, including the αA helix, is crucial for activity. The conformation of this segment is stabilized through interactions with a helix in the poorly conserved C-terminal segment of Rat1 or the D1 domain of Xrn1. Finally, domains D2–D4 in Xrn1 may be involved in substrate binding and/or interactions with other proteins.

The recent structures of three XRNs and two FEN superfamily proteins have greatly expanded our knowledge on substrate recognition and catalytic mechanism of these enzymes. At the same time, additional structural information is needed to fully understand how the two metal ions are coordinated by the enzyme and participate in catalysis. Moreover, the activities of the XRNs are likely regulated through interactions with other proteins. Structural studies of such complexes should provide further insights into these important and ubiquitous enzymes.

ACKNOWLEDGMENTS

This research is supported in part by a grant from the NIH to L. T. (GM090059).

REFERENCES

[1] Chang JH, Xiang S, Tong L. 5'-3' Exoribonucleases. In: Nicholson AW, editor. Ribonucleases. Berlin: Springer-Verlag; 2011. p. 167–92.

[2] Stevens A. An exoribonuclease from Saccharomyces cerevisiae: effect of modifications of 5' end groups on the hydrolysis of substrates to 5' mononucleotides. Biochem Biophys Res Commun 1978;81:656–61.

[3] Stevens A, Poole TL. 5'-exonuclease-2 of Saccharomyces cerevisiae. Purification and features of ribonuclease activity with comparison to 5'-exonuclease-1. J Biol Chem 1995;270:16063–9.

[4] Yang W. Nucleases: diversity of structure, function and mechanism. Quart Rev Biophys 2011;44:1–93.

[5] Till DD, et al. Identification and developmental expression of a 5'-3' exoribonuclease from Drosophila melanogaster. Mech Develop 1998;79:51–5.

[6] Newbury S, Woollard A. The 5'-3' exoribonuclease xrn-1 is essential for ventral epithelial enclosure during C. elegans embryogenesis. RNA 2004;10:59–65.

[7] Kastenmayer JP, Green PJ. Novel features of the XRN-family in Arabidopsis: evidence that AtXRN4, one of several orthologs of nuclear Xrn2p/Rat1p, functions in the cytoplasm. Proc Natl Acad Sci USA 2000;97:13985–90.

[8] Xiang S, Cooper-Morgan A, Jiao X, Kiledjian M, Manley JL, Tong L. Structure and function of the 5'→3' exoribonuclease Rat1 and its activating partner Rai1. Nature 2009;458:784–8.

[9] Chang JH, Xiang S, Xiang K, Manley JL, Tong L. Structural and biochemical studies of the 5'→3' exoribonuclease Xrn1. Nature Struct Mol Biol 2011;18:270–6.

[10] Jinek M, Coyle SM, Doudna JA. Coupled 5' nucleotide recognition and processivity in Xrn1-mediated mRNA decay. Mol Cell 2011;41:600–8.

[11] Solinger JA, Pascolini D, Heyer W-D. Active-site mutations in the Xrn1p exoribonuclease of Saccharomyces cerevisiae reveal a specific role in meiosis. Mol Cell Biol 1999;19:5930–42.

[12] Page AM, Davis K, Molineux C, Kolodner RD, Johnson AW. Mutational analysis of exoribonuclease I from Saccharomyces cerevisiae. Nucl Acid Res 1998;26:3707–16.

[13] Johnson AW. Rat1p and Xrn1p are functionally interchangeable exoribonucleases that are restricted to and required in the nucleus and cytoplasm, respectively. Mol Cell Biol 1997;17:6122–30.

[14] Shobuike T, Tatebayashi K, Tani T, Sugano S, Ikeda H. The dhp1+ gene, encoding a putative nuclear 5′ → 3′ exoribonuclease, is required for proper chromosome segregation in fission yeast. Nucl Acid Res 2001;29:1326–33.

[15] Xue Y, et al. Saccharomyces cerevisiae RAI1 (YGL246c) is homologous to human DOM3Z and encodes a protein that binds the nuclear exoribonuclease Rat1p. Mol Cell Biol 2000;20:4006–15.

[16] Jiao X, Xiang S, Oh C-S, Martin CE, Tong L, Kiledjian M. Identification of a quality-control mechanism for mRNA 5′-end capping. Nature 2010;467:608–11.

[17] Cavalli G, Paro R. Chromo-domain proteins: iinking chromatin structure to epigenetic regulation. Curr Opin Cell Biol 1998;10:354–60.

[18] Kyrpides NC, Woese CR, Ouzounis CA. KOW: a novel motif linking a bacterial transcription factor with ribosomal proteins. Trends Biochem Sci 1996;21:425–6.

[19] Mayer BJ. SH3 domains: complexity in moderation. J Cell Sci 2001;114:1253–63.

[20] Thomson T, Lasko P. Tudor and its domains: germ cell formation from a Tudor perspective. Cell Res 2005;15:281–91.

[21] le Maire A, et al. A tandem of SH3-like domains participates in RNA binding in KIN17, a human protein activated in response to genotoxics. J Mol Biol 2006;364:764–76.

[22] Ponting CP. Tudor domains in proteins that interact with RNA. Trends Biochem Sci 1997;22:51–2.

[23] Selenko P, Sprangers R, Stier G, Buhler D, Fischer U, Sattler M. SMN tudor domain structure and its interaction with the Sm proteins. Nat Struct Biol 2001;8:27–31.

[24] Corsini L, Sattler M. Tudor hooks up with DNA repair. Nat Struct Mol Biol 2007;14:98–9.

[25] Nielsen PR, et al. Structure of the chromo barrel domain from the MOF acetyltransferase. J Biol Chem 2005;280:32326–31.

[26] Gajiwala KS, et al. Structure of the winged-helix protein hRFX1 reveals a new mode of DNA binding. Nature 2000;403:916–21.

[27] Tsutakawa SE, et al. Human flap endonuclease structures, DNA double-base flipping, and a unified understanding of the FEN1 superfamily. Cell 2011;145:198–211.

[28] Orans J, et al. Structures of human exonuclease 1 DNA complexes suggest a unified mechanism for nuclease family. Cell 2011;145:212–23.

[29] Ceska TA, Sayers JR, Stier G, Suck D. A helical arch allowing single-stranded DNA to thread through T5 5′-exonuclease. Nature 1996;382:90–3.

[30] Mueser TC, Nossal NG, Hyde CC. Structure of bacteriophage T4 RNase H, a 5′ to 3′ RNA-DNA and DNA-DNA exonuclease with sequence similarity to the RAD2 family of eukaryotic proteins. Cell 1996;85:1101–12.

[31] Devos JM, Tomanicek SJ, Jones CE, Nossal NG, Mueser TC. Crystal structure of bacteriophage T4 5′ nuclease in complex with a branched DNA reveals how Flap endonuclease-1 family nucleases bind their substrates. J Biol Chem 2007;282:31713–24.

[32] Hwang KY, Baek K, Kim H-Y, Cho Y. The crystal structure of flap endonuclease-1 from Methanococcus jannaschii. Nat Struct Biol 1998;5:707–13.

[33] Sayers JR, Artymiuk PJ. Flexible loops and helical arches. Nat Struct Biol 1998;5:668–70.

[34] Chapados BR, et al. Structural basis for FEN-1 substrate specificity and PCNA-mediated activation in DNA replication and repair. Cell 2004;116:39–50.

[35] Sakurai S, et al. Structural basis for recruitment of human flap endonuclease 1 to PCNA. EMBO J 2005;24:683–93.

[36] Patel N, et al. Flap endonucleases pass 5′-flaps through a flexible arch using a disorder-thread-order mechanism to confer specificity for free 5′-ends. Nucl Acid Res 2012;40:4507–19.

[37] Kim Y, Eom SH, Wang J, Lee D-S, Suh SW, Steitz TA. Crystal structure of Thermus aquaticus DNA polymerase. Nature 1995;376:612–6.

[38] Murali R, Sharkey DJ, Daiss JL, Murthy HMK. Crystal structure of Taq DNA polymerase in complex with an inhibitory Fab: the Fab is directed against an intermediate in the helix-coil dynamics of the enzyme. Proc Natl Acad Sci USA 1998;95:12562–7.

[39] Clissold PM, Ponting CP. PIN domains in nonsense-mediated mRNA decay and RNAi. Curr Biol 2000;10:R888–90.

[40] Glavan F, Behm-Ansmant I, Izaurralde E, Conti E. Structures of the PIN domains of SMG6 and SMG5 reveal a nuclease within the mRNA surveillance complex. EMBO J 2006;25:5117–25.

[41] Grasby JA, Finger LD, Tsutakawa SE, Atack JM, Tainer JA. Unpairing and gating: sequence-independent substrate recognition by FEN superfamily nucleases. Trends Biochem Sci 2012;37:74–84.

[42] Bar-Nahum G, Epshtein V, Ruckenstein AE, Rafikov R, Mustaev A, Nudler E. A ratchet mechanism of transcription elongation and its control. Cell 2005;120:183–93.

[43] Guo Q, Sousa R. Translocation by T7 RNA polymerase: a sensitively poised Brownian ratchet. J Mol Biol 2006;358:241–54.

[44] Houdusse A, Sweeney HL. Myosin motors: missing structures and hidden springs. Curr Opin Struct Biol 2001;11:182–94.

[45] Sinturel F, Pellegrini O, Xiang S, Tong L, Condon C, Benard L. Real-time fluorescence detection of exoribonucleases. RNA 2009;15:2057–62.

[46] Armon A, Graur D, Ben-Tal N. ConSurf: an algorithmic tool for the identification of functional regions in proteins by surface mapping of phylogenetic information. J Mol Biol 2001;307:447–63.

CHAPTER SEVEN

Rat1 and Xrn2: The Diverse Functions of the Nuclear Rat1/Xrn2 Exonuclease

Michal Krzyszton, Monika Zakrzewska-Placzek, Michal Koper, Joanna Kufel[1]

Faculty of Biology, Institute of Genetics and Biotechnology, University of Warsaw, Pawinskiego 5a, Warsaw, Poland

[1]Corresponding author: e-mail address: kufel@ibb.waw.pl

Contents

1. Introduction	132
2. The Life Story of Rat1/Xrn2	133
3. Sequence and Structural Data	135
4. One Protein, Many Functions	138
4.1 RNA processing	138
4.2 Transcription termination: The torpedo model	141
4.3 RNA surveillance	149
5. Overlapping Functions of Rat1 and Xrn1	154
6. Xrn2 as a Silencing Suppressor	155
7. Perspectives	156
Acknowledgments	157
References	157

Abstract

The role of the nucleus of a eukaryotic cell during gene expression is not only limited to transcription and RNA processing but also includes the initial stages of RNA surveillance. All of these processes, and more precisely, transcription elongation and termination, 5′-end RNA maturation, and the removal of processing intermediates and aberrant molecules, require the activity of the nuclear 5′-3′ exoribonuclease Rat1/Xrn2. This protein, together with its cytoplasmic counterpart, Xrn1, constitutes a highly conserved eukaryotic family of nucleases, whose roles exceed participation in RNA metabolism alone. Despite many years of extensive research and recent findings related to the structure and function of these enzymes revealed almost every year, several aspects are yet to be discovered.

The Enzymes, Volume 31
ISSN 1874-6047
http://dx.doi.org/10.1016/B978-0-12-404740-2.00007-0

1. INTRODUCTION

The nuclear Rat1/Xrn2, together with the cytoplasmic Xrn1, constitutes the XRN family of conserved exonucleases that is unique for eukaryotic organisms. Both enzymes display highly processive 5′-3′ activity toward 5′ monophosphorylated RNA substrates and control the majority of cellular processes which require truncating or removing 5′ RNA sequences. Together, they are responsible for the 5′ processing or turnover of a broad range of substrates and, as a consequence, impact on different developmental processes and cellular metabolic pathways, including RNA interference, stress response, and hormone signaling.

RNA turnover pathways proceed in either a 5′-3′ or a 3′-5′ direction with participation of the XRN nucleases or the exosome complex, respectively (reviewed in Ref. [1]). Usually, both mechanisms contribute to the execution of this process to varying extent and sometimes cooperate to achieve the most efficient effect, but there seems to be no clear rules dictating the preferred mode, which may differ depending on the organism or be substrate specific. Both pathways are subject to complex regulatory strategies and require the involvement of auxiliary factors, such as RNA-binding proteins, deadenylases, decapping enzymes, and helicases, which may encourage the choice of the more productive mechanism.

As the structure and function of Xrn1 is covered in previous chapters, here the most relevant aspects concerning the nuclear Rat1/Xrn2 will be described. Generally, due to its localization to the nucleus, Rat1 activities are biased toward RNA processing rather than degradation. Rat1 also operates more often on RNA precursors, excised intermediates of processing reactions, nascent (i.e., polymerase-associated) transcripts, and a broad range of unstable, intergenic, noncoding species. Nevertheless, most of the functions of Rat1 in yeast can be fulfilled by Xrn1, especially when it is targeted to the nucleus, so it is apparent that the diverse roles of both enzymes in RNA metabolism result from their different subcellular locations. Curiously enough, the yeast *RAT1* gene is essential for viability while *XRN1* can be deleted, but the temperature-sensitive *rat1* mutant phenotype is rescued by Xrn1, suggesting that the essential function of Rat1 can be substituted by Xrn1. However, despite decades of extensive research this Rat1 activity has not been identified, probably owing to the extensive redundancies between both nucleases. The only Rat1 activity described to date that cannot be replaced by Xrn1 is in transcription termination, as this process requires not only the exonuclease activity but also the interactions with other key factors.

2. THE LIFE STORY OF RAT1/XRN2

The gene, whose product was later demonstrated to be responsible for the majority of the nuclear 5′-3′ exoribonuclease activity in *Saccharomyces cerevisiae*, was identified in at least five independent genetic screens. The first one, aimed at discerning temperature-sensitive (*ts*) mutations causing the nuclear retention of polyadenylated RNAs, was performed with oligo(dT) FISH (fluorescence *in situ* hybridization) [2] and provided an essential gene encoding a 116 kDa protein, named *RAT1* (ribonucleic acid trafficking 1) with a homology to the previously cloned *XRN1*. At the nonpermissive temperature, the *rat1-1* mutant accumulated poly(A)+ polymerase II (Pol I)-dependent RNAs in nuclear spots [2], a phenotype which later was also associated with mutations in components of the cleavage factor I (CF I) complex involved in mRNA 3′-end formation [3] and in the Mtr4 subunit of the TRAMP and exosome complexes [4]. In addition, the *rat1-1* strain showed defects in 5′-end processing of the 5.8S rRNA at nonpermissive conditions (see below) and had decreased levels of actin mRNA [2]. The second genetic screen, designed to find factors engaged in Pol III transcription initiation, identified a *TAP1* (transcription activation protein 1) gene, whose temperature-sensitive mutation resulted in a decrease in tRNA and 5S rRNA synthesis and subsequent inhibition of rRNA biogenesis [5]. Despite the fact that *TAP1* is identical to *RAT1*, poly(A)+ RNA retention in the nucleus was not observed in the *tap1-1* strain. How *tap1-1* mutation affects tRNA expression is unclear, but it is possibly through the involvement of Rat1 in tRNA surveillance [6]. In turn, the temperature-sensitive mutation in *HKE1* (homology to *KEM1/XRN1*) was found in a screen for genes involved in the targeting of proteins to mitochondria, even though it was not directly linked to this process [7].

Purification of Rat1/Tap1/Hke1 from yeast cells enabled the demonstration of its 5′-3′ exoribonuclease activity and suggested that it is identical to the previously observed exo-2 activity, which is separate from exo-1 (Xrn1) [7–9]. Finally, the *rsf11* (requiring SWI4 11) temperature-sensitive mutant was identified in a screen for genes synthetically lethal with the deletion of *SWI4* [10]. The basis of this interaction is not known, but it can be suppressed by overexpression of the *RME1*-encoded transcription repressor, which prevents cells from entering meiosis. Based on the homology between Rat1 and Xrn1, it was predicted that they can be at least partially interchangeable, but the first attempts to suppress the *hke1-1/rat1-1* phenotype by overexpression of Xrn1 were unsuccessful, most likely due to the nuclear localization of Rat1 [7,11]. Indeed, mutations in the nuclear localization sequence (NLS) of Rat1, found in a genetic screen for suppressors of the *xrn1Δ ski2Δ* double-mutant lethality, that led to Rat1 mislocalization to the cytoplasm, rescued all known *xrn1Δ* phenotypes, including cytoplasmic RNA turnover defects, sensitivity to the microtubule-destabilizing drug benomyl, and lower sporulation [11]. Even a simple overexpression of Rat1 can complement *xrn1Δ* defects, probably by saturating nuclear import [12]. On the other hand, *rat1-1* *ts*-lethality can be suppressed by targeting catalytically active Xrn1 to the nucleus, which supports the notion that Xrn1 and Rat1 can perform the same functions under some circumstances and confirms that the nuclear 5′-3′ exoribonuclease activity is essential for cell viability [11]. Interestingly, Rat1 lacking the NLS still complemented *rat1-1* defects, possibly due to additional nuclear targeting mechanisms.

Studies of the catalytic properties of Rat1 revealed that it is a processive, Mg^{2+}-dependent and ATP-independent 5′-3′ exoribonuclease, generating 5′-NMPs as a product, with no discernible endonucleolytic activity and a poor capacity to hydrolyze RNAs with a cap or triphosphate group at the 5′-end [7,9]. Similarly to Xrn1, Rat1 also possesses some activity toward single-stranded DNA, but this is only about 5% of its ribonuleolytic capacity. Mg^{2+} can be substituted by Mn^{2+}, but the enzyme does not achieve optimal rates even at high concentrations of manganese ions [9].

In yeast cells, the abundance of Rat1 reaches only approximately 6% of Xrn1 levels [9]. Both Xrn1 and Rat1 can be inhibited by the deletion of *MET22/HAL2* [13], coding for a phosphatase that is able to convert pAp (adenosine 5′,3′ bisphosphate) to phosphate and Ap. As pAp blocks the 5′-3′ exoribonuclease activity of Xrn1 and Rat1, and Met22/Hal2 is inhibited by lithium, both nucleases are also repressed in the presence of this ion.

Rat1 homologues were identified in many eukaryotic organisms. *Schizosaccharomyces pombe* Dhp1p is at least partially able to suppress both the *xrn1Δ* phenotypes and *rat1-1* *ts*-lethality in *S. cerevisiae*. The gene encoding this protein is essential for *S. pombe* viability [14], and the *ts* *dhp1-1* mutant retains poly(A)+ RNAs in the nucleus and has aberrant chromosome segregation during mitosis at restrictive conditions [15]. Interestingly, the latter defect is shared with the mutation in *dis3+* gene encoding the exosome component [16]. Dhp1 was proposed to interact with and to be stabilized by the Rai1 (Rat1-interacting protein 1) homologue, Din1, which was able to suppress the *dhp1-1 ts* phenotype when overexpressed [15]. The lethality of *DHP1* deletion in *S. pombe* is rescued by expression of the mouse Rat1 homologue, Dhm1 [17], illustrating the conserved functions of this protein. The human counterpart of Rat1, Xrn2, was also identified and extensively studied [18], and genes encoding Rat1 homologues are present in other higher eukaryotes, including *Drosophila melanogaster* and *Caenorhabditis elegans* [19].

Interestingly, higher plants lack XRN1 homologues and possess only orthologues of Rat1/Xrn2 [20]. *Arabidopsis thaliana* has three XRN proteins: nuclear XRN2 and XRN3, which are able to suppress *rat1-1* lethality in *S. cerevisiae*, and cytoplasmic XRN4, which lacks the NLS and acts as a functional homologue of Xrn1, except that it is not responsible for the overall mRNA decay, but only acts on selected substrates [21]. The single-cell algae *Chlamydomonas reinhardtii* is also thought to have three Rat1 homologues, XRN1/2/3, but none of them seem to contain the NLS, and XRN1 may actually be targeted to mitochondria [22].

Initially, Rat1 was shown to copurify with a 45 kDa protein [12], which was later identified as Rai1 [23]. *S. pombe* Rai1 was demonstrated to have two enzymatic activities: pyrophosphohydrolase, which is able to remove pyrophosphate from pppNpRNA and phosphodiesterase-decapping endonuclease, which releases GpppN from RNAs with unmethylated caps [24,25]. Yeast cells lacking Rai1 are slow growing and are synthetic lethal with a *rat1-1 ts* mutation even at the permissive conditions [23]. Rai1 activates Rat1 and assists in several Rat1-mediated processes, including 5'-end rRNA processing and rRNA degradation, nuclear mRNA decay, and the torpedo mechanism of Pol I and Pol II transcription termination [23,26–29]. In addition, Rai1 stabilizes Rat1 and stimulates its exoribonuclease activity *in vitro*, especially when RNA substrates have strong secondary structures [23,25].

3. SEQUENCE AND STRUCTURAL DATA

According to the present phylogenetic and structural classification, exoribonucleases encompass six superfamilies [30]. Rat1/Xrn2 and Xrn1 constitute the XRN superfamily, which is exclusive to eukaryotes and, until recently, was believed to be the only group capable of RNA degradation in the 5′-3′ direction. All XRN members share a common conserved N-terminal domain designated XRN_N [20,31,32]. Despite exhibiting vestigial conservation of the determining motif, they belong to the PIN (PilT N-terminus) domain-containing clan of proteins [33,34]. More recently, it appeared that the 5′-3′ exoribonuclease activity is not a unique feature of the XRN family. The yeast alternative 5′-3′ exonuclease Rrp17 and its higher eukaryotic homologues are not similar to known exonucleases but nevertheless contain a conserved domain that is important for their catalytic activity [35]. Mammalian CPSF-73, in addition to its endonuclease function in histone mRNA 3′-end formation, may also act as a 5′-3′ exonuclease in the degradation of the resulting downstream cleavage products [36]. Finally, the first example of 5′-3′ exonuclease activity in prokaryotes was demonstrated for the *Bacillus subtilis* endoribonuclease RNase J1, which is structurally related to CPSF-73 [37,38]. However, the XRN nuclease family is still considered to be responsible for the majority of the cellular 5′-3′ exoribonuclease activity and is one of the key players in RNA metabolism.

The amino acid sequence of Rat1/Xrn2 and Xrn1 displays a high level of homology in the XRN_N domain, which is formed of two conserved regions, CR1 and CR2. These motifs are separated by a divergent and less-structured spacer that is more sensitive to protease treatment [39]. The N-terminal domain has an overall higher content of acidic residues, of which seven in CR1 are strongly conserved and are responsible for the nuclease activity [40,41]. Xrn1 proteins are longer than Rat1/Xrn2 (1528 aa vs. 1006 aa in *S. cerevisiae*) and the extended C-terminal part of Xrn1 is composed of more basic amino acids and shows the lowest degree of conservation.

The crystal structures of the fission yeast Rat1, in complex with its activating partner Rai1, or of Xrn1 from *Kluyveromyces lactis* and *D. melanogaster* offered a detailed view of the architecture of XRN proteins [25,42,43]. According to the 2.2 Å resolution, crystallographic data obtained for the *S. pombe* full-length Rai1 (41 kDa) either alone or in a complex with a C-terminally truncated Rat1 (1–885 aa; 101 kDa), which still supports cell viability, CR1 and CR2 of Rat1 form a large, well-knit domain with the most conserved acidic residues of CR1 in the active site of the enzyme [25]. CR1 is built of seven β-sheets sealed by α-helices, of which the longest αD-helix forms the tower domain, which is extended for 30 Å. As Xrn1 proteins do not contain the αD-helix, the tower domain is probably unique for Rat1/Xrn2 and has been proposed to be responsible for Rat1-specific functions, such as transcription termination [25]. In turn, CR2 is composed of a few helices and long loops and contributes to the overall Rat1 structure by providing additional active-site residues and, more importantly, by creating a pocket that surrounds the acidic residues directly involved in catalysis. Such a conformation is more closed compared to the active sites of other known nucleases, and provides a convincing structural explanation for the absence of an endonucleolytic activity in the XRN-family enzymes.

The less-conserved C-terminal region of Rat1 is also less structured and is composed of several loops, which locate mainly at the bottom of the structure, on the opposite side of the active site. It is important that the C-terminal part of the protein constitutes the interface for interaction with its partner Rai1, which substantially stabilizes Rat1. The interacting surfaces of both proteins include ~ 800 Å^2 and show a high degree of shape compatibility. Although Rai1 has no exoribonuclease activity itself, binding of Rai1 greatly enhances Rat1 processivity by helping to unwind RNAs with stable secondary structures, which causes Rat1 stalling. These results are fully consistent with previous biochemical observations for the *S. cerevisiae* Rat1–Rai1 complex [23]. Surprisingly, only fungal Rai1 proteins contain conserved residues responsible for the interaction with Rat1, and consistently, the mammalian Rai1 homologue, DOM3Z, most likely does not associate with Xrn2 and its depletion does not affect Pol II transcription termination or the cotranscriptional removal of aberrant transcripts by Xrn2 [25,44].

Structural studies demonstrated that Rai1 and DOM3Z may have a completely new protein fold and also suggested a catalytic capacity, which resides in a large pocket, with four conserved amino acids at the bottom. The enzymatic activity, that is independent of the interaction with Rat1, was confirmed by biochemical studies to be 5′ RNA pyrophosphohydrolase [25]. An additional, also Rat1-independent, phosphodiesterase-decapping endonuclease activity of Rai1 was recently identified, primarily toward RNAs carrying an unmethylated cap structure [24].

4. ONE PROTEIN, MANY FUNCTIONS

4.1. RNA processing

4.1.1 rRNA maturation

The majority of events leading to the formation of mature and functional rRNA molecules occur in the nucleolus, where pre-rRNA is encoded by multiple tandemly repeated genes. Three of the four rRNAs, 18S, 5.8S, and 25/28S, are derived from the large polycistronic pre-rRNA precursor (35S in yeast, 47S in mouse and human cells) transcribed by RNA Pol I, which contains external transcribed spacers (5′ and 3′ETS) at both ends and internal transcribed spacers (ITS1 and ITS2) separating the mature sequences. The 5S rRNA is independently transcribed as a precursor by Pol III.

The maturation of rRNA has been the most extensively studied in *S. cerevisiae*, for which the detailed processing pathway composed of multiple steps was established (Fig. 7.1A), but some information is also available for *Xenopus laevis*, plant, human, and mouse cells (reviewed in Refs. [45–48]). The primary pre-rRNA transcript is processed by endonucleolytic cleavages, exonucleolytic trimming, and covalent nucleotide modifications (2′-O–methylation and pseudouridylation of uridine residues) at evolutionarily conserved sites. It is well established that XRN-family proteins are involved in ribosomal RNA processing not only in yeast but also in multicellular eukaryotes. Consistent with its 5′-3′ exonucleolytic activity, yeast and mammalian Rat1/Xrn2 were shown to be responsible for generating the 5′-ends of the mature 25/28S and 5.8S rRNAs, following the release of their immediate precursors by endonucleolytic cleavages [49–52]. In fact, the 5′ extensions of 25 and 5.8S rRNAs in yeast are trimmed by at least three 5′-3′ exoRNases: Rat1, probably in combination with its activator Rai1, Xrn1, acting in the absence of Rat1, and the atypical, but highly conserved, nuclease Rrp17 [23,29,35,49]. The defects in 25S/5.8S rRNA processing intensify when both Rat1 and Xrn1 are missing, but are not detected when only Xrn1 is deleted, whereas Rrp17 acts in parallel to the activity of Rat1, indicated by a distinct pre-rRNA processing phenotype that is observed upon Rrp17 depletion. Moreover, functional analyses of *rat1*, *rai1*, and *rrp17* mutants suggest the existence of a link between the 5′ and 3′ processing of 5.8S rRNA by Rat1 or Rrp17 and the exosome [23,29,35,53].

The role of Rat1 in rRNA maturation was recently confirmed with the CRAC (crosslinking and analysis of cDNAs) approach [54]. Consistent with

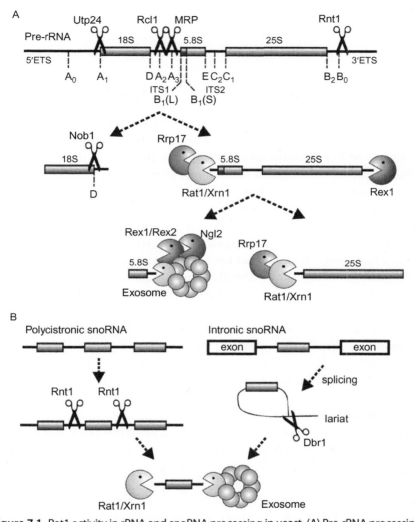

Figure 7.1 Rat1 activity in rRNA and snoRNA processing in yeast. (A) Pre-rRNA processing pathway. The primary transcript is covalently modified by snoRNPs and cotranscriptionally cleaved by Rnt1 endonuclease in the 3′ETS at site B_0. Released 35S pre-rRNA undergoes U3 snoRNP-dependent processing at sites A_0, A_1, and A_2, which leads to the removal of 5′ETS and separation of 35S pre-rRNA into 20S and 27S. The putative endonucleases responsible for the early cleavages, A_0 and A_2, are Utp24 and Rcl1, respectively. The 20S pre-rRNA is exported into the cytoplasm and cleaved endonucleolytically at site D by Nob1, whereas processing of the 27S pre-rRNA proceeds via two alternative pathways. The majority (∼85%) is cleaved at site A_3 by RNase MRP, which is followed by 5′-end exoribonucleolytic trimming by Rat1–Rai1, Xrn1, and Rrp17 and 3′-end maturation by Rex1. The 27SB pre-rRNAs are subsequently cleaved at site C_2 in ITS2, and the resulting intermediates are trimmed by Rat1—Rai1 at the 5′-end and the exosome, Rex1/Rex2 and Ccr4-like RNase Ngl2 at the 3′-end, leading to the formation of the mature 25S rRNAs and two forms of 5.8S rRNAs. (B) The mature 5′- and 3′-ends of polycistronic and intronic snoRNAs are processed by exonucleolytic trimming by Rat1 and Xrn1, and the exosome, following their release from precursors by Rnt1 cleavage or intron lariat debranching, respectively. (See color plate section in the back of the book.)

Rat1 involvement in the degradation of excised precursor fragments, this analysis revealed Rat1 associations at multiple sites in the pre-rRNA, particularly over the 5′ and 3′ regions of the 5′ETS. High levels of Rat1 crosslinking were also observed across the regions flanking other pre-rRNA processing intermediates, known as Rat1 substrates in ITS1 and ITS2, and also over the 5′ regions of mature 5.8S and 25S rRNAs, confirming the Rat1 activity in the 5′-end formation of these species. In addition, the dependence of Rat1 binding at the regions surrounding cleavage sites in ITS1 and ITS2 on some components of pre-60S ribosome particles that associate around the 3′-end of 5.8S, further confirmed that 5.8S pre-rRNA processing at both ends is coordinated. It has been proposed that this cooperation is achieved by the folding of the pre-rRNA, particularly the structural reorganization of ITS2, which brings both termini into close proximity [53,54].

Surprisingly, the plant nuclear Rat1 homologues, *Arabidopsis* AtXRN2 and AtXRN3 are apparently not involved in the 5′ maturation of either 5.8S or 25S rRNAs, but AtXRN2 was demonstrated to contribute to the early endonucleolytic processing of pre-rRNA at site P via a U3-containing ribonucleoprotein (RNP) complex [55].

The role of Rat1 in the degradation of discarded pre-rRNAs and pre-rRNA maturation by-products is described in Section 6.3.

4.1.2 snoRNA processing

Rat1 and Xrn1 are involved in the regulation of rRNA levels not only by their activity in pre-rRNA processing but also through the 5′ end formation of small nucleolar RNAs (snoRNAs) that are essential for correct rRNA biogenesis. In yeast, early cleavages leading to the formation of the mature 18S rRNA require U3, U14, snR10 and snR30 snoRNAs, and the remaining box C/D and box H/ACA snoRNAs participate in pre-rRNA modification, by 2′-O-methylation and pseudouridylation, respectively.

The genomic organization of genes encoding snoRNAs varies substantially among different organisms (reviewed in Ref. [56]). In vertebrates, the majority are localized within introns of protein-coding genes, but some, including U3, are transcribed independently from their own promoter by Pol II. On the other hand, the majority of snoRNA-encoding genes in *S. cerevisiae* form independent transcription units, with only a few being enclosed within introns or transcribed as polycistronic precursors. The latter type of genomic organization is widespread in plants, where the majority of snoRNAs are derived from polycistronic transcripts that undergo enzymatic processing.

In yeast, polycistronic precursors and the majority of independently transcribed box C/D snoRNAs, as well as a few box H/ACA snoRNAs, are endonucleolytically cleaved by a double-strand-specific endonuclease Rnt1 (Fig. 7.1B). These are then subsequently trimmed exonucleolytically by Rat1 and Xrn1 at their 5'-ends and by the exosome at their 3'-ends to generate mature snoRNAs [50,57–59]. It is thought that plant polycistronic snoRNAs are processed through a similar mechanism; however, the enzymes involved in this process remain unknown, with the exception of the plant exosome complex, and neither the *Arabidopsis* Rnt1 homologue, AtRtl2, nor the nuclear 5'-3' exonucleases, AtXRN2 and AtXRN3, participate in the 5'-end processing of snoRNAs [55,60,61].

Also, the 5'-end formation of the majority of intronic snoRNAs requires the activity of Rat1 in yeast (Fig. 7.1B) and an unspecified 5'-3' exonuclease in vertebrates [50,62,63]. This mode of maturation is largely splicing dependent, since the exonucleolytic trimming follows pre-mRNA processing and debranching of the intron lariat. However, there are some examples of snoRNAs that can be processed independently of splicing [63–66].

4.2. Transcription termination: The torpedo model

4.2.1 Polymerase II

Although transcription termination has been much less explored than its initiation, the past 25 years has revealed that the well timed and accurate release of the nascent transcript and RNA polymerases from the DNA template is crucial for the fate of the RNA as well as for overall genome maintenance.

Two mechanisms have been proposed for this process in the case of mRNA coding genes, allosteric (antiterminator) and torpedo (Fig. 7.2A; reviewed in Refs. [67–69]). The allosteric model postulates that 3' processing factors trigger conformational changes in the transcription elongation complex, which facilitates the dissociation of antiterminator factors and/or binding of termination factors [70]. The torpedo model was first proposed by Connelly and Manley [71] and Proudfoot [72], who suggested that the 3' pre-mRNA product generated by the cleavage and polyadenylation apparatus is attacked by 5'-3' exoribonuclease activity, which degrades the Pol II-associated RNA faster than it is synthesized, while removing the polymerase from the template. In fact, both termination pathways are often orchestrated and, with some exceptions, can be unified in one allosteric-torpedo model [73].

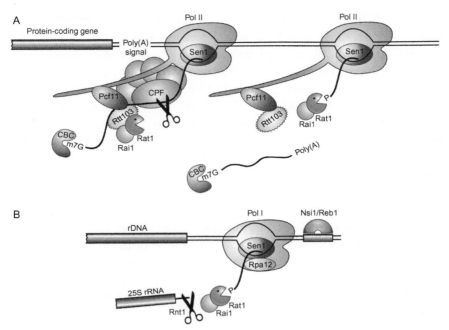

Figure 7.2 Function of Rat1 in the torpedo mechanism of transcription termination. (A) The nascent Pol II mRNA transcript is cleaved cotranscriptionally by the Ysh1 component of the cleavage and polyadenylation specificity factor (CPSF). The resulting exposed monophosphorylated 5'-end of the downstream RNA is a substrate for rapid degradation by Rat1/Rai1, which had been recruited to the Pol II complex via Pcf11 and Rtt103 proteins interacting with the Ser2-phosphorylated CTD. The elongation complex is destabilized upon Rat1/Rai1 catching up with the polymerase. Rat1-dependent termination is stimulated by the Sen1 helicase that interacts with the Rpb1 large Pol II subunit and facilitates Rat1 recruitment. (B) The newly synthesized pre-rRNA is cleaved cotranscriptionally by the Rnt1 endonuclease and the resulting 3' product is degraded by Rat1/Rai1, which torpedoes Pol I. The efficient Pol I termination also requires the binding of Nsi1 and/or Reb1 at the T1 terminator (green box), causing polymerase pausing, the Pol I-specific subunit Rpa12 and helicase Sen1, which associates with rDNA and directly interacts with Rnt1. (See color plate section in the back of the book.) The efficient Pol I termination also requires the binding of Nsi1 and/or Reb1 at the T1 terminator (green box), causing polymerase pausing, the Pol I-specific subunit Rpa12 and helicase Sen1, which associates with rDNA and directly interacts with Rnt1.

Later work revealed that Rat1 in yeast and Xrn2 in humans were responsible for the torpedo activity [27,74,75] and that lack of either enzyme, as well as the Rat1 activator Rai1 in yeast, leads to a termination defect, which manifests as an extensive transcription read-through. It is also possible that a plant Rat1 homologue, *Arabidopsis* AtXRN3, acts as a transcription termination factor similarly to Rat1/Xrn2. High-throughput approaches revealed an accumulation of noncoding transcripts from the 3′-end of mRNA and microRNA (miRNA) genes in the *xrn3* knockdown mutant, which may correspond to transcriptional read-through molecules [76]. Thus, AtXRN3 may participate in a global RNA 3′-end surveillance as a result of its torpedo termination activity.

The torpedo mechanism depends on the exonuclease catalytic activity of Rat1, and not merely on its presence [27]. The monophosphorylated RNA substrate for Rat1/Xrn2 during mRNA synthesis is generated as a result of a cotranscriptional cleavage (CoTC) at the polyadenylation site by the mRNA 3′-end formation machinery. More precisely, this process is performed by the Ysh1 component of the CF II in yeast or the corresponding CPSF-73 subunit of the human cleavage and polyadenylation specificity factor (CPSF) (reviewed in Ref. [77]). Two factors were proposed to contribute to efficient termination: the strength of the poly(A) signal and the action of the exonuclease degrading the 5′ cleavage product, which facilitates pausing of the polymerase downstream of the poly(A) site and promotes its release [75]. On the other hand, impaired Rat1–dependent termination does not abolish proper poly(A)-site recognition and mRNA cleavage [27].

Additional Rat1/Xrn2 entry sites may arise through autocatalytic CoTC downstream of the poly(A) site in mammals [74] or endonucleolytic cleavage by Rnt1 in yeast [78]. The latter case was reported as a fail-safe termination mechanism for protein-coding genes, which, together with the pathway mediated by the Nrd1/Nab3/Sen1 complex (NRD), provides a backup mode for Pol II release.

Although Rat1/Xrn2 is required for efficient Pol II termination, Rat1 fails to trigger termination *in vitro* and its 5′-3′ degradation activity is not sufficient to promote polymerase dissociation *in vivo* [73,79]. Consistently, yeast Xrn1 targeted to the nucleus is capable of the cotranscriptional degradation of nascent RNA but does not rescue termination defects caused by a Rat1 deficiency. These observations suggest that the exonuclease catching up with the polymerase is necessary but not sufficient to destabilize the elongation complex, and that an additional element of the torpedo mechanism operates during termination. It has been postulated that a Rat1 unique feature, the extended tower domain, not present in Xrn1 proteins, may be responsible for its termination properties [25]. It is conceivable that the tower domain undergoes certain modifications or serves as an interface for interactions with termination-enhancing factors. One of the possible candidates for factors stimulating Rat1-dependent termination is the RNA helicase Sen1, which is a key component of the termination pathway for noncoding RNAs in yeast (see below) that also contributes to the termination of protein-coding genes, with the strongest effect on shorter mRNAs [80–82]. Sen1 interacts via its N-terminal domain with the largest Pol II subunit, Rpb1 [83] and impairing Sen1 helicase activity impairs genome-wide Pol II distribution over coding and noncoding genes [81]. The human Sen1 homologue, Senataxin, was shown to directly promote Xrn2-mediated transcription termination by resolving the RNA:DNA hybrids (R-loops) formed behind the elongating Pol II, predominantly at G-rich pause sites downstream of the poly(A) site [84]. This activity facilitates the access of Xrn2 to poly(A) site $3'$ cleavage products and thus contributes to the recruitment of the torpedo exonuclease. The mode of action of yeast Sen1 is thought to be similar [85].

An unanswered question is how Rat1 is recruited to the elongating polymerase. Several observations support Rat1 association via proteins interacting with the Pol II C-terminal domain (CTD) through their CTD-interacting domain (CID), namely the Pcf11 subunit of the yeast cleavage factor IA (CFIA) and Rtt103 (regulator of Ty1 transposition 103). Rat1 and Rai1 are present at the promoters and coding regions with a strong enrichment at the $3'$-ends of genes. The functional interaction between Rat1 and Pcf11 was shown to facilitate the mutual corecruitment of both factors: Rat1 links termination to the $3'$-end formation by stimulating the recruitment of $3'$-end processing factors, especially the CFIA subunits Pcf11 and Rna15, while Pcf11 contributes to Rat1 association over the poly(A) site [73]. Also, human Pcf11 was shown to enhance degradation of nascent RNA and promote transcription termination [86]. In turn, Rtt103, which also associates near the $3'$-ends of genes, copurifies with Rat1, Rai1, and Pcf11, and together with Pcf11 cooperatively recognizes the Ser2-phosphorylated CTD of the elongating Pol II [87,88]. On the other hand, Rat1 distribution over genes is not altered in the absence of Rtt103 [87]. In addition, Rat1 also terminates transcription by Pol I [29,82], which lacks a CTD, suggesting that Rat1 recruitment may also occur via other mechanisms. Indeed, the association of hXrn2 was reported to be mediated by p54nrb/PFS (protein-associated splicing factor), which are multifunctional proteins involved in transcription, splicing, and polyadenylation [89]. The role of p54nrb/PFS in Pol II transcription termination was also demonstrated in *C. elegans* [90].

In addition to mRNAs, Pol II synthesizes a broad range of noncoding transcripts, including small nuclear RNAs and snoRNAs and a variety of unstable RNAs. In yeast, the transcription of these relatively short species is terminated mainly by an alternative NRD complex, composed of the RNA-binding proteins Nrd1 and Nab3 and the ATP-dependent helicase Sen1 [80,81,91–93]. Since this mechanism most likely does not entail a cotranscriptional endonucleolytic cleavage [92], it is the Sen1 RNA:DNA hybrid-unwinding activity that is thought to be essential for polymerase dissociation, perhaps in a manner similar to bacterial Rho DNA–RNA helicase. Rho probably destabilizes the elongation complex by translocating along and removing the nascent RNA from the polymerase [94]. An alternative, allosteric model posits that Rho loads onto polymerase and induces rearrangements in the enzyme active site at termination sites [95]. Both modes of action may well apply for Sen1 helicase.

Although Rat1 is recruited to snoRNA genes, particularly intronic ones, NRD-dependent termination of snoRNA transcripts is not impaired when it is missing, and it was, therefore, suggested that it participates in some premature termination events [92,96]. It is still possible, however, that the Rat1 torpedo mechanism may contribute to transcription termination of snoRNAs, such as U3 or snR40, whose 3′-end is released by Rnt1 cleavage that exposes the remaining nascent transcript 5′-end for Rat1 attack [97,98].

Surprisingly, Rat1 was shown to colocalize with Pcf11 at telomeres and at the Pol III-transcribed tRNA, 5S rRNA, and *SCR1* genes [96]. The functional significance of these observations is currently unclear, but alternative termination modes of these RNAs, and the degradation of aberrant transcripts or the processing of unstable noncoding RNAs, such as telomeric TERRA [99], are a possibility (Table 7.1).

4.2.2 Polymerase I

A strong presence of Rat1 is also observed at Pol I rRNA genes [96]. This is due to the recently reported involvement of Rat1 torpedo in the transcription termination of Pol I transcription (Fig. 7.2B; [82,29]). rRNA genes are arranged in the nucleolus as tandem rDNA repeats. Approximately, 150 transcriptional units in yeast up to several hundred in mammals undergo a complicated and independent regulation of expression. In yeast, each rDNA repeat carrying the 35S rRNA gene is flanked by intergenic spacers 1 and 2 (IGS1 and IGS2). Approximately, 90% of rRNA transcription terminates at the primary T1 located in IGS1, ∼93 nt downstream of the mature 25S rRNA 3′-end, while the remaining transcription stops at the fail-safe T2 region, ∼160 nt further

Table 7.1 Yeast Rat1-cooperating proteins

Factor	Interaction	Activity or function	References
rRNA and snoRNA processing			
Las1	Physical	Modulates Rat1–Rai1	[53]
Nop15	Physical	60S ribosomal subunit biogenesis	[35]
Rai1	Physical	Rat1 stabilization and activation, pyrophosphohydrolase, and phosphodiesterase-decapping endonuclease	[12,23,25]
Rnt1	Functional	RNAase III double-strand-specific endoribonuclease	[50,57,59]
Rrp17	Physical	5′–3′ Exoribonuclease, nuclear	[35]
Xrn1	Genetic	5′-3′ Exoribonuclease, cytoplasmic	[11]
Transcription termination			
Npl3	Physical	Stimulates transcription termination	[100]
Nrd1	Genetic	RNA-binding protein, part of the NRD complex	[78]
Pcf11	Functional	Subunit of the cleavage factor IA, scaffolding protein	[73]
Rai1	Physical		[12,23,25]
Rna15	Functional	Subunit of the cleavage factor IA	[73]
Rnh2	Genetic	Ribonuclease H2	[99]
Rnt1	Functional		[78,82,101]
Rpo21	Genetic	Large subunit of RNA PolII	[102]
Rtt103	Physical	Recruitment of Rat1–Rai1	[87]
Sen1	Genetic	ATP-dependent helicase, part of the NRD complex	[78,82,85]
RNA surveillance			
Pap1	Genetic	Poly(A) polymerase, mRNA polyadenylation	[99]
Pcf11	Functional		[96]
Rai1	Physical		[12,23,25]
Rpb1	Genetic	Large subunit of RNA Pol II	[102]
Ski2	Genetic	RNA helicase, component of the Ski complex, exosome cofactor	[11]

Table 7.1 Yeast Rat1-cooperating proteins—cont'd

Factor	Interaction	Activity or function	References
Tan1	Genetic	$tRNA^{Ser}$ ac^4C_{12} acetyltransferase	[6]
Trm44	Genetic	$tRNA^{Ser}$ Um_{44} 2'-O-methyltransferase	[6]
Trm8	Genetic	Subunit of a tRNA methyltransferase complex	[6]
Trf4	Genetic	Poly(A) polymerase of the TRAMP complex	[99]
Xrn1	Genetic		[11]

downstream (reviewed in Ref. [103]). Proteins responsible for efficient termination at T1 are Nsi1 and, probably to a lesser extent, transcription factor Reb1, which bind specific sequences within the rDNA terminator and presumably form a steric hindrance that stops elongating Pol I [101, 104, 105]. Termination at T2 depends on the T-stretch element, which might form a weak RNA:DNA heteroduplex promoting Pol I destabilization. In addition to Nsi1 and Reb1, several other factors, linking transcription and pre-RNA processing, were demonstrated to be involved in Pol I termination. These include the Pol I-specific subunit Rpa12 and the endoribonuclease Rnt1 that carries out CoTC at the 3'-end of the primary pre-rRNA transcript [101]. It was found that this cleavage, similarly to pre-mRNA processing at the poly(A) site, exposes a free monophosphate group at the 5'-end of the Pol I-bound transcript that is used by the torpedo activity of Rat1 [82,29]. Depletion of Rat1 or the absence of Rai1 resulted in a severe termination defect, manifested by the increased Pol I occupancy downstream of terminators and a strong accumulation of read-through transcripts. As is the case for Pol II termination by the torpedo, the mechanism for Pol I also requires the catalytic activity of Rat1. The other analogy with the Pol II model regards the involvement of the Sen1 helicase, which is associated with rDNA near the terminators and most likely cooperates with Rat1 to ensure efficient Pol I termination in a similar manner to its action on Pol II [82]. The mechanism, which recruits Rat1 or Sen1 to the elongating Pol I and to rDNA genes, is unknown, but Sen1 directly interacts with Rnt1 and this may contribute to its association with rDNA [83]. There is no direct evidence for Xrn2-mediated Pol I torpedo termination in higher eukaryotes, but human Xrn2 was shown to associate with Pol I transcription termination and chromatin remodeling factors, TTF-I and Rsf1, suggesting that Xrn2 may be also involved in this mechanism [106].

4.2.3 Cooperation between termination and RNA surveillance

A vast number of genes in higher organisms are regulated via promoter-proximal Pol II pausing (reviewed in Refs. [107,108]). A paused polymerase may remain in the arrested mode, resume RNA synthesis or cause its premature termination. It is not entirely surprising that Xrn2 is a key factor in the premature termination of nascent transcription at polymerase pause sites. Human Xrn2 was demonstrated to interact with the transcription termination and mRNA decapping factors, TTF2, Dcp1a, Dcp2, and Edc3, which colocalize at the 5′-ends of genes, suggesting that termination is connected with nuclear RNA decapping and 5′-3′ degradation [106]. Depletion of these proteins, as well as polymerase negative elongation factors that induce promoter-proximal pausing, leads to redistribution of Pol II away from promoters, in line with the Pol II release from arrest. Together, this data is consistent with a model, whereby cotranscriptional decapping of nascent transcripts near the sites of Pol II pausing triggers their degradation by Xrn2 and, as a consequence, polymerase dissociation by the action of the torpedo. The torpedo premature termination of Pol II transcription could be important in the regulation of so-called pervasive transcription, at least for a large population of mammalian genes that display promoter-proximal pausing. Consequently, Xrn2 is engaged in the control of regulated decapping and may also limit divergent antisense transcription from bidirectional promoters.

Another example of hXrn2 activity, possibly also associated with its function in transcription termination, is the production of transcription start site-associated RNAs. These short ($\sim$38 nt) noncoding RNAs probably arise as a result of nucleolytic trimming of nascent transcripts by Xrn2 that are protected by a Pol II that has stalled at transcriptional pause sites [109].

In addition to mRNA genes, hXrn2 is recruited to Pol II independently transcribed miRNA genes, and degrades exposed nascent transcripts following the cotranscriptional Drosha cleavage of long precursors. As a result of this torpedo termination of miRNA transcripts, Xrn2 contributes to the Drosha-mediated transcriptional attenuation of downstream genes [110]. Furthermore, hXrn2 was shown to take part in the biosynthesis pathway of miRNA. More specifically, together with the exosome complex, Xrn2 is responsible for the clearance of intronic by-product sequences that remain after cotranscriptional splicing-independent processing of intronic miRNAs by Drosha, a Microprocessor RNase-III like enzyme [111].

A transcription termination-related mechanism of the surveillance of uncapped nascent transcripts by Rat1 was also reported in yeast [102]. Here, Rat1-dependent premature termination was observed to occur evenly throughout the genes upstream of poly(A) sites when capping activity was impaired or reduced by decreased Pol II processivity. Rat1 was postulated not only to degrade uncapped transcripts and to promote transcription termination of futile polymerase molecules generating aberrant RNAs but also to play a positive role in the polymerase transition from the initiation to the elongation mode. These examples underline the importance of Rat1 activity in establishing interconnections between all stages of transcription and RNA surveillance.

4.3. RNA surveillance

The plethora of RNA molecules produced by the activity of RNA polymerases requires precise quality control mechanisms that are responsible for the elimination of aberrant or superfluous RNA species and also take part in the normal regulation of RNA metabolism. Some of these processes take place in the nucleus and involve the function of Rat1/Xrn2 (Fig. 7.3).

4.3.1 mRNA

mRNA biogenesis in the nucleus is a highly orchestrated process involving the addition of the cap structure, splicing, 3'-end processing, such as cleavage and polyadenylation, assembly into RNP particles, and finally, export. Incorrectly processed transcripts are potentially detrimental to the cell, particularly unspliced pre-mRNAs with an altered reading frame, which can be a source of aberrant or toxic proteins. The most well characterized pathway removing defective mRNAs, named nonsense-mediated decay (NMD), acts in the cytoplasm on transcripts containing premature stop codons and is tightly linked to translation (reviewed in Ref. [112]). Many aberrant transcripts can also be degraded in the nucleus by Rat1 and the nuclear exosome; however, what rules dictate the choice of the preferred mechanism is unclear.

The nuclear decay pathway was demonstrated to act on unspliced pre-mRNAs and incompletely spliced intermediates in a splicing deficient *prp2-1* helicase mutant, where splicing is blocked prior to the first catalytic step and spliceosome assembly is not affected [113]. In this case, the major activity is provided by the exosome and the Rat1-dependent pathway appears to have a minor impact, yet at the nonpermissive temperature the

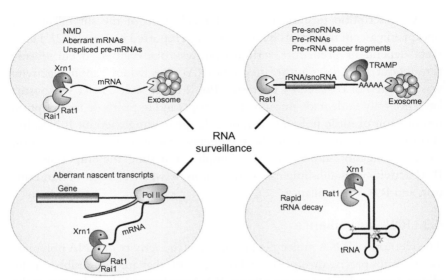

Figure 7.3 Rat1 contribution to RNA surveillance mechanisms. mRNA quality control pathways at the posttranscriptional (upper left) or transcriptional (bottom left) level require the 5′-3′ exoribonucleases, Rat1 and Xrn1, and the exosome. These activities remove aberrantly synthesized or spliced mRNAs, and mRNAs containing premature termination codons (nonsense-mediated decay, NMD). The 5′-3′ degradation of nascent transcripts, which are still linked to Pol II, leads to the premature transcription termination. The nuclear polyadenylation-mediated rRNA/snoRNA surveillance pathway (upper right) employs both Rat1 and the exosome/TRAMP complexes. The oligo(A) tails added by TRAMP to the 3′-ends of the nuclear decay substrates stimulate the activity of the exosome. The Rapid tRNA decay (bottom right) involves both Xrn1 and Rat1 that eliminate improperly modified tRNAs. Known hypomodifications, which target tRNAs for Rapid tRNA decay, are indicated as asterisks: yellow for tRNASer^{Ser} (ac^{4}C$_{12}$ and Um$_{44}$) and green for tRNA$^{Val(AAC)}$ (m^{7}G$_{46}$ and m^{5}C). (See color plate section in the back of the book.)

rat1-1 mutant accumulates both types of unprocessed transcripts. Nuclear exosome and Rat1 also eliminate pre-mRNAs carrying a 3′ splice site mutation, but not those with 5′ splice site or branch point mutations, as these are efficiently exported into the cytoplasm [114,115]. Indeed, the latter pre-mRNAs that fail to assemble the spliceosome, as well as lariat-containing intermediates that cannot complete the second step of splicing, are targeted for cytoplasmic degradation mainly by Xrn1 or the cytoplasmic exosome [115]. Notably, defects in nuclear RNA surveillance often lead to an increase in the level of mature mRNAs, which points to a competition between pre-mRNA degradation and splicing [113].

The degradation of unspliced pre-mRNAs in the nucleus may simply be a result of their prolonged nuclear retention. In fact, the existence of the DRN (decay of RNA in the nucleus) pathway was reported, which removes RNAs retained in the nucleus due to mutations in RNA export factors, including the nucleoporins Nup116 or Rat7, as well as in the Hpr1 component of the THO complex [26,116]. This pathway involves $3'$-$5'$ degradation by the nuclear exosome cofactor, exoribonuclease Rrp6, and probably also the $5'$-$3'$ decay by Rat1, as DRN is suppressed by the lack of the cap binding complex subunit Cbp80 or the Rat1 activator Rai1.

For some pre-mRNAs, a specific fail-safe mechanism exists removing unspliced and partially spliced transcripts and lariat-containing introns, that also regulates mature mRNA levels as a consequence. This pathway entails the presence of the endonuclease Rnt1 cleavage site, a specific stem-loop structure, in pre-mRNA introns. Rnt1 cleavage can create an entry site for the exoribonucleases, Rrp6 from the exosome complex or Rat1/Xrn1 [117], with the relative contribution of Rat1 and Xrn1 depending on a particular pre-mRNA. A similar mechanism, involving Rnt1, Rat1, and Rrp6, was also reported for iron-uptake protein mRNAs [118]. Here, the stem-loop structure recognized by Rnt1 is located within the sequence of the mature mRNA. Interestingly, *rnt1Δ* cells are both hypersensitive to iron toxicity and have a delayed response to iron starvation, probably due to elevated levels of iron-uptake proteins. The same set of factors, Rnt1, Rat1, and Xrn1, contributes to the removal of polycistronic and $5'$- or $3'$-extended forms of iron-uptake pre-mRNAs that are produced as a result of aberrant transcription initiation and termination [118]. Also, unspliced and partially spliced *MATa1* pre-mRNAs containing the noncanonical Rnt1 stem-loop structure located in the second exon–intron boundary are degraded in a similar fashion [119]. Rat1 and the nuclear exosome are also responsible for the elimination of the aberrantly spliced exon2-skipped *MATa1* mRNA and of three other yeast genes containing two introns.

Unexpectedly, the decay of the AU–rich (ARE)-containing *CTH2* mRNA in yeast was reported to be Rat1 dependent [120], while normally, at least in mammalian cells, ARE mRNAs are targeted for cytoplasmic degradation by the exosome or Xrn1 (reviewed in Ref. [121]). The basis for this atypical behavior is not known, but it may be connected to a noncanonical *CTH2* mRNA $3'$-end formation involving the NRD and exosome complexes [120].

In addition, mammalian Xrn2 nuclease was shown to participate in nuclear RNA surveillance. In human cells, Xrn2, with a less important contribution of the nuclear exosome, is involved in the removal of pre-mRNAs, whose splicing or termination is disturbed [44]. As mRNA maturation is mainly cotranscriptional, it is not surprising that the quality control and degradation of aberrantly processed transcripts occurs for nascent, chromatin-bound RNAs. In particular, molecules with mutations in the 3′ splice site or cleavage and polyadenylation signal, which are not easily released from the transcriptional machinery, are principal targets of this mechanism. Given that the degradation of some Xrn2 substrates is compromised when the DCP2 decapping enzyme is knocked down, they are probably decapped prior to removal. The cotranscriptional nature of this process is supported by the general association of Xrn2 and DCP2 with chromatin, which suggests their widespread involvement in quality control mechanisms.

As can be expected, it is not only pre-mRNA molecules that are subject to surveillance. In mammalian cells, some pri–miRNA precursors are encoded in pre-mRNA introns. Cotranscriptional pre-miRNA release, resulting from the cleavage by Drosha, creates entry sites for the exosome and Xrn2, enabling the rapid removal of introns without the degradation of exons that are protected by the binding of the Pol II complex [111]. This situation strongly resembles yeast intron-encoded snoRNAs or yeast introns with the stem–loop structure that is recognized by Rnt1.

Recently, it has been reported that the 5′-end of mRNAs and the cap structure can also undergo surveillance conducted by the Rat1-interacting protein, pyrophosphohydrolase, and the decapping endonuclease Rai1 [24]. The second Rai1 activity is stimulated by the interaction with Rat1, so it is thought that Rat1 participates in the degradation of uncapped 5′-monophosphorylated substrates generated by Rai1.

These examples illustrate the general impact of nuclear quality control pathways involving the contribution of Rat1 on cellular metabolism.

4.3.2 Rapid tRNA decay
Mature tRNAs are heavily modified, which contributes to their stability and ensures their proper aminoacylation and the correct interactions with the translation machinery. In yeast strains defective in tRNA modifications, these molecules are rapidly deacylated and degraded by a novel pathway independent of the nuclear exosome [6,122]. On the contrary, the exosome is implicated in the removal of aberrant pre-tRNAs rather than mature species [123]. The rapid tRNA decay (RTD) pathway employs

Xrn1 and Rat1, with their relative contribution depending on the tRNA molecule [6,124]. The involvement of both enzymes suggests that RTD may occur either in the nucleus or in the cytoplasm. This, in turn, implies that hypomodified tRNAs are reimported into the nucleus or vouch for the cytoplasmic pool of Rat1. The feature of the aberrant tRNA resulting from hypomodifications or point mutations, which is recognized by Rat1 and Xrn1, is a less stable acceptor- and T-stem, which exposes the tRNA 5′-end, making it accessible for exonucleases [43,124]. In addition to the RTD pathway, *rat1-1* cells accumulate unspliced but end-matured precursors of some tRNAs that are extended by 1–2 nt [13], but the rationale behind this phenotype is unclear.

4.3.3 rRNA and snoRNA

The correct amount of ribosomes, and therefore their correct surveillance, is a crucial point for the whole-cell metabolism. The inhibition of ribosome biogenesis in *S. cerevisiae* targets pre-rRNA species for degradation by the exosome preceded by their polyadenylation [28]. As the accumulation of poly(A)+ pre-rRNAs observed in the *rrp6Δ* mutant is enhanced by the inactivation of Rat1, as well as the fact that their shortened and heterogeneous 5′-ends become more homogenous, Rat1 is likely to play a minor role in their removal. In contrast, mammalian Xrn2 appears to have a more profound impact in a related process, as it constitutes a basic activity that is responsible for the elimination of abortive Pol I transcripts as well as aberrant pre-rRNA processing intermediates [52]. In addition, both Rat1 and Xrn2 degrade some pre-rRNA spacer fragments from 5′ETS, ITS1, and ITS2 regions [13,50,125]. In agreement with this function, Rat1 was found to directly bind to these spacer RNAs [54]. In a somehow parallel process, misfolded or misassembled yeast pre-snoRNAs are degraded by Rat1 or/and Xrn1 [126].

4.3.4 TERRA and others

In eukaryotic cells telomeric and subtelomeric regions are transcribed by Pol II, giving rise to a class of noncoding RNAs: the sense TERRA and the antisense ARRET (reviewed in Ref. [127]). In yeast, these ncRNAs are polyadenylated and they affect telomere length, probably by creating RNA:DNA hybrids inhibiting telomerase functions or interacting with the RNA component of telomerase. It appears that Rat1 is the main activity responsible for removing TERRA, probably together with the TRAMP complex component, alternative poly(A) polymerase Trf4 [99].

As a consequence *rat1-1* cells have shortened telomeres. Interestingly, only TERRA RNAs generated from one of the two yeast telomere types that contain X-elements are silenced posttranscriptionally due to the exonucleolytic activity of Rat1, while the Y'-containing telomeres are silenced transcriptionally, mainly by proteins recruited by the repressor/activator Rap1, with only a minor contribution of Rat1 [128].

It has been recently reported that the level of a large number of long noncoding RNAs (lncRNA) is regulated by a decapping-dependent pathway of 5'-3' degradation [129]. The majority of these lncRNAs are degraded by Xrn1 following the removal of the cap structure by Dcp2, but a significant fraction is removed by Rat1. For example, the defective nuclear decay of *GAL10* lncRNA caused by the *rat1* mutation is probably a major source of *GAL1* mRNA downregulation. Interestingly, in this case, Rai1 does not appear to assist Rat1 activity.

Similarly, plant nuclear Rat1 homologues are involved in RNA surveillance mechanisms. In *Arabidopsis*, AtXRN2 and AtXRN3 were shown to participate in the polyadenylation-mediated decay of mature and precursor rRNAs and pre-rRNA spacers [55], as well as in the degradation of miRNA processing by-products, the so-called MIRNA loops [130].

5. OVERLAPPING FUNCTIONS OF RAT1 AND XRN1

As described above, all phenotypes in yeast cells lacking Xrn1 can be rescued by the expression of Rat1 mutated in the NLS and mislocalized to the cytoplasm. Conversely, the *ts*-lethality of the *rat1-1* mutant can be suppressed by Xrn1 targeted to the nucleus [11]. These observations suggest at least partial interchangeability of the two nucleases. Indeed, in several cases, Rat1 and Xrn1 show a high degree of functional redundancy. Xrn1 partially substitutes for a nonfunctional Rat1 in some RNA processing and degradation processes, including rRNA and snoRNA 5'-end trimming and the removal of rRNA processing intermediates [49–51,57]; the cotranscriptional degradation of downstream products during transcription termination [29,73]; and RNA surveillance of pre-mRNAs, hypomodified tRNAs, and polyadenylated pre-rRNAs [6,28,113,124]. Usually, Rat1 is considered to be the major player in these processes, as the deletion of *XRN1* alone has little effect, but intensifies molecular phenotypes observed in Rat1-deficient cells. Surprisingly, Xrn1 fused to an NLS fails to rescue Pol II transcription termination defects in the *rat1* mutant [73]. This is most likely due to the ability of Rat1 to enhance the recruitment of 3′-end formation factors, a task that cannot be carried out by Xrn1.

Some redundancies between Rat1 and Xrn1 activities are also observed in the cytoplasm, where Xrn1 is the major player in mRNA turnover. However, in yeast cells lacking Xrn1, transcripts targeted for degradation have shortened, heterogeneous 5′-ends and this phenotype is suppressed by the additional *rat1* mutation, supporting the notion that Rat1 has the ability to degrade cytoplasmic mRNAs, at least in the absence of Xrn1 [131,132].

The fact that Xrn1 and Rat1 are interchangeable when mislocalized to the other compartment indicates that their diverse cellular activities in RNA metabolism largely result from their different subcellular localizations, but even in normal conditions they may have partially overlapping functions in the nucleus. In turn, the other 5′-3′ exonuclease Rrp17 appears to have an essential, independent role in the 5′ processing of 60S subunit rRNAs [35].

6. XRN2 AS A SILENCING SUPPRESSOR

Small RNAs are well-established regulators of gene expression. Their biogenesis and functions are relatively well characterized, but little is known about their degradation pathways. Even so, in some cases Xrn2 has been reported to have a role in this process. In *C. elegans*, a screen for suppressors of a temperature-sensitive *let7* miRNA allele pointed to XRN-2 [19]. It was found that XRN-2 knockdown increased the level of some, but not all, miRNAs as well as certain passenger miR★ strands, with an apparent preference for those that can be loaded onto AGO proteins [19,133]. Similar effects were observed for the depletion of XRN-1, showing that both nucleases are involved in miR or miR★ strand turnover. On the other hand, pri- or pre-miRNAs or discarded precursor fragments are not XRN-2 substrates. In addition to its function in degradation, XRN-2 was proposed to have some unclear role in releasing miRNAs from the complex with AGO prior to their decay. These observations led to a proposal of a "proof-reading" mechanism for strand selection from the miR:miR★ duplex based on the XRN-mediated elimination of a molecule devoid of and not stabilized by a target. The fact, that XRN-1 and -2 have an impact on AGO-associated miRNAs, and not on free species, suggests that AGO proteins may somehow assist the recruitment of the nucleases.

Recently, an unexpected role of Xrn2 in specific cytoplasmic foci related to the elimination of nontargeted siRNAs has been described [134]. The persistent presence of siRNA–RISC complexes possessing no cognate target is toxic to mammalian cells, probably due to off-targeting effects or saturation of the free RISC pool. High levels of nontargeted siRNAs stimulate expression of the ER-resident protein NPGPx (nonselenocysteine-containing phospholipid hydroperoxide glutathione peroxidase), the depletion of which leads to apoptosis induced by nontargeted siRNA stress. NPGPx normally recruits nuclear Xrn2 to cytoplasmic foci, which may correspond to P-bodies, where these two proteins form a complex via disulfide bonds, an interaction that is essential for the removal of nontargeted siRNAs by Xrn2. Consistently, human Xrn2 was reported to coimmunoprecipitate with three cytoplasmic NMD factors, UPF1, UPF2, and UPF3X [135] and several cysteine residues in Xrn2, some of which are important for the formation of disulfide bonds with NPGPx and are conserved between humans and yeast [134], suggesting a mechanism of Rat1 translocation to the cytoplasm.

In addition, *Arabidopsis* Rat1 counterparts have functions related to RNAi. In plants, a complicated small RNA-based pathway of transgene silencing is triggered by aberrant transgene transcription or simply by the overaccumulation of transgene RNAs. It has been reported that AtXRN2/3/4 and, indirectly, FIERY1, a homologue of yeast Met22/Hal2, act as silencing suppressors, possibly by decreasing the pool of transgene transcripts via their degradation [130].

7. PERSPECTIVES

Through its 5′-3′ exoribonucleolytic activity, Rat1/Xrn2 not only acts in the processing and degradation of different classes of nuclear RNAs but also contributes to transcription termination by the torpedo mechanism and participates in the RNA surveillance pathways of both coding and noncoding transcripts. It will be important to establish the precise mechanisms of its recruitment to the elongating polymerase and accurate rules that dictate how this enzyme discriminates between aberrant and correct RNA molecules. Another question is whether Xrn2 activity in higher eukaryotes requires any activators or the support of additional interacting factors. Finally, an element that is still missing is the existence of a connection between the molecular functions of Rat1/Xrn2 in RNA metabolism and outcomes at the cellular and organism level. Although there is one example of a link between the expression of Xrn2 and human lung cancer [136], several other correlations are expected to emerge in the future, as has been the case for other RNA enzymatic activities. Notably, mutations in the exosome components were demonstrated to be associated with developmental, neurodegenerative, and autoimmune disorders [137-139].

ACKNOWLEDGMENTS

This work was supported by the Foundation for Polish Science grant [TEAM/2008-2/] and the National Science Centre grant [N N301 314537].

REFERENCES

[1] Ghosh S, Jacobson A. RNA decay modulates gene expression and controls its fidelity. Wiley Interdiscip Rev RNA 2010;1:351–61.

[2] Amberg DC, Goldstein AL, Cole CN. Isolation and characterization of RAT1: an essential gene of *Saccharomyces cerevisiae* required for the efficient nucleocytoplasmic trafficking of mRNA. Genes Dev 1992;6:1173–89.

[3] Hammell CM, et al. Coupling of termination, 3′ processing, and mRNA export. Mol Cell Biol 2002;22:6441–57.

[4] Kadowaki T, Chen S, Hitomi M, Jacobs E, Kumagai C, Liang S, et al. Isolation and characterization of *Saccharomyces cerevisiae* mRNA transport-defective (*mtr*) mutants. J Cell Biol 1994;126:649–59.

[5] Di Segni G, McConaughy BL, Shapiro RA, Aldrich TL, Hall BD. TAP1, a yeast gene that activates the expression of a tRNA gene with a defective internal promoter. Mol Cell Biol 1993;13:3424–33.

[6] Chernyakov I, Whipple JM, Kotelawala L, Grayhack EJ, Phizicky EM. Degradation of several hypomodified mature tRNA species in *Saccharomyces cerevisiae* is mediated by Met22 and the 5′-3′ exonucleases Rat1 and Xrn1. Genes Dev 2008;22:1369–80.

[7] Kenna M, Stevens A, McCammon M, Douglas MG. An essential yeast gene with homology to the exonuclease-encoding *XRN1/KEM1* gene also encodes a protein with exoribonuclease activity. Mol Cell Biol 1993;13:341–50.

[8] Stevens A. Purification and characterization of a *Saccharomyces cerevisiae* exoribonuclease which yields 5′-mononucleotides by a 5′-3′ mode of hydrolysis. J Biol Chem 1980;255:3080–5.

[9] Stevens A, Poole TL. 5′-exonuclease-2 of *Saccharomyces cerevisiae*. Purification and features of ribonuclease activity with comparison to 5′-exonuclease-1. J Biol Chem 1995;270:16063–9.

[10] Toone WM, et al. Rme1, a negative regulator of meiosis, is also a positive activator of G1 cyclin gene expression. EMBO J 1995;14:5824–32.

[11] Johnson AW. Rat1p and Xrn1p are functionally interchangeable exoribonucleases that are restricted to and required in the nucleus and cytoplasm, respectively. Mol Cell Biol 1997;17:6122–30.

[12] Poole TL, Stevens A. Comparison of features of the RNase activity of 5′-exonuclease-1 and 5′-exonuclease-2 of *Saccharomyces cerevisiae*. Nucleic Acids Symp Ser 1995;33:79–81.

[13] Dichtl B, Stevens A, Tollervey D. Lithium toxicity in yeast is due to the inhibition of RNA processing enzymes. EMBO J 1997;16:7184–95.

[14] Sugano S, Shobuike T, Takeda T, Sugino A, Ikeda H. Molecular analysis of the dhp1+ gene of *Schizosaccharomyces pombe*: an essential gene that has homology to the DST2 and RAT1 genes of *Saccharomyces cerevisiae*. Mol Gen Genet 1994;243:1–8.

[15] Shobuike T, Tatebayashi K, Tani T, Sugano S, Ikeda H. The dhp1+ gene, encoding a putative nuclear 5′-3′ exoribonuclease, is required for proper chromosome segregation in fission yeast. Nucleic Acids Res 2001;29:1326–33.

[16] Kinoshita N, Goebl M, Yanagida M. The fission yeast dis3+ gene encodes a 110-kDa essential protein implicated in mitotic control. Mol Cell Biol 1991;11:5839–47.

[17] Shobuike T, Sugano S, Yamashita T, Ikeda H. Characterization of cDNA encoding mouse homolog of fission yeast dhp1+ gene: structural and functional conservation. Nucleic Acids Res 1995;23:357–61.

[18] Zhang M, et al. Cloning and mapping of the XRN2 gene to human chromosome 20p11.1-p11.2. Genomics 1999;59:252–4.

[19] Chatterjee S, Großhans H. Active turnover modulates mature microRNA activity in Caenorhabditis elegans. Nature 2009;461:546–9.

[20] Kastenmayer JP, Green PJ. Novel features of the XRN-family in Arabidopsis: evidence that AtXRN4, one of several orthologs of nuclear Xrn2p/Rat1p, functions in the cytoplasm. Proc Natl Acad Sci USA 2000;97:13985–90.

[21] Souret FF, Kastenmayer JP, Green PJ. AtXRN4 degrades mRNA in Arabidopsis and its substrates include selected miRNA targets. Mol Cell 2004;15:173–83.

[22] Zimmer SL, Fei Z, Stern DB. Genome-based analysis of Chlamydomonas reinhardtii exoribonucleases and poly(A) polymerases predicts unexpected organellar and exosomal features. Genetics 2008;179:125–36.

[23] Xue Y, et al. Saccharomyces cerevisiae RAI1 (YGL246c) is homologous to human DOM3Z and encodes a protein that binds the nuclear exoribonuclease Rat1p. Mol Cell Biol 2000;20:4006–15.

[24] Jiao X, Xiang S, Oh C-S, Martin CE, Tong L, Kiledjian M. Identification of a quality-control mechanism for mRNA 5′-end capping. Nature 2010;467:608–11.

[25] Xiang S, Cooper-Morgan A, Jiao X, Kiledjian M, Manley JL, Tong L. Structure and function of the 5′-3′ exoribonuclease Rat1 and its activating partner Rai1. Nature 2009;458:784–8.

[26] Das B, Butler JS, Sherman F. Degradation of normal mRNA in the nucleus of Saccharomyces cerevisiae. Mol Cell Biol 2003;23:5502–15.

[27] Kim M, et al. The yeast Rat1 exonuclease promotes transcription termination by RNA polymerase II. Nature 2004;432:517–22.

[28] Fang F, Phillips S, Butler JS. Rat1p and Rai1p function with the nuclear exosome in the processing and degradation of rRNA precursors. RNA 2005;11:1571–8.

[29] El Hage A, Koper M, Kufel J, Tollervey D. Efficient termination of transcription by RNA polymerase I requires the 5′ exonuclease Rat1 in yeast. Genes Dev 2008;22:1069–81.

[30] Zuo Y, Deutscher MP. Exoribonuclease superfamilies: structural analysis and phylogenetic distribution. Nucleic Acids Res 2001;29:1017–26.

[31] Szankasi P, Smith GR. Requirement of S. pombe exonuclease II, a homologue of S. cerevisiae Sep1, for normal mitotic growth and viability. Curr Genet 1996;30:284–93.

[32] Till DD, et al. Identification and developmental expression of a 5′-3′ exoribonuclease from Drosophila melanogaster. Mech Dev 1998;79:51–5.

[33] Glavan F, Behm-Ansmant I, Izaurralde E, Conti E. Structures of the PIN domains of SMG6 and SMG5 reveal a nuclease within the mRNA surveillance complex. EMBO J 2006;25:5117–25.

[34] Anantharaman V, Aravind L. The NYN domains: novel predicted RNAses with a PIN domain-like fold. RNA Biol 2006;3:18–27.

[35] Oeffinger M, et al. Rrp17p is a eukaryotic exonuclease required for 5′ end processing of Pre-60S ribosomal RNA. Mol Cell 2009;36:768–81.

[36] Dominski Z, Yang XC, Marzluff WF. The polyadenylation factor CPSF-73 is involved in histone-pre-mRNA processing. Cell 2005;123:37–48.

[37] Mathy N, Bénard L, Pellegrini O, Daou R, Wen T, Condon C. 5′-to-3′ exoribonuclease activity in bacteria: role of RNase J1 in rRNA maturation and 5′ stability of mRNA. Cell 2007;129:681–92.

[38] Condon C. What is the role of RNase J in mRNA turnover? RNA Biol 2010;7:316–21.

[39] Tishkoff DX, Johnson AW, Kolodner RD. Molecular and genetic analysis of the gene encoding the *Saccharomyces cerevisiae* strand exchange protein Sep1. Mol Cell Biol 1991;11:2593–608.

[40] Page AM, Davis K, Molineux C, Kolodner RD, Johnson AW. Mutational analysis of exoribonuclease I from *Saccharomyces cerevisiae*. Nucleic Acids Res 1998;26:3707–16.

[41] Solinger JA, Pascolini D, Heyer WD. Active-site mutations in the Xrn1p exoribonuclease of *Saccharomyces cerevisiae* reveal a specific role in meiosis. Mol Cell Biol 1999;19:5930–42.

[42] Chang JH, Xiang S, Xiang K, Manley JL, Tong L. Structural and biochemical studies of the 5′-3′ exoribonuclease Xrn1. Nat Struct Mol Biol 2011;18:270–6.

[43] Jinek M, Coyle SM, Doudna JA. Coupled 5′ nucleotide recognition and processivity in Xrn1-mediated mRNA decay. Mol Cell 2011;41:600–8.

[44] Davidson L, Kerr A, West S. Co-transcriptional degradation of aberrant pre-mRNA by Xrn2. EMBO J 2012;31:2566–78.

[45] Venema J, Tollervey D. Ribosome synthesis in *Saccharomyces cerevisiae*. Annu Rev Gen 1999;33:261–311.

[46] Gerbi SA, Borovjagin AV. Pre-ribosomal RNA processing in multicellular organisms. In: Olson MOJ, editor. The nucleolus. New York: Kluwer Academic/Plenum Publishers; 2004. p. 170–98.

[47] Henras AK, et al. The post-transcriptional steps of eukaryotic ribosome biogenesis. Cell Mol Life Sci 2008;65:2334–59.

[48] Phipps KR, Charette JM, Baserga SJ. The small subunit processome in ribosome biogenesis—progress and prospects. Wiley Interdiscip Rev RNA 2011;2:1–21.

[49] Henry Y, Wood H, Morrissey JP, Petfalski E, Kearsey S, Tollervey D. The 5′ end of yeast 5.8S rRNA is generated by exonucleases from an upstream cleavage site. EMBO J 1994;13:2452–63.

[50] Petfalski E, Dandekar T, Henry Y, Tollervey D. Processing of the precursors to small nucleolar RNAs and rRNAs requires common components. Mol Cell Biol 1998;18:1181–9.

[51] Geerlings TH, Vos JC, Raue HA. The final step in the formation of 25S rRNA in *Saccharomyces cerevisiae* is performed by 5′-3′ exonucleases. RNA 2000;6:1698–703.

[52] Wang M, Pestov DG. 5′-end surveillance by Xrn2 acts as a shared mechanism for mammalian pre-rRNA maturation and decay. Nucleic Acids Res 2011;39:1811–22.

[53] Schillewaert S, Wacheul L, Lhomme F, Lafontaine DL. The evolutionarily conserved protein Las1 is required for pre-rRNA processing at both ends of ITS2. Mol Cell Biol 2011;32:430–44.

[54] Granneman S, Petfalski E, Tollervey D. A cluster of ribosome synthesis factors regulate pre-rRNA folding and 5.8S rRNA maturation by the Rat1 exonuclease. EMBO J 2011;30:4006–19.

[55] Zakrzewska-Placzek M, Souret FF, Sobczyk GJ, Green PJ, Kufel J. *Arabidopsis thaliana* XRN2 is required for primary cleavage in the pre-ribosomal RNA. Nucleic Acids Res 2010;38:4487–502.

[56] Filipowicz W, Pogacic V. Biogenesis of small nucleolar ribonucleoproteins. Curr Opin Cell Biol 2002;14:319–27.

[57] Qu LH, et al. Seven novel methylation guide small nucleolar RNAs are processed from a common polycistronic transcript by Rat1p and RNase III in yeast. Mol Cell Biol 1999;19:1144–58.

[58] Allmang C, Kufel J, Chanfreau G, Mitchell P, Petfalski E, Tollervey D. Functions of the exosome in rRNA, snoRNA and snRNA synthesis. EMBO J 1999;18:5399–410.

[59] Lee CY, Lee A, Chanfreau G. The roles of endonucleolytic cleavage and exonucleolytic digestion in the 5′-end processing of S. cerevisiae box C/D snoRNAs. RNA 2003;9:1362–70.

[60] Chekanova JA, et al. Genome-wide high-resolution mapping of exosome substrates reveals hidden features in the Arabidopsis transcriptome. Cell 2007;131:1340–53.

[61] Comella P, et al. Characterization of a ribonuclease III-like protein required for cleavage of the pre-rRNA in the 3′ETS in Arabidopsis. Nucleic Acids Res 2008;36:1163–75.

[62] Kiss T, Filipowicz W. Exonucleolytic processing of small nucleolar RNAs from pre-mRNA introns. Genes Dev 1995;9:1411–24.

[63] Caffarelli E, Fatica A, Prislei S, De Gregorio E, Fragapane P, Bozzoni I. Processing of the intron-encoded U16 and U18 snoRNAs: the conserved C and D boxes control both the processing reaction and the stability of the mature snoRNA. EMBO J 1996;15:1121–31.

[64] Villa T, Ceradini F, Presutti C, Bozzoni I. Processing of the intron-encoded U18 small nucleolar RNA in the yeast Saccharomyces cerevisiae relies on both exo- and endonucleolytic activities. Mol Cell Biol 1998;18:3376–83.

[65] Ooi SL, Samarsky D, Fournier M, Boeke JD. Intronic snoRNA biosynthesis in Saccharomyces cerevisiae depends on the lariat-debranching enzyme: intron length effects and activity of a precursor snoRNA. RNA 1998;4:1096–110.

[66] Giorgi C, Fatica A, Nagel R, Bozzoni I. Release of U18 snoRNA from its host intron requires interaction of Nop1p with the Rnt1p endonuclease. EMBO J 2001;20:6856–65.

[67] Rosonina E, Kaneko S, Manley JL. Terminating the transcript: breaking up is hard to do. Genes Dev 2006;20:1050–6.

[68] Richard P, Manley JL. Transcription termination by nuclear RNA polymerases. Genes Dev 2009;23:1247–69.

[69] Kuehner JN, Pearson EL, Moore C. Unravelling the means to an end: RNA polymerase II transcription termination. Nat Rev Mol Cell Biol 2011;12:283–94.

[70] Logan J, Falck-Pedersen E, Darnell JEJ, Shenk T. A poly(A) addition site and a downstream termination region are required for efficient cessation of transcription by RNA polymerase II in the mouse beta maj-globin gene. Proc Natl Acad Sci USA 1987;84:8306–10.

[71] Connelly S, Manley JL. A functional mRNA polyadenylation signal is required for transcription termination by RNA polymerase II. Genes Dev 1988;2:440–52.

[72] Proudfoot NJ. How RNA polymerase II terminates transcription in higher eukaryotes. Trends Biochem Sci 1989;14:105–10.

[73] Luo W, Johnson AW, Bentley DL. The role of Rat1 in coupling mRNA 3′-end processing to transcription termination: implications for a unified allosteric-torpedo model. Genes Dev 2006;20:954–65.

[74] West S, Gromak N, Proudfoot NJ. Human 5′-3′ exonuclease Xrn2 promotes transcription termination at co-transcriptional cleavage sites. Nature 2004;432:522–5.

[75] Gromak N, West S, Proudfoot NJ. Pause sites promote transcriptional termination of mammalian RNA polymerase II. Mol Cell Biol 2006;26:3986–96.

[76] Kurihara Y, et al. Surveillance of 3′ noncoding transcripts requires FIERY1 and XRN3 in Arabidopsis. G3 (Bethesda) 2012;2:487–98.

[77] Chan S, Choi EA, Shi Y. Pre-mRNA 3′-end processing complex assembly and function. Wiley Interdisc Rev RNA 2011;2:321–35.

[78] Rondón AG, Mischo HE, Kawauchi J, Proudfoot NJ. Fail-safe transcriptional termination for protein-coding genes in S. cerevisiae. Mol Cell 2009;36:88–98.

[79] Dengl S, Cramer P. Torpedo nuclease Rat1 is insufficient to terminate RNA polymerase II in vitro. J Biol Chem 2009;284:21270–9.

[80] Steinmetz EJ, Conrad NK, Brow DA, Corden JL. RNA-binding protein Nrd1 directs poly(A) independent 3'-end formation of RNA polymerase II transcripts. Nature 2001;413:327–31.

[81] Steinmetz EJ, Warren CL, Kuehner JN, Panbehi B, Ansari AZ, Brow DA. Genome-wide distribution of yeast RNA polymerase II and its control by Sen1 helicase. Mol Cell 2006;24:735–46.

[82] Kawauchi J, Mischo H, Braglia P, Rondon A, Proudfoot NJ. Budding yeast RNA polymerases I and II employ parallel mechanisms of transcriptional termination. Genes Dev 2008;22:1082–92.

[83] Ursic D, Chinchilla K, Finkel JS, Culbertson MR. Multiple protein/protein and protein/RNA interactions suggest roles for yeast DNA/RNA helicase Sen1p in transcription, transcription-coupled DNA repair and RNA processing. Nucleic Acids Res 2004;32:2441–52.

[84] Skourti-Stathaki K, Proudfoot NJ, Gromak N. Human senataxin resolves RNA/DNA hybrids formed at transcriptional pause sites to promote Xrn2-dependent termination. Mol Cell 2011;42:794–805.

[85] Mischo HE, et al. Yeast Sen1 helicase protects the genome from transcription-associated instability. Mol Cell 2011;41:21–32.

[86] West S, Proudfoot NJ. Human Pcf11 enhances degradation of RNA polymerase II-associated nascent RNA and transcriptional termination. Nucleic Acids Res 2008;36:905–14.

[87] Kim M, Ahn SH, Krogan NJ, Greenblatt JF, Buratowski S. Transitions in RNA polymerase II elongation complexes at the 3' ends of genes. EMBO J 2004;23:354–64.

[88] Lunde BM, et al. Cooperative interaction of transcription termination factors with the RNA polymerase II C-terminal domain. Nat Struct Mol Biol 2010;17:1195–201.

[89] Kaneko S, Rozenblatt-Rosen O, Meyerson M, Manley JL. The multifunctional protein p54nrb/PSF recruits the exonuclease XRN2 to facilitate pre-mRNA 3' processing and transcription termination. Genes Dev 2007;21:1779–89.

[90] Cui M, Allen MA, Larsen A, Macmorris M, Han M, Blumenthal T. Genes involved in pre-mRNA 39-end formation and transcription termination revealed by a lin-15 operon Muv suppressor screen. Proc Natl Acad Sci USA 2008;105:16665–70.

[91] Arigo JT, Eyler DE, Carroll KL, Corden JL. Termination of cryptic unstable transcripts is directed by yeast RNA-binding proteins Nrd1 and Nab3. Mol Cell 2006;23:841–51.

[92] Kim M, Vasiljeva L, Rando OJ, Zhelkovsky A, Moore C, Buratowski S. Distinct pathways for snoRNA and mRNA termination. Mol Cell 2006;24:723–34.

[93] Vasiljeva L, Kim M, Mutschler H, Buratowski S, Meinhart A. The Nrd1-Nab3-Sen1 termination complex interacts with the Ser5-phosphorylated RNA polymerase II C-terminal domain. Nat Struct Mol Biol 2008;15:795–804.

[94] Richardson JP. How Rho exerts its muscle on RNA. Mol Cell 2006;22:711–2.

[95] Epshtein V, Dutta D, Wade J, Nudler E. An allosteric mechanism of Rho-dependent transcription termination. Nature 2010;463:245–9.

[96] Kim H, et al. Gene-specific RNA polymerase II phosphorylation and the CTD code. Nat Struct Mol Biol 2010;17:1279–86.

[97] Chanfreau G, Legrain P, Jacquier A. Yeast RNase III as a key processing enzyme in small nucleolar RNAs metabolism. J Mol Biol 1998;248:975–88.

[98] Kufel J, Allmang C, Chanfreau G, Petfalski E, Lafontaine DLJ, Tollervey D. Precursors to the U3 snoRNA lack snoRNP proteins but are stabilized by La binding. Mol Cell Biol 2000;20:5415–24.

[99] Luke B, Panza A, Redon S, Iglesias N, Li Z, Lingner J. The Rat1p 5' to 3' exonuclease degrades telomeric repeat-containing RNA and promotes telomere elongation in *Saccharomyces cerevisiae*. Mol Cell 2008;32:465–77.

[100] Hurt E, Luo MJ, Rother S, Reed R, Strasser K. Cotranscriptional recruitment of the serine-arginine-rich (SR)-like proteins Gbp2 and Hrb1 to nascent mRNA via the TREX complex. Proc Natl Acad Sci USA 2004;101:1858–62.

[101] Prescott EM, et al. Transcriptional termination by RNA polymerase I requires the small subunit Rpa12p. Proc Natl Acad Sci USA 2004;101:6068–73.

[102] Jimeno-González S, Haaning LL, Malagon F, Jensen TH. The yeast 5'-3' exonuclease Rat1p functions during transcription elongation by RNA polymerase II. Mol Cell 2010;37:580–7.

[103] Reeder RH, Lang WH. Terminating transcription in eukaryotes: lessons learned from RNA polymerase I. Trends Biochem Sci 1997;22:473–7.

[104] Reeder RH, Guevara P, Roan JG. *Saccharomyces cerevisiae* RNA polymerase I terminates transcription at the Reb1 terminator *in vivo*. Mol Cell Biol 1999;19:7369–76.

[105] Reiter A, et al. The Reb1-homologue Ydr026c/Nsi1 is required for efficient RNA polymerase I termination in yeast. EMBO J 2012;31:3480–93.

[106] Brannan K, et al. mRNA decapping factors and the exonuclease Xrn2 function in widespread premature termination of RNA polymerase II transcription. Mol Cell 2012;46:311–24.

[107] Nechaev S, Adelman K. Pol II waiting in the starting gates: regulating the transition from transcription initiation into productive elongation. Biochim Biophys Acta 2011;1809:34–45.

[108] Li J, Gilmour DS. Promoter proximal pausing and the control of gene expression. Curr Opin Genet Dev 2011;21:231–5.

[109] Valen E, et al. Biogenic mechanisms and utilization of small RNAs derived from human protein-coding genes. Nat Struct Mol Biol 2011;18:1075–82.

[110] Ballarino M, et al. Coupled RNA processing and transcription of intergenic primary microRNAs. Mol Cell Biol 2009;29:5632–8.

[111] Morlando M, Ballarino M, Gromak N, Pagano F, Bozzoni I, Proudfoot NJ. Primary microRNA transcripts are processed co-transcriptionally. Nat Struct Mol Biol 2008;15:902–9.

[112] Chang YF, Imam JS, Wilkinson MF. The nonsense-mediated decay RNA surveillance pathway. Annu Rev Biochem 2007;76:51–74.

[113] Bousquet-Antonelli C, Presutti C, Tollervey D. Identification of a regulated pathway for nuclear pre-mRNA turnover. Cell 2000;102:765–75.

[114] Legrain P, Rosbash M. Some cis- and trans-acting mutants for splicing target pre-mRNA to the cytoplasm. Cell 1989;57:573–83.

[115] Hilleren PJ, Parker R. Cytoplasmic degradation of splice-defective pre-mRNAs and intermediates. Mol Cell 2003;12:1453–65.

[116] Kuai L, Das B, Sherman F. A nuclear degradation pathway controls the abundance of normal mRNAs in *Saccharomyces cerevisiae*. Proc Natl Acad Sci USA 2005;102:13962–7.

[117] Danin-Kreiselman M, Lee CY, Chanfreau G. RNAse III-mediated degradation of unspliced pre-mRNAs and lariat introns. Mol Cell 2003;11:1279–89.

[118] Lee A, Henras AK, Chanfreau G. Multiple RNA surveillance pathways limit aberrant expression of iron uptake mRNAs and prevent iron toxicity in *S. cerevisiae*. Mol Cell 2005;19:39–51.

[119] Egecioglu DE, Kawashima TR, Chanfreau GF. Quality control of MATa1 splicing and exon skipping by nuclear RNA degradation. Nucleic Acids Res 2012;40:1787–96.

[120] Ciais D, Bohnsack MT, Tollervey D. The mRNA encoding the yeast ARE-binding protein Cth2 is generated by a novel 3′ processing pathway. Nucleic Acids Res 2008;36:3075–84.

[121] von Roretz C, Di Marco S, Mazroui R, Gallouzi IE. Turnover of AU-rich-containing mRNAs during stress: a matter of survival. Wiley Interdiscip Rev RNA 2011;2:336–47.

[122] Alexandrov A, et al. Rapid tRNA decay can result from lack of nonessential modifications. Mol Cell 2006;21:87–96.

[123] Kadaba S, Wang X, Anderson JT. Nuclear RNA surveillance in Saccharomyces cerevisiae: Trf4p-dependent polyadenylation of nascent hypomethylated tRNA and an aberrant form of 5S rRNA. RNA 2006;12:508–21.

[124] Whipple JM, Lane EA, Chernyakov I, D'Silva S, Phizicky EM. The yeast rapid tRNA decay pathway primarily monitors the structural integrity of the acceptor and T-stems of mature tRNA. Genes Dev 2011;25:1173–84.

[125] Schillewaert S, Wacheul L, Lhomme F, Lafontaine DLJ. The evolutionarily conserved protein LAS1 is required for pre-rRNA processing at both ends of ITS2. Mol Cell Biol 2012;32:430–44.

[126] Villa T, Ceradini F, Bozzoni I. Identification of a novel element required for processing of intron-encoded box C/D small nucleolar RNAs in Saccharomyces cerevisiae. Mol Cell Biol 2000;20:1311–20.

[127] Feuerhahn S, Iglesias N, Panza A, Porro A, Lingner J. TERRA biogenesis, turnover and implications for function. FEBS Lett 2010;584:3812–8.

[128] Iglesias N, Redon S, Pfeiffer V, Dees M, Lingner J, Luke B. Subtelomeric repetitive elements determine TERRA regulation by Rap1/Rif and Rap1/Sir complexes in yeast. EMBO Rep 2011;12:587–93.

[129] Geisler S, Lojek L, Khalil AM, Baker KE, Coller J. Decapping of long noncoding RNAs regulates inducible genes. Mol Cell 2012;45:279–91.

[130] Gy I, et al. Arabidopsis FIERY1, XRN2, and XRN3 are endogenous RNA silencing suppressors. Plant Cell 2007;19:3451–61.

[131] Muhlrad D, Decker CJ, Parker R. Deadenylation of the unstable mRNA encoded by the yeast MFA2 gene leads to decapping followed by 5′-3′ digestion of the transcript. Genes Dev 1994;8:855–66.

[132] He F, Jacobson A. Upf1p, Nmd2p, and Upf3p regulate the decapping and exonucleolytic degradation of both nonsense-containing mRNAs and wild-type mRNAs. Mol Cell Biol 2001;21:1515–30.

[133] Chatterjee S, Fasler M, Bussing I, Großhans H. Target-mediated protection of endogenous microRNAs in C. elegans. Dev Cell 2011;20:388–96.

[134] Wei P-C, Lo W-T, Su M-I, Shew J-Y, Lee W-H. Non-targeting siRNA induces NPGPx expression to cooperate with exoribonuclease XRN2 for releasing the stress. Nucleic Acids Res 2011;40:323–32.

[135] Lejeune F, Li X, Maquat LE. Nonsense-mediated mRNA decay in mammalian cells involves decapping, deadenylating, and exonucleolytic activities. Mol Cell 2003;12:675–87.

[136] Lu Y, et al. Genetic variants cis-regulating Xrn2 expression contribute to the risk of spontaneous lung tumor. Oncogene 2010;29:1041–9.

[137] Wan J, et al. Mutations in the RNA exosome component gene EXOSC3 cause pontocerebellar hypoplasia and spinal motor neuron degeneration. Nat Genet 2012;44:704–8.

[138] Brouwer R, Pruijn GJ, van Venrooij WJ. The human exosome: an autoantigenic complex of exoribonucleases in myositis and scleroderma. Arthritis Res 2001;3:102–6.

[139] Astuti D, et al. Germline mutations in DIS3L2 cause the Perlman syndrome of overgrowth and Wilms tumor susceptibility. Nat Genet 2012;44:277–84.

CHAPTER EIGHT

Normal and Aberrantly Capped mRNA Decapping

Megerditch Kiledjian[1], Mi Zhou, Xinfu Jiao

Department of Cell Biology and Neuroscience, Rutgers University, Piscataway, New Jersey, USA
[1]Corresponding author: e-mail address: kiledjian@biology.rutgers.edu

Contents

1. Introduction   166
2. mRNA-Decapping Proteins in the Exonucleolytic Pathway of mRNA Decay   167
    2.1 Dcp2-decapping enzyme   167
    2.2 Nudt16-decapping protein   169
3. Presence of an Aberrant Cap-Decapping Protein in *S. cerevisiae*   170
    3.1 Rai1 is a pyrophosphohydrolase that removes the 5′-terminal
        diphosphate from an uncapped RNA   170
    3.2 Identification of Rai1 as a decapping endonuclease protein that
        selectively functions on aberrantly capped mRNAs   171
    3.3 Rai1 stimulates Rat1 activity   172
    3.4 Generation of aberrantly capped mRNAs   173
    3.5 mRNA capping under nutrient stress   173
4. Additional Potential Functions of Rai1   174
    4.1 Rai1 in transcription termination and 5′-end mRNA capping quality control   174
    4.2 Rai1 in rRNA processing   175
5. Future Directions   176
    5.1 Mammalian rai1 homolog   176
    5.2 Regulation of capping   176
Acknowledgment   177
References   177

Abstract

Messenger RNAs transcribed by RNA polymerase II are modified at their 5′-end by the cotranscriptional addition of a 7-methylguanosine (m^7G) cap. The cap is an important modulator of gene expression and the mechanism and components involved in its removal have been extensively studied. At least two decapping enzymes, Dcp2 and Nudt16, and an array of decapping regulatory proteins remove the m^7G cap from an mRNA exposing the 5′-end to exonucleolytic decay. In contrast, relatively less is known about the decay of mRNAs that may be aberrantly capped. The recent demonstration that the *Saccharomyces cerevisiae* Rai1 protein selectively hydrolyzes aberrantly capped mRNAs provides new insights into the modulation of mRNA that lack a canonical m^7G

The Enzymes, Volume 31
ISSN 1874-6047
http://dx.doi.org/10.1016/B978-0-12-404740-2.00008-2

cap 5'-end. Whether an mRNA is uncapped or capped but missing the N7 methyl moiety, Rai1 hydrolyzes its 5'-end to generate an mRNA with a 5' monophosphate. Interestingly, Rai1 heterodimerizes with the Rat1 5'–3' exoribonuclease, which subsequently degrades the 5'-end monophosphorylated mRNA. Importantly, Rat1 stimulates the 5'-end hydrolysis activities of Rai1 to generate a 5'-end unprotected mRNA substrate for Rat1 and, in turn, Rai1 stimulates the activity of Rat1. The Rai1–Rat1 heterodimer functions as a molecular motor to detect and degrade mRNAs with aberrant caps and defines a novel quality control mechanism that ensures mRNA 5'-end integrity. The increase in aberrantly capped mRNA population following nutritional stress in S. cerevisiae demonstrates the presence of aberrantly capped mRNAs in cells and further reinforces the functional significance of the Rai1 in ensuring mRNA 5'-end integrity.

1. INTRODUCTION

The stability and translational efficiency of eukaryotic mRNAs are significantly influenced by the 5'-end 7-methylguanosine (m^7G) cap of eukaryotic mRNAs [1–3]. The cap is cotranscriptionally added and consists of a guanine nucleoside methylated at the N7 position attached to the terminal nucleoside of the RNA by an unusual 5'–5' pyrophosphate linkage [4–6]. Capping is carried out by the combination of three enzymatic activities consisting of a triphosphatase, guanylyltransferase, and methyltransferase [5,7]. Three different proteins carry out the distinct activities in yeast, while the triphosphatase and guanylyltransferase activities are carried out by a single bifunctional capping enzyme in mammals [8]. The presence of the methyl group on the cap is essential for recognition by the cap-binding complex, CBC and eIF4E [9–11]. The capping enzymes are recruited to the nascent transcript generated by RNA polymerase II through its carboxyl terminal domain (CTD) [12,13].

Historically, cap addition has been perceived as a default process that lacks regulation and proceeds to completion. In contrast, removal of the cap is where the regulation is believed to occur and catalyzed by the Dcp2 [14–16] and Nudt16 [17,18] decapping enzymes to release m^7GDP and 5'-monophosphate RNA. The exposed 5' monophosphate RNA is subsequently subjected to degradation by the Xrn1 5'–3' exoribonuclease to clear the mRNA body [19,20]. Interestingly, Dcp2 and Nudt16 function on an N7 methyl cap substrate and minimally function on unmethylated cap [15,18] raising an intriguing question regarding the fate of mRNAs aberrantly lacking a cap or the N7 methyl moiety. Recent reports have shown that at least in yeast cells, aberrantly capped mRNAs are generated upon exposure to nutrient stress and a heterodimeric complex consisting of Rai1–Rat1 degrades these mRNAs [21]. Rai1 is an aberrant cap-decapping enzyme that preferentially hydrolyzes mRNAs with noncanonical 5′-ends to generate a 5′-end monophosphate mRNA that can be utilized as substrate for the nuclear Rat1 exoribonuclease. Importantly, the two subunits mutually stimulate one another's activity whereby Rai1 enhances Rat1 activity and Rat1 stimulates Rai1 activity. Here, we discuss the contribution of Rai1 to aberrant cap decapping to initiate decay of mRNAs with aberrant 5′-ends in a 5′-end quality control mechanism.

2. mRNA-DECAPPING PROTEINS IN THE EXONUCLEOLYTIC PATHWAY OF mRNA DECAY

In eukaryotes, mRNA decay is generally triggered by shortening of the 3′ poly(A) tail to deadenylate the mRNA (please see Chapter 10 for a review of mRNA deadenylation). The deadenylated mRNA is subsequently shunted into one of two different directional decay pathways to undergo either 5′–3′ or 3′–5′ decay. In the 3′–5′ pathway, the RNA is degraded from the 3′end by the RNA exosome, which consists of a multisubunit complex with a 3′–5′ exoribonuclease component (please see Chapter 4 for a review of the exosome). The resulting cap structure is hydrolyzed by a scavenger-decapping enzyme, DcpS, to release m^7GMP and nucleotide diphosphate [22,23]. In the 5′–3′ decay pathway, the mRNA 5′-end cap structure is initially cleaved by the 5′ decapping enzymes, Dcp2 or Nut16, to release m^7GDP and a 5′-end monophosphorylated RNA [18,24–27] (Fig. 8.1). Following removal of the cap, the resulting 5′ monophosphorylated RNA serves as a substrate for one of two major 5′–3′ exoribonucleases, the nuclear yeast Rat1 exoribonuclease (Xrn2 in mammals) or cytoplasmic Xrn1 protein [20,28].

2.1. Dcp2-decapping enzyme

Dcp2 was the first decapping enzyme identified and is highly conserved in eukaryotes. It is a member of the Nudix family of proteins with a central domain consisting of a Nudix fold structure, which catalyzes the decapping step. Biochemical analyses have shown that Dcp2 can hydrolyze both monomethyl (m^7G) and trimethyl ($m^{2,2,7}G$) capped RNA with poor activity on

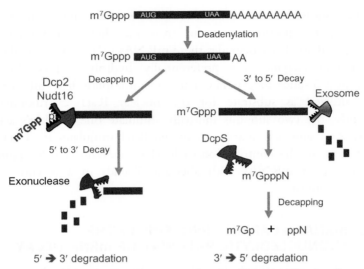

Figure 8.1 Eukaryotic exonucleolytic mRNA decay pathways. Following shortening of the poly(A) tail, eukaryotic mRNAs can undergo either 5'–3' or 3'–5' decay. In the 5' decay pathway, the mRNA 5'-end is decapped by Dcp2 or Nudt16 to release m^7GDP and generates a 5'-end monophosphorylated mRNA that is subsequently degraded by exoribonucleases Xrn1 or Xrn2. In the 3' decay pathway, the deadenylated mRNA is continued to be degraded from the 3'-end by the RNA exosome complex to degrade the RNA body resulting in the generation of a residual cap structure. The cap structure is subsequently degraded by the scavenger-decapping enzyme, DcpS which hydrolyzes the cap structure to release m^7GMP. (For color version of this figure, the reader is referred to the online version of this chapter.)

unmethylated capped (G–cap) RNA [15,29–31]. Dcp2 is an RNA-binding protein and requires a capped RNA substrate that is longer than 25 nucleotides and the RNA body contributes to the substrate specificity for Dcp2 decapping [31,32]. Structural analyses of Dcp2 revealed that it directly interacts with the cap and RNA body to recognize its substrate where the C-terminus of the NUDIX domain forms a conserved RNA-binding channel to accommodate the RNA [33]. The RNA-binding property of Dcp2 is consistent with the observed transcript specificity. Dcp2 preferentially binds a 5' terminal stem loop structure termed Dcp2 binding and decapping element (DBDE), which promotes recruitment of Dcp2 and subsequent decapping [34]. The DBDE is not restricted to a specific primary sequence and consists of at least an 8 base-pair-long stem with an intervening loop positioned within the 5' terminal 10 nucleotides of an mRNA for optimal Dcp2-mediated decapping.

Dcp2 is not ubiquitously expressed in all tissues and is developmentally regulated. Robust levels of Dcp2 protein are detected in the mouse embryonic brain, heart, liver, and kidney, while it can only be detected in the corresponding adult brain but not in the latter three tissues [18]. The selective function of Dcp2 in transcript decapping is further substantiated by genome-wide profiling of Dcp2-responsive mRNAs, which identified approximately 200 mRNAs that increased upon reduction in Dcp2 protein levels. Interestingly, a disproportional number of these mRNAs are involved in innate immunity and reduced Dcp2 levels correlate with enhanced resistance to viral challenge [35]. These findings demonstrate that Dcp2 is not a default decapping enzyme that functions on all mRNAs.

2.2. Nudt16-decapping protein

The transcript specificity of Dcp2 indicated that additional decapping enzymes are present in mammalian cells. Consistent with this premise, a second decapping protein, Nudt16, was also shown to possess mRNA-decapping activity [18]. Similar to Dcp2, Nudt16 is a member of the Nudix family of hydrolases. It was initially identified as a 29-kD nuclear protein in Xenopus (X29) that selectively bound and decapped the U8 snoRNA *in vitro* [36]. It was subsequently shown to be conserved across metazoans and thought to be a nuclear snoRNA-decapping protein [37]. More recently, human Nudt16 was shown to be localized to both the nuclear and cytoplasmic compartments [18,38] and regulate the stability of a subset of mRNAs in cells including the angiomotin-like 2 mRNA [18]. Intriguingly, in addition to regulating the stability of distinct mRNAs, Dcp2 and Nudt16 coordinately function on at least one mRNA. Both enzymes are involved in the decay of the c-myc mRNA where the half-life incrementally increases with the reduction in each protein, but most significantly increases when both are simultaneously reduced [17]. Conversely, the two decapping proteins function redundantly on the c-jun mRNA. A reduction in either decapping protein had no adverse consequence on stability of the c-jun mRNA, while an increased half-life was only detected upon reduction in both [17].

Dcp2 and Nudt16 each influencing the stability of a small subset of mRNAs suggests there are as yet additional mRNA-decapping enzymes that remain unidentified. Mammalian genomes contain 22 Nudt family proteins, two of which are Dcp2 and Nudt16. Whether any of the remaining 20 Nudt proteins also function in mRNA decapping remains to be determined.

Interestingly, yeast contains six putative Nudix proteins, one of which is Dcp2. Although a homolog of Nudt16 is not evident in *Saccharomyces cerevisiae*, whether the remaining five Nudix proteins possess intrinsic mRNA-decapping activity is not yet known.

3. PRESENCE OF AN ABERRANT CAP-DECAPPING PROTEIN IN *S. CEREVISIAE*

The selective decapping of m^{7}G-capped mRNAs by Dcp2 and Nudt16 and the requirement of a 5′ monophosphate substrate RNA for Rat1 and Xrn1 raises an interesting question regarding the decay of mRNAs containing either an unmethylated cap or lacking a cap altogether. These latter two mRNAs with an aberrant 5′-end would be precluded from Dcp2 or Nudt16 decapping and would be protected from 5′-end exonucleolytic decay. The recent identification of the Rat1-interacting protein, Rai1 as a protein with both pyrophosphohydrolase and decapping endonuclease activities [21,39] revealed the existence of a novel quality control mechanism that initiates the decay of aberrantly capped mRNAs.

3.1. Rai1 is a pyrophosphohydrolase that removes the 5′-terminal diphosphate from an uncapped RNA

Rai1 was initially identified as an uncharacterized polypeptide in highly purified preparations of Rat1 [28] and subsequently shown to be a potent stimulator of Rat1 exonuclease activity [40]. Based on the intrinsic unstable nature of Rat1 *in vitro* [28] and the lack of detectable biochemical activity, Rai1 was proposed to stimulate Rat1 activity by stabilization of the Rat1 protein. Structural analysis of the Rat1–Rai1 heterodimer as well as Rai1 alone revealed a conserved octahedral-coordinated sphere of a cation within Rai1 and the putative mammalian homolog, Dom3Z [39]. Biochemical analysis demonstrated that Rai1 possesses pyrophosphohydrolytic activity, which specifically cleaves within the 5′ triphosphate to release the terminal two phosphates and generates a monophosphorylated 5′-end RNA (pppRNA → pp + pRNA) that can subsequently serve as a substrate for Rat1 [39]. Importantly, a mutually stimulatory association exists between Rat1 and Rai1. In addition to the stimulation of the Rat1 exonuclease activity by Rai1, Rat1 in turn stimulates the pyrophospholytic activity of Rai1 [39] thus comprising a mutually complementary degradation complex that can detect and degrade mRNAs lacking a 5′-end cap (Fig. 8.2).

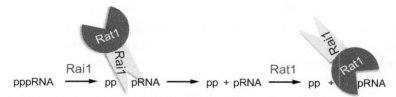

Figure 8.2 Rai1 possesses RNA 5′-end phosphohydrolase activity. In yeast cells, Rai1 cleaves the triphosphorylated 5′-end of an mRNA within the α and β phosphates to generate pyrophosphate and release a 5′-end monophosphorylated RNA that is subsequently degraded by 5′–3′ exoribonuclease Rat1. Rai1 and Rat1 form a stable heterodimer and mutually enhance each other's respective activities. (For color version of this figure, the reader is referred to the online version of this chapter.)

Figure 8.3 Rai1 possesses decapping endonuclease. Rai1 possesses a decapping endonuclease activity that removes the entire cap structure from an mRNA containing a 5′cap lacking the N7-methyl moiety. As shown in Fig. 8.2, the generated 5′-end monophosphorylated RNA is degraded by Rat1. (For color version of this figure, the reader is referred to the online version of this chapter.)

3.2. Identification of Rai1 as a decapping endonuclease protein that selectively functions on aberrantly capped mRNAs

The ability of Rai1 to specifically cleave a triphosphate linkage between the α and β phosphates and generate a 5′-end monophosphorylated mRNA is reminiscent of the Dcp2 and Nudt16 mRNA–decapping enzymes. This similarity indicates that Rai1 may also function on capped RNAs. Addressing this question yielded two surprising findings. First, Rai1 can function on capped RNAs, but preferentially functions on mRNAs possessing an unmethylated cap [21]. Second, the mode of hydrolysis shifts from a pyrophosphohydrolase activity within the triphosphate linkage to a phosphodiesterase-decapping endonuclease activity exclusively following the penultimate nucleotide to remove the entire cap structure (GpppN-RNA → GpppN + pRNA) [21] (Fig. 8.3). Importantly, despite the altered site of cleavage, the 5′-end of the resulting RNA contains a monophosphate and can be degraded by Rat1. This unexpected activity of Rai1 *in vitro* was also evident in cells harboring the temperature-sensitive ABD1 N7 methyltransferase mutant allele (*abd1-5*) which generates capped but not methylated primary transcripts at the non-permissive temperature [41]. Steady state levels and stability of mRNAs increased in cap methylation-deficient strains containing a mutant *RAI1* gene (*abd1-5 rai1Δ*), implicating Rai1 in the regulation of aberrantly capped mRNAs lacking an N7-methyl moiety [21]. The combined activities of Rai1 in hydrolyzing an uncapped RNA retaining a 5'-end triphosphate or an RNA containing an unmethylated cap strongly implicates Rai1 in a quality control mechanism that initiates the demise of mRNAs containing an aberrant 5'-end. Moreover, the presence of such an activity directed against aberrant, but not normal, caps further supports the growing body of evidence that, like other steps of pre-mRNA processing, capping may also be a regulated step.

3.3. Rai1 stimulates Rat1 activity

Rai1 forms within a heterodimeric complex with Rat1 and stimulates its exonucleolytic activity [28,39,40]. Conversely, both the pyrophosphohydrolase and decapping endonuclease activities of Rai1 are stimulated by Rat1 [21,39], demonstrating the significance of the heterodimer formation for their respective functions. Cocrystal structure of the Rat1–Rai1 heterodimer reveals an interaction interface involving the β8–αE segment and β4 strand of Rai1 [39] (please see Chapter 7 and this chapter for structural parameters of Rat1 and Rai1). Mutations within the protein–protein interaction interface ablate the stimulatory activities of the two proteins without altering their respective basal level of catalytic activity [21,39]. Through the formation of a Rat1–Rai1 heterodimer, the complex functions as a molecular motor that detects and degrades mRNAs with aberrant 5'-ends, regardless of whether it is uncapped or capped without the N7 methyl group. Rat1 will stimulate the hydrolase activities of Rai1 to generate a 5'-end monophosphorylated RNA that is degraded by the Rat1 subunit, which in turn is stimulated by Rai1 (Figs. 8.2 and 8.3).

Rai1 appears to contribute to the enhanced activity of Rat1 by stabilization of the Rat1 structure [39]. The Rat1–Rai1 heterodimer is more stable *in vitro* [40] compared to Rat1 alone [28] and the CTD of Rat1 is essential for its activity [42]. The interaction of Rai1 with a loop contained within the critical CTD of Rat1 that is on the opposite side of the Rat1 active site is proposed to stabilize the C-terminal loop structure which, in turn, indirectly coordinates and stabilizes the Rat1 active site [39].

3.4. Generation of aberrantly capped mRNAs

Upon exposure to environmental stress, mRNAs are primarily released from polysomes and sequestered into structures termed stress granules as an adaptive response in eukaryotic cells [43–45]. Stress granules are cytoplasmic mRNP structures that contain an array of RNA-binding proteins, translation initiation factors, large and small ribosomal subunit protein components, and mRNAs (for reviews, see Refs. [46–48]). Following return to an unstressed state, stress granules dissociate and the silenced mRNAs return to the polysomal pool of translating mRNAs [47]. Although a concomitant decrease in overall transcription occurs following stress, a complete transcriptional inhibition does not proceed and a diverse array of genes are either up- or downregulated [49–51]. Such a situation presents the cell with an apparent paradox, at a time when cytoplasmic mRNAs are being sequestered, the nucleus is transcribing mRNAs to be transported to the cytoplasm only to be sequestered. Yeast cells have evolved an intriguing mechanism to address this apparent dilemma by regulating the extent to which the nascent transcripts generated upon exposure to stress are aberrantly capped. mRNAs that are either uncapped or incompletely capped by a guanosine lacking the N7 methyl moiety are subjected to 5′-end hydrolysis by the Rai1 protein to generate a 5′ monophosphorylated mRNA that would be degraded by the Rat1 protein [21].

3.5. mRNA capping under nutrient stress

Exposure of S. cerevisiae to either glucose or amino acid starvation leads to rapid decay of two mRNAs, PGK1 and ACT1, which are relatively stable in cells grown under normal growth conditions [21]. It should be noted that the increased instability is observed only when the studies were carried out with the incorporation of a lag phase to enable clearing of preexisting mRNAs generated prior to the onset of the stress condition. Importantly, a similar instability was not detected in an isogenic strain disrupted for the RAI1 gene where the mRNAs remained significantly more stable [21]. These findings were interpreted to indicate that the addition of the m^7G cap is inefficient when cells are deprived of nutrients and Rai1 selectively hydrolyzes their 5′-end to expose a 5′ monophosphorylated mRNA for Rat1-directed decay (Figs. 8.2 and 8.3). Immunoprecipitation of capped mRNAs under conditions that separate m^7G-capped mRNA from aberrantly unmethylated capped or uncapped mRNAs [21] confirmed the

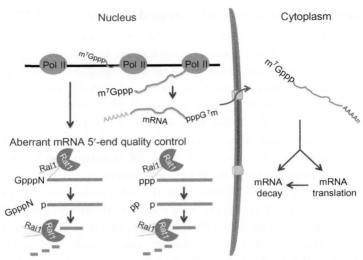

Figure 8.4 Schematic diagram of Rat1–Rai1 heterodimer functions on mRNA 5′-end capping quality control. RNA polymerase II (Pol II) nascent transcripts are initially capped at the 5′-end with an N7 methyl cap structure (m7GpppN) cotranscriptionally. The premRNA is further processed by splicing (not shown) and polyadenylation and the resulting mature mRNA is transported from the nucleus to the cytoplasm to undergo mRNA translation and/or decay. Transcripts lacking an m^7G cap are targeted by the Rai1 phosphohydrolase or decapping endonuclease activity to generate a 5′-end monophosphated mRNA that is cleared by the exoribonuclease activity of Rat1. (For color version of this figure, the reader is referred to the online version of this chapter.)

aberrantly capped nature of the newly synthesized mRNAs under the stress parameters. The aberrant capped RNA-decapping activity of Rai1 and 5–3′ exoribonuclease activity of Rat1 indicate that Rai1–Rat1 heterodimer plays an essential role in clearing mRNAs with aberrant 5′-end caps. Rai1–Rat1 appear to be involved in a quality control mechanism that ensures mRNA 5′-end integrity by an aberrant cap-mediated mRNA decay mechanism (Fig. 8.4).

4. ADDITIONAL POTENTIAL FUNCTIONS OF RAI1

4.1. Rai1 in transcription termination and 5′-end mRNA capping quality control

Addition of the 5′ cap is cotranscriptional and occurs at the initial stages of transcription following synthesis of the first 20–40 nt when both the capping and methyltransferase enzymes are associated with Ser5-phosphorylated

CTD of RNAP II [12,13]. This corresponds to the checkpoint pause stage when capping is believed to occur and RNAP II enters an elongation phase upon Ser2 phosphorylation and subsequently Ser7 phosphorylation [13,52–56]. Several lines of recent evidence indicates that this coupling may promote the synthesis of only properly capped mRNAs and has a built-in quality control mechanism to clear aberrantly capped mRNAs. Disruption of the *CEG1*-capping enzyme gene or the *ABD1* N7-methyltransferase gene results in nonviable yeast cells demonstrating that the capping and methylation steps are both essential functions. Interestingly, disruption of the normal timing of transcription yields aberrantly capped pre-mRNAs. An RNA polymerase II with a catalytic site mutation that leads to a reduction in the rate of transcription (*rpb1-N488D*) [57] generates mRNAs lacking a cap [58]. Moreover, similar to observations in *ceg1* mutants harboring a temperature-sensitive *rat1* allele that results in increased mRNA levels [59], mRNA levels also increase in *rpb1-N488D-rat1-1* cells with a corresponding increase in uncapped mRNAs [58]. The decrease in capping was attributed to a reduction in promoter occupancy of the Ceg1 protein due to less-efficient CTD Ser-5 phosphorylation. These findings demonstrate the importance of Rat1 in the cotranscriptional clearing of uncapped pre-mRNAs in a process perhaps analogous to the "torpedo" model of termination where Rat1 facilitates transcription termination following cleavage/polyadenylation at the 3′-end of a gene [60–62]. However, one important distinction is that the downstream RNA fragment following cleavage at the polyadenylation site contains a 5′ monophosphate that is an effective substrate for Rat1. In contrast, the 5′-end of an uncapped primary transcript would be expected to contain a 5′ triphosphate, which would be resistant to the Rat1 5′-end monophosphate RNA-specific exonuclease [28,63]. Although not yet directly demonstrated, it is highly likely that Rai1 is also present early in the transcription cycle and cleaves either the triphosphate or unmethylated capped 5′-end to enable Rat1 access to the pre-mRNA.

4.2. Rai1 in rRNA processing

Rat1 has long been shown to be involved in the proper 5′-end processing of 5.8S and 25S rRNAs [54–56]. The 5.8S rRNA consists of two species, $5.8S_S$ and $5.8S_L$ that differ in their 5′ length. The $5.8S_S$ is the predominant form and is generated by an endonucleolytic cleavage at site A3 followed by exonucleolytic trimming by Rat1 [40,64]. Yeast strains disrupted for Rai1 are compromised in their production of $5.8S_s$ and predominantly shift to generating the $5.8S_L$ form, which is reversed with Rat1 overexpression [40] indicating that the significance of Rai1 is to stimulate Rat1 activity. Surprisingly, an apparent Rat1-independent function of Rai1 was also noted in 5.8S rRNA processing where defects in 3′-end processing were detected [40]. Importantly, the defect in 5.8S 3′-end processing was analogous to defects observed in strains disrupted of the nuclear 3′–5′ exonuclease, Rrp6 suggestive of a functional interplay between Rai1 and Rrp6 [65] in addition to the interaction with Rat1. At least in yeast 5.8S rRNA maturation, Rai1 influences both Rat1 and Rrp6 exonucleolytic activity. Consistent with the absence of the Rat1 interaction domain within the putative mammalian homolog of Rai1, Dom3Z [39], the interaction of Dom3Z with Xrn2 could not be detected by two hybrid analyses [66]. However, yeast two hybrid interactions are detected between Dom3Z and the mammalian homolog of Rrp6, PM/Scl-100 [67], suggesting that Dom3Z may influence 3′-end decay in mammals.

5. FUTURE DIRECTIONS

5.1. Mammalian rai1 homolog

Rai1 shares weak primary sequence identity but significant structural identity with a putative mammalian homolog, Dom3Z [39,40]. The conserved structure of Dom3Z and Rai1 at the active site indicates that Dom3Z may also possess catalytic activity. Dom3Z is a nuclear protein that can also be localized to a subset of cytoplasmic P bodies detected by GFP-Dcp1a [68]. Further, knockdown of Dom3Z in mammalian U2OS cells results in a marked reduction in cytoplasmic P bodies [68]. These observations suggest that Dom3Z may have a role in mRNA metabolism in both the nucleus and cytoplasm. Future studies on the role of Dom3Z in potential aberrant cap regulation and mRNA metabolism are necessary to begin delineating the potential role of Dom3Z in mRNA catabolism.

5.2. Regulation of capping

The overwhelming perception has generally been that the addition of the 5′-end cap is an unregulated process that always proceeds to completion. The demonstration that capping is inefficient upon exposure of yeast cells to nutrient starvation indicates that this perception may be more an indication of our naiveté and opens a new realm of potential regulation previously unforeseen. It is likely that cells exploit, and accentuate, an already existing regulatory mechanism following exposure to nutrient stress. Whether additional stress conditions also elicit the generation of aberrantly capped mRNAs remains to be determined, although early indications are that at least heat shock also leads to aberrantly capped mRNA in yeast (Xinfu Jiao and Megerditch Kiledjian, unpublished observations). A more pressing question is whether capping is normally inefficient and fulfills a modulatory process to maintain functional mRNA homeostasis. Further studies will begin to address these pressing questions in addition to the mechanism by which any mode of regulation is controlled and the factors involved.

ACKNOWLEDGMENT

This work was supported by NIH grant GM67005 to M. K.

REFERENCES

[1] Liu H, Kiledjian M. Decapping the message: a beginning or an end. Biochem Soc Trans 2006;34:35–8.
[2] Meyer S, Temme C, Wahle E. Messenger RNA turnover in eukaryotes: pathways and enzymes. Crit Rev Biochem Mol Biol 2004;39:197–216.
[3] Merrick WC. Cap-dependent and cap-independent translation in eukaryotic systems. Gene 2004;332:1–11.
[4] Furuichi Y, Shatkin AJ. Viral and cellular mRNA capping: past and prospects. Adv Virus Res 2000;55:135–84.
[5] Ghosh A, Lima CD. Enzymology of RNA cap synthesis. Wiley Interdiscip Rev RNA 2010;1:152–72.
[6] Shatkin AJ. Capping of eucaryotic mRNAs. Cell 1976;9:645–53.
[7] Shuman S. Capping enzyme in eukaryotic mRNA synthesis. Prog Nucleic Acid Res Mol Biol 1995;50:101–29.
[8] Yue Z, Maldonado E, Pillutla R, Cho H, Reinberg D, Shatkin AJ. Mammalian capping enzyme complements mutant Saccharomyces cerevisiae lacking mRNA guanylyltransferase and selectively binds the elongating form of RNA polymerase II. Proc Natl Acad Sci USA 1997;94:12898–903.
[9] Fischer PM. Cap in hand: targeting eIF4E. Cell Cycle 2009;8:2535–41.
[10] Gingras AC, Raught B, Sonenberg N. eIF4 initiation factors: effectors of mRNA recruitment to ribosomes and regulators of translation. Annu Rev Biochem 1999;68:913–63.
[11] Goodfellow IG, Roberts LO. Eukaryotic initiation factor 4E. Int J Biochem Cell Biol 2008;40:2675–80.
[12] Ho CK, Shuman S. Distinct roles for CTD Ser-2 and Ser-5 phosphorylation in the recruitment and allosteric activation of mammalian mRNA capping enzyme. Mol Cell 1999;3:405–11.
[13] Mandal SS, Chu C, Wada T, Handa H, Shatkin AJ, Reinberg D. Functional interactions of RNA-capping enzyme with factors that positively and negatively regulate promoter escape by RNA polymerase II. Proc Natl Acad Sci USA 2004;101:7572–7.

[14] Dunckley T, Parker R. The DCP2 protein is required for mRNA decapping in Saccharomyces cerevisiae and contains a functional MutT motif. EMBO J 1999;18:5411–22.

[15] Wang Z, Jiao X, Carr-Schmid A, Kiledjian M. The hDcp2 protein is a mammalian mRNA decapping enzyme. Proc Natl Acad Sci USA 2002;99:12663–8.

[16] Lykke-Andersen J. Identification of a human decapping complex associated with hUpf proteins in nonsense-mediated decay. Mol Cell Biol 2002;22:8114–21.

[17] Li Y, Song M, Kiledjian M. Differential utilization of decapping enzymes in mammalian mRNA decay pathways. RNA 2011;17:419–28.

[18] Song MG, Li Y, Kiledjian M. Multiple mRNA decapping enzymes in mammalian cells. Mol Cell 2010;40:423–32.

[19] Decker CJ, Parker R. A turnover pathway for both stable and unstable mRNAs in yeast: evidence for a requirement for deadenylation. Genes Dev 1993;7:1632–43.

[20] Hsu CL, Stevens A. Yeast cells lacking 5′–>3′ exoribonuclease 1 contain mRNA species that are poly(A) deficient and partially lack the 5′ cap structure. Mol Cell Biol 1993;13:4826–35.

[21] Jiao X, Xiang S, Oh C, Martin CE, Tong L, Kiledjian M. Identification of a quality-control mechanism for mRNA 5′-end capping. Nature 2010;467:608–11.

[22] Liu H, Rodgers ND, Jiao X, Kiledjian M. The scavenger mRNA decapping enzyme DcpS is a member of the HIT family of pyrophosphatases. EMBO J 2002;21:4699–708.

[23] Wang Z, Kiledjian M. Functional link between the mammalian exosome and mRNA decapping. Cell 2001;107:751–62.

[24] Coller J, Parker R. Eukaryotic mRNA decapping. Annu Rev Biochem 2004;73:861–90.

[25] Franks TM, Lykke-Andersen J. The control of mRNA decapping and P-body formation. Mol Cell 2008;32:605–15.

[26] Garneau NL, Wilusz J, Wilusz CJ. The highways and byways of mRNA decay. Nat Rev Mol Cell Biol 2007;8:113–26.

[27] Li Y, Kiledjian M. Regulation of mRNA decapping. Wiley Interdiscip Rev RNA 2010;1:253–65.

[28] Stevens A, Poole TL. 5′-exonuclease-2 of Saccharomyces cerevisiae. Purification and features of ribonuclease activity with comparison to 5′-exonuclease-1. J Biol Chem 1995;270:16063–9.

[29] Cohen LS, Mikhli C, Jiao X, Kiledjian M, Kunkel G, Davis RE. Dcp2 Decaps m2,2,7G pppN-capped RNAs, and its activity is sequence and context dependent. Mol Cell Biol 2005;25:8779–91.

[30] Piccirillo C, Khanna R, Kiledjian M. Functional characterization of the mammalian mRNA decapping enzyme hDcp2. RNA 2003;9:1138–47.

[31] Steiger M, Carr-Schmid A, Schwartz DC, Kiledjian M, Parker R. Analysis of recombinant yeast decapping enzyme. RNA 2003;9:231–8.

[32] Li Y, Ho ES, Gunderson SI, Kiledjian M. Mutational analysis of a Dcp2-binding element reveals general enhancement of decapping by 5′-end stem-loop structures. Nucleic Acids Res 2009;37:2227–37.

[33] Deshmukh MV, et al. mRNA decapping is promoted by an RNA-binding channel in Dcp2. Mol Cell 2008;29:324–36.

[34] Li Y, Song MG, Kiledjian M. Transcript-specific decapping and regulated stability by the human Dcp2 decapping protein. Mol Cell Biol 2008;28:939–48.

[35] Li Y, Dai J, Song M, Fitzgerald-Bocarsly P, Kiledjian M. Dcp2 decapping protein modulates mRNA stability of the critical interferon regulatory factor (IRF) IRF-7. Mol Cell Biol 2012;32:1164–72.

[36] Ghosh T, Peterson B, Tomasevic N, Peculis BA. Xenopus U8 snoRNA binding protein is a conserved nuclear decapping enzyme. Mol Cell 2004;13:817–28.

[37] Taylor MJ, Peculis BA. Evolutionary conservation supports ancient origin for Nudt16, a nuclear-localized, RNA-binding, RNA-decapping enzyme. Nucleic Acids Res 2008;36:6021–34.

[38] Lu G, et al. hNUDT16: a universal decapping enzyme for small nucleolar RNA and cytoplasmic mRNA. Protein Cell 2011;2:64–73.

[39] Xiang S, Cooper-Morgan A, Jiao X, Kiledjian M, Manley JL, Tong L. Structure and function of the 5′−>3′ exoribonuclease Rat1 and its activating partner Rai1. Nature 2009;458:784–8.

[40] Xue Y, et al. Saccharomyces cerevisiae RAI1 (YGL246c) is homologous to human DOM3Z and encodes a protein that binds the nuclear exoribonuclease Rat1p. Mol Cell Biol 2000;20:4006–15.

[41] Schwer B, Saha N, Mao X, Chen HW, Shuman S. Structure-function analysis of yeast mRNA cap methyltransferase and high-copy suppression of conditional mutants by AdoMet synthase and the ubiquitin conjugating enzyme Cdc34p. Genetics 2000;155:1561–76.

[42] Shobuike T, Tatebayashi K, Tani T, Sugano S, Ikeda H. The dhp1(+) gene, encoding a putative nuclear 5′−>3′ exoribonuclease, is required for proper chromosome segregation in fission yeast. Nucleic Acids Res 2001;29:1326–33.

[43] Anderson P, Kedersha N. Stressful initiations. J Cell Sci 2002;115:3227–34.

[44] Anderson P, Kedersha N. RNA granules. J Cell Biol 2006;172:803–8.

[45] Anderson P, Kedersha N. Stress granules: the Tao of RNA triage. Trends Biochem Sci 2008;33:141–50.

[46] Anderson P, Kedersha N. RNA granules: post-transcriptional and epigenetic modulators of gene expression. Nat Rev Mol Cell Biol 2009;10:430–6.

[47] Buchan JR, Parker R. Eukaryotic stress granules: the ins and outs of translation. Mol Cell 2009;36:932–41.

[48] Erickson SL, Lykke-Andersen J. Cytoplasmic mRNP granules at a glance. J Cell Sci 2011;124:293–7.

[49] Brauer MJ, et al. Coordination of growth rate, cell cycle, stress response, and metabolic activity in yeast. Mol Biol Cell 2008;19:352–67.

[50] Causton HC, et al. Remodeling of yeast genome expression in response to environmental changes. Mol Biol Cell 2001;12:323–37.

[51] Gasch AP, et al. Genomic expression programs in the response of yeast cells to environmental changes. Mol Biol Cell 2000;11:4241–57.

[52] Gao L, Gross DS. Sir2 silences gene transcription by targeting the transition between RNA polymerase II initiation and elongation. Mol Cell Biol 2008;28:3979–94.

[53] Chiu YL, Ho CK, Saha N, Schwer B, Shuman S, Rana TM. Tat stimulates cotranscriptional capping of HIV mRNA. Mol Cell 2002;10:585–97.

[54] Hong SW, et al. Phosphorylation of the RNA polymerase II C-terminal domain by TFIIH kinase is not essential for transcription of Saccharomyces cerevisiae genome. Proc Natl Acad Sci USA 2009;106:14276–80.

[55] Pei Y, Schwer B, Shuman S. Interactions between fission yeast Cdk9, its cyclin partner Pch1, and mRNA capping enzyme Pct1 suggest an elongation checkpoint for mRNA quality control. J Biol Chem 2003;278:7180–8.

[56] Schwer B, Mao X, Shuman S. Accelerated mRNA decay in conditional mutants of yeast mRNA capping enzyme. Nucleic Acids Res 1998;26:2050–7.

[57] Malagon F, Kireeva ML, Shafer BK, Lubkowska L, Kashlev M, Strathern JN. Mutations in the Saccharomyces cerevisiae RPB1 gene conferring hypersensitivity to 6-azauracil. Genetics 2006;172:2201–9.

[58] Jimeno-Gonzalez S, Haaning LL, Malagon F, Jensen TH. The yeast 5′-3′ exonuclease Rat1p functions during transcription elongation by RNA polymerase II. Mol Cell 2010;37:580–7.

[59] Kim HJ, et al. mRNA capping enzyme activity is coupled to an early transcription elongation. Mol Cell Biol 2004;24:6184–93.

[60] Connelly S, Manley JL. A functional mRNA polyadenylation signal is required for transcription termination by RNA polymerase II. Genes Dev 1988;2:440–52.

[61] Kim M, et al. The yeast Rat1 exonuclease promotes transcription termination by RNA polymerase II. Nature 2004;432:517–22.

[62] West S, Gromak N, Proudfoot NJ. Human 5′ → 3′ exonuclease Xrn2 promotes transcription termination at co-transcriptional cleavage sites. Nature 2004;432:522–5.

[63] Kenna M, Stevens A, McCammon M, Douglas MG. An essential yeast gene with homology to the exonuclease-encoding XRN1/KEM1 gene also encodes a protein with exoribonuclease activity. Mol Cell Biol 1993;13:341–50.

[64] Henry Y, Wood H, Morrissey JP, Petfalski E, Kearsey S, Tollervey D. The 5′ end of yeast 5.8S rRNA is generated by exonucleases from an upstream cleavage site. EMBO J 1994;13:2452–63.

[65] Fang F, Phillips S, Butler JS. Rat1p and Rai1p function with the nuclear exosome in the processing and degradation of rRNA precursors. RNA 2005;11:1571–8.

[66] Lehner B, Semple JI, Brown SE, Counsell D, Campbell RD, Sanderson CM. Analysis of a high-throughput yeast two-hybrid system and its use to predict the function of intracellular proteins encoded within the human MHC class III region. Genomics 2004;83:153–67.

[67] Lehner B, Sanderson CM. A protein interaction framework for human mRNA degradation. Genome Res 2004;14:1315–23.

[68] Zheng D, Chen CY, Shyu AB. Unraveling regulation and new components of human P-bodies through a protein interaction framework and experimental validation. RNA 2011;17:1619–34.

CHAPTER NINE

Activity and Function of Deadenylases

Christiane Harnisch, Bodo Moritz, Christiane Rammelt, Claudia Temme, Elmar Wahle[1]

Martin-Luther-University of Halle–Wittenberg, Institute of Biochemistry and Biotechnology, Kurt-Mothes-Strasse 3, Halle, Germany
[1]Corresponding author: e-mail address: ewahle@biochemtech.uni-halle.de

Contents

1.	Introduction	182
2.	The Poly(A) Nuclease (PAN)	183
	2.1 Discovery, structure, and catalytic properties	183
	2.2 Biological function	184
3.	The Poly(A)-Specific Ribonuclease	185
	3.1 Discovery, structure, and catalytic properties	185
	3.2 Biological function	188
4.	The CCR4–NOT Complex	190
	4.1 Discovery, structure, and catalytic properties	190
	4.2 Biological function	198
	Acknowledgments	203
	References	203

Abstract

Shortening of the poly(A) tail is the first and often rate-limiting step in mRNA degradation. Three poly(A)-specific 3′ exonucleases have been described that can carry out this reaction: PAN, composed of two subunits; PARN, a homodimer; and the CCR4–NOT complex, a heterooligomer that contains two catalytic subunits and may have additional functions in the cell. Current evidence indicates that all three enzymes use a two-metal ion mechanism to release nucleoside monophosphates in a hydrolytic reaction. The CCR4–NOT is the main deadenylase in all organisms examined, and mutations affecting the complex can be lethal. The contribution of PAN, apparently an initial deadenylation preceding the activity of CCR4–NOT, is less important, whereas the activity of PARN seems to be restricted to specific substrates or circumstances, for example, stress conditions. Rapid deadenylation and decay of specific mRNAs can be caused by recruitment of both PAN and the CCR4–NOT complex. This function can be carried out by RNA-binding proteins, for example, members of the PUF family. Alternatively, miRNAs can recruit the deadenylase complexes with the help of their associated GW182 proteins.

The Enzymes, Volume 31
ISSN 1874-6047
http://dx.doi.org/10.1016/B978-0-12-404740-2.00009-4

1. INTRODUCTION

Poly(A) tails in eukaryotic cells come in two flavors, differing in their mode of synthesis, their length, and their function. The first type of poly(A) tail is a characteristic modification of mRNAs. Polyadenylation is an obligatory feature of nuclear pre-mRNA processing and is catalyzed by a "canonical" poly(A) polymerase, that is, an orthologue of *Saccharomyces cerevisiae* Pap1p. Such poly(A) tails are long, more than 200 nucleotides in mammalian cells, and around 70 nucleotides in *S. cerevisiae*. They are degraded in the cytoplasm by continuous shortening throughout the lifetime of the mRNA. As subsequent steps of mRNA decay are delayed until deadenylation has proceeded beyond a certain limit, deadenylation can be considered a timer for mRNA decay and the poly(A) tail itself a stabilizing modification [1,2]. Typically, the enzymes catalyzing mRNA deadenylation are poly(A)-specific and do not invade the mRNA body to any significant extent [3]. These enzymes will be the subject of this review.

Poly(A) tails of the second type are added to a large collection of RNAs which have in common that they are to be degraded quickly (e.g., intergenic transcripts, defective tRNAs) or have to undergo 3′ trimming (e.g., intermediates of rRNA processing or precursors to snoRNAs). These poly(A) tails are generated by one of the "noncanonical" poly(A) polymerases, relatives of *S. cerevisiae* Trf4p, *S. pombe* Cid1, or metazoan GLD2, and are normally just a few nucleotides long. Their function is to promote 3′ shortening or degradation, presumably both by providing an unstructured landing pad for the processive exosome and by interactions between this nuclease and the poly(A) polymerase complex generating the tail [1]. Thus, poly(A) tails of this sort have a very transient existence, being degraded the moment they are synthesized, and can be considered destabilizing RNA modifications, similar to the function of poly(A) tails in prokaryotes. While the exosome digests the poly(A) tail, it also proceeds further into the RNA; it is thus not considered a deadenylase and is treated in a separate chapter.

Although the distinction between the two types of poly(A) tails is fundamental, there are connections between them: poly(A) tails generated by one type of poly(A) polymerase can serve as primers to be elongated by another [4–6], and we will discuss one example of a mammalian deadenylase participating in the type of 3′ trimming reaction carried out by the exosome in yeast [7].

This review is focussed on the three widely conserved deadenylases that have been described: in the order of their discovery, the poly(A) nuclease (PAN2–3), the poly(A)-specific ribonuclease (PARN), and the CCR4–NOT complex. The rates of degradation contribute to controlling the steady state levels of different mRNAs, and changes in stability can be used to up- or downregulate mRNAs [2]. As the rate of deadenylation is a major factor determining the overall rate of degradation of an mRNA, deadenylases are controlled by mRNA-specific factors, and some of these will also be discussed.

2. THE POLY(A) NUCLEASE (PAN)

2.1. Discovery, structure, and catalytic properties

Discovery of the PAN was prompted by the observation that an *S. cerevisiae* mutant deficient for the cytoplasmic poly(A) binding protein (Pab1p; PABPC is the orthologous mammalian protein) had longer poly(A) tails. Biochemical assays in crude extracts revealed a poly(A) degrading ribonuclease activity that was dependent on its substrate being complexed with Pab1p [8,9]. The purified enzyme contained two subunits, the products of the PAN2 and PAN3 genes [10,11]. The subunit stoichiometry of the PAN complex has not been defined, but two-hybrid experiments suggest an oligomerization of Pan3p [12]. PAN is present in most eukaryotes but has apparently been lost in a number of unicellular organisms and in *Arabidopsis* [13].

PAN2 (127 kDa in yeast), the catalytic subunit of the complex, is a member of the DEDD superfamily of 3′ exonucleases [10,14,15]. These enzymes use a two-metal ion mechanism to hydrolyze the phosphodiester bond by an in-line nucleophilic substitution, and the four acidic side chains from which the family name is derived coordinate the two catalytic Mg^{2+} ions. PAN releases 5′ AMP from its substrate in a distributive reaction and depends on a 3′ hydroxyl group [15,16]. Efficient deadenylation requires PAN3 (76 kDa in yeast), which binds tightly to PAN2 [11,15]. A small C-terminal fragment of yeast Pan3p is sufficient for the interaction with Pan2p [12].

The dependence on PABPC is conserved in the human enzyme [15]. The poly(A) binding protein presumably increases the substrate affinity of PAN as suggested by its direct interaction with PAN3 [12,15]. However, as PABPC can be partially replaced by spermidine or high salt concentrations, screening of phosphate charges is probably also involved. Even under conditions of spermidine-dependent activity, PAN prefers poly(A), demonstrating an inherent, PABPC-independent specificity of the enzyme [15,16]. Under the same conditions, the stimulatory effect of PAN3 appears modest at most [15]; it thus is currently unclear whether PAN3 has any effect on PAN2 beyond mediating the interaction with PABPC. In its Pabp1p-dependent activity, PAN does not remove poly(A) tails completely, leaving about 20–25 nt [8,17]. Presumably, this reflects the length of poly(A) covered by the last molecule of Pab1p; its departure prevents further degradation of the substrate.

The C-terminal helical domain (PABC domain) of Pab1p [18,19] and a preceding linker sequence are sufficient for binding to an N-terminal region of Pan3p [12]. Like many other interactions of the PABC domain, binding is mediated by the ∼15 amino acid PAM2 motif also found in other proteins interacting with PABPC [20,21]. Mutations decreasing the Pan3p—Pab1p interaction result in longer poly(A) tails in vivo, consistent with reduced PAN activity [12,20,21].

In contrast to the function of Pab1p as a global activator of PAN, the yeast protein Pbp1p has been identified as an inhibitor of PAN [22]. The inhibitory activity may involve an association with Pab1p [12], but the mechanism and biological function are not further defined.

2.2. Biological function

Deletions of yeast PAN2 or PAN3 do not cause a significant growth defect, but bulk poly(A) tails are ∼20 nucleotides longer than in control cells [10,11]. A major reason for this effect seems to be that, in wild-type cells, there is an initial PAN-dependent shortening of newly made poly(A) tails [23]. A role of PAN in further poly(A) tail shortening did not become apparent until a pan2Δ mutation was combined with a ccr4Δ mutation: whereas the single mutations had a mild (ccr4Δ) or no phenotype (pan2Δ), the double mutant displayed a complete block in deadenylation [17]. Initial deadenylation by PAN followed by further shortening predominantly by the CCR4–NOT complex appears to be conserved in mammalian cells [24]. In Drosophila S2 cells, PAN is also of minor importance compared to CCR4-NOT-dependent deadenylation, but a temporal order has not been established [25]. The fact that, after induction of a regulated promoter in yeast, the earliest detectable transcripts have already undergone PAN-dependent shortening implies that this reaction occurs early, at the very beginning of the life of an mRNA [23]. Genetic analysis in yeast further suggested that early PAN- and Pab1p-dependent deadenylation may play a role in nuclear export of mRNA and, thus, in permitting its translation and initiating its decay [26–29]. If this is true, initial PAN-dependent trimming should be a nuclear event. However, direct localization studies have found PAN to be mostly cytoplasmic in yeast and mammalian cells [15,24,30] (http://yeastgfp.yeastgenome.org).

One important mechanism by which miRNAs affect gene expression is through an acceleration of mRNA deadenylation. While miRNA-dependent deadenylation functions mainly through the CCR4–NOT complex, PAN also appears to play a modest role [31]. GW182 proteins function as adapters to recruit PAN through the PAN3 subunit [32–34]. PAN is also involved, together with CCR4–NOT, in mammalian nonsense-mediated decay [24].

In summary, PAN appears to share many of its substrates with the CCR4–NOT complex; the division of labor between the two deadenylase complexes may be mostly temporal, PAN acting before CCR4–NOT.

3. THE POLY(A)-SPECIFIC RIBONUCLEASE

3.1. Discovery, structure, and catalytic properties

In hindsight, PARN was first described as the main deadenylating activity in HeLa cell extracts [35,36]. The enzyme was then purified to homogeneity from calf thymus [37,38] and *Xenopus* oocytes [39], and cloning was based on these preparations [38,40]. PARN is widely but not universally conserved, being absent from *S. cerevisiae* and *Drosophila*.

Human PARN is a homodimer of 74 kDa subunits [41]. X-ray structures show an N-terminal nuclease domain interrupted by an R3H domain and followed by an RNA recognition motif (RRM) domain [41,42] (Fig. 9.1). The major dimerization surface is formed by the nuclease domain, and mutations disrupting it strongly reduce enzymatic activity [41]. The C-terminal ~ 130 amino acids are predicted to be unstructured but may contain a second dimerization surface [43]. As a member of the DEDD superfamily of $3'$ exonucleases described earlier [40,41,44,45], PARN degrades RNA from the $3'$ end, releasing $5'$-NMPs. The $3'$ -OH group required for activity [36,37] forms a hydrogen bond to Glu30, one of the side chains also involved in metal ion binding [41] (Fig. 9.2).

Among the homopolymers, poly(A) is the preferred PARN substrate. Depending on the reaction conditions, poly(U) is degraded less effectively, and poly(C) and poly(G) are largely resistant. When polyadenylated RNA is used as a substrate, the poly(A) tail is degraded, the deadenylated RNA body accumulates transiently and is then also hydrolyzed [37–39]. The RRM domain specifically binds poly(A) [47], but a PARN mutant lacking the

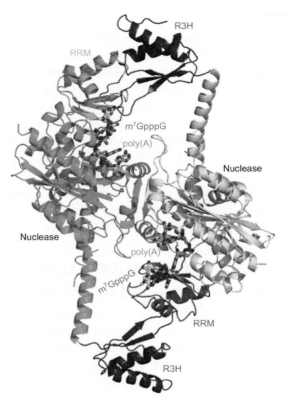

Figure 9.1 A model for homodimeric PARN. The structure has been reconstructed from three independently determined X-ray structures (see text). The nuclase domains of the two monomers are green and yellow. The monomer on the left is in the closed conformation, the one on the right in the open conformation. Contributions of both RRM and nuclease domain to cap binding and the clash between cap and poly(A) in the closed conformation are visible. From Ref. [42], with permission. (See color plate section in the back of the book.)

domain, while less active, retains its specificity for poly(A) [41]. Likewise, a deletion of the R3H domain reduces RNA affinity but not substrate specificity [41]. The active site of PARN itself is specific for A residues [48], but, surprisingly, the structure of the cocrystal does not reveal any A-specific hydrogen bonds. The basis of poly(A) recognition thus remains to be elucidated.

The activity of PARN is modestly stimulated by its interaction with the m^7G cap found at the 5' ends of mRNAs [38,49,50]. The cap increases the processivity of PARN, which is low on noncapped substrates, and also

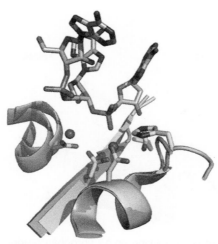

Figure 9.2 Alignment of the catalytic centers of PARN and *S. pombe* Pop2p. Human PARN (PDB 2A1R) [41] is in yellow and aligned with POP2 (PDB 3G0Z) [46] in lime; active site residues (DEDD) are in stick representation and colored by element. The oligo(rA) substrate (light blue) is from the PARN structure, the two catalytic metal ions are from the Pop2p structure. The 3′-OH of the nucleotide in the upper right is in hydrogen bonding distance to an active site glutamate residue. (See color plate section in the back of the book.)

facilitates progress of the enzyme into heteropolymeric RNA sequences [37,49,51]. The cap is bound by the RRM domain with an affinity in the low micromolar range, much weaker than the nanomolar affinities of "canonical" cap-binding proteins [47,52,53]. Whereas the surface of the β-sheet of the RRM domain is the conventional site of nucleic acid interactions, the cap binds to the "edge" of the RRM in PARN. The methylated base is stacked on top of a single tryptophan residue, in contrast to its insertion between two aromatic side chains in high–affinity cap-binding proteins [42,52,53] (Fig. 9.3). In a cocrystal structure of PARN with m^{7}GpppG, each of the two RRM domains with their bound caps adopts a different position with respect to the corresponding nuclease domain. In particular, in the "closed" conformation, the nuclease domain contributes to cap binding [42] (Fig. 9.1). Interestingly, in this conformation, the position of the cap on the nuclease domain overlaps with the position of oligo(A) determined in a different cocrystal, explaining why high concentrations of free cap inhibit the degradation even of noncapped substrates [51]. A reasonable model assumes that, in the interaction with a capped substrate RNA, PARN may use one subunit to bind the cap and the other to attack the 3′ end of the same

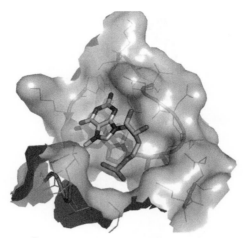

Figure 9.3 Binding of m^{7}G by stacking on a tryptophan residue in the RRM of PARN. Based on PDB 3D45 and in Ref. [42]. (See color plate section in the back of the book.)

RNA molecule. Communication between the two subunits is suggested by the stimulatory effect of low concentrations of free cap on the degradation of uncapped substrates [51].

3.2. Biological function

The preference of PARN for poly(A) and its stimulation by a 5' cap strongly suggest mRNAs as substrates. However, mammalian PARN is localized in the cell nucleus with a strong enrichment in the nucleolus and in Cajal bodies [7,24]. *Xenopus* PARN is also predominantly localized in oocyte nuclei [40]. Although the experiments certainly do not exclude the presence of low concentrations of PARN in the cytoplasm, the localization appears at odds with a predominant function of PARN in cytoplasmic mRNA decay.

A deletion of the PARN orthologue in *S. pombe* had no obvious phenotype, and the same was true for simultaneous knockdown of both orthologues in *Caenorhabditis elegans* [54]. However, RNA analyses were not reported in either organism. In contrast, T-DNA insertions into the single *Arabidopsis* gene encoding a functional PARN orthologue caused delayed embryo development and lethality [54,55]. Among five mRNAs examined, one had a significantly longer poly(A) tail in a homozygous PARN mutant [54]. In contrast, examination of dozens of transcripts in a hypomorphic mutant revealed no differences in poly(A) tail lengths [56]. *Arabidopsis* PARN is induced by various stress conditions [57]. In comparing functions of *Arabidopsis* PARN and of its vertebrate orthologues, one has to keep in mind that their overall sequence identity is only ~20% with higher similarities limited to the nuclease and other domains, and that *Arabidopsis* PARN is entirely [54] or partially [55] localized in the cytoplasm.

In *Xenopus* oocytes, PARN is responsible for the so-called default mRNA deadenylation, a general, sequence-independent deadenylation reaction triggered by progesterone-induced oocyte maturation, the resumption of meiosis resulting in the differentiation of the oocyte into a fertilizable egg. Deadenylation coincides with breakdown of the nuclear envelope, which allows nuclear PARN to meet its cytoplasmic substrates. However, additional activation of PARN by posttranslational modifications seems likely [39,40,49]. Poly(A) tail lengths of many RNAs important during the development of *Xenopus* oocytes and early embryos are regulated by an RNA sequence dubbed the cytoplasmic polyadenylation element (CPE). The CPE, via the CPE-binding protein (CPEB), initially causes shortening of the long poly(A) tail added in the nucleus. Upon maturation, the CPE, together with the "canonical" polyadenylation signal AAUAAA, directs cytoplasmic poly(A) tail elongation, overriding default deadenylation (reviewed in [58]). It has been reported that PARN associates with CPEB and fulfills two functions, first catalyzing CPE-dependent poly(A) tail shortening and then counteracting the activity of the associated poly(A) polymerase GLD2 to keep poly(A) tails short. Upon oocyte maturation, CPEB phosphorylation releases PARN and allows poly(A) tail extension [59]. Note, though, that others have not confirmed an association between CPEB and PARN [60].

In mammalian cells, the prevailing evidence indicates that mRNA deadenylation is carried out mostly by the CCR4–NOT complex, and the role of PARN is reserved for special cases or circumstances. For example, PARN has been reported to be induced and activated by UV irradiation [61], in agreement with stress-induced PARN expression in *Arabidopsis* (see above). A role of PARN in mRNA deadenylation promoted by AU-rich elements (AREs) has been proposed [62–64]. While some of these claims were supported by increased stability of relevant RNAs upon PARN depletion, other studies did not detect a role of PARN in mRNA decay promoted by AREs or other destabilizing elements [24,65–67]. These discrepancies need to be resolved. A human host factor restricting the replication of HIV apparently makes use of PARN to induce degradation of specific viral mRNAs [68].

An unexpected role of PARN in the processing of small nuclear RNAs (snoRNAs) of the H/ACA type has recently been uncovered. In mammals, these RNAs are derived from intron-encoded precursors by splicing, intron debranching, and 5′ and 3′ exonuclease digestion. Upon PARN knockdown, late processing intermediates of these snoRNAs accumulated with short oligo(A) tails attached to a few remaining intron nucleotides. The data suggested that PARN not only degrades the oligo(A) tails but may also remove the last few intron nucleotides [7]. It appears that, in this process, PARN has taken over a role that is played by the exosome and its associated nuclease RRP6 in yeast.

4. THE CCR4–NOT COMPLEX
4.1. Discovery, structure, and catalytic properties

Most subunits of the CCR4–NOT complex were initially identified in various genetic screens aiming at the identification of transcription factors (recently reviewed in [69]). Evidence supporting a role of the complex in transcriptional regulation will not be discussed in this review. Experiments to examine a function of the CCR4–NOT complex in deadenylation were prompted by sequence similarities of the CCR4 and POP2/CAF1 subunits with exonucleases, and the analysis of deletion mutants uncovered clear defects in mRNA turnover and deadenylation [17,70]. The composition of the CCR4–NOT complex was first determined in yeast, and the subunits Caf1p (=Pop2p), Ccr4p, Not1–5p, Caf40p, and Caf130p were identified [71,72]. Similar but not identical results were then obtained in *Drosophila* [73,74], trypanosomes [75], and humans [72,76–78] (Table 9.1 and Fig. 9.4; recently reviewed in [69,80]).

CCR4 (94 kDa in yeast) is a member of the EEP class of exo- and endonucleases related to bovine DNase I and *E. coli* exonuclease III. CCR4 was not found in trypanosomes and several other unicellular eukaryotes [75]. The human and mouse genomes encode two CCR4 orthologues called CNOT6 (hCCR4a) and CNOT6L (hCCR4b) [24,78,81]. These appear to be integrated into the CCR4–NOT complex in a mutually exclusive manner [77]. Integration of CCR4 into the complex is mediated primarily by a conserved leucine-rich repeat (LRR), which precedes the C-terminal catalytic domain and associates with CAF1 [76,81–83]. *In vitro* deadenylase activity has been demonstrated for yeast Ccr4p [17,84–86] and for both mammalian homologues [78,85,87]. Consistent with its *in vivo* function, CCR4 is a Mg^{2+}-dependent, poly

Table 9.1 Subunits of the CCR4–NOT complex

Saccharomyces cerevisiae (baker's yeast)		Drosophila melanogaster (fruit fly)			Homo sapiens (human)		
Protein	Alternative names[a]	Homolog protein	Related proteins	Annotation ID[b]; gene and alternative names	Closest homolog(es)	Related proteins	Names[c]
NOT1	CDC39, YCR093W	*Not1*		CG34407; *Not1*	CNOT1		CCR4–NOT transcription complex, subunit 1, CDC39, NOT1
NOT2	CDC36, YDL165W	*Not2*		CG2161; *Regena (Rga)*	CNOT2		CCR4–NOT transcription complex, subunit 2, CDC36, NOT2
NOT3	YIL038C	*Not3/5*		CG8426; *L(2)NC136 (lethal (2) NC136)*	CNOT3		CCR4–NOT transcription complex, subunit 3, NOT3
NOT5	YPR072W	n.e.			n.e.		
CCR4	YAL021C	*twin*		CG31137; *twin, CCR4*	CNOT6		CCR4–NOT transcription complex, subunit 6
					CNOT6L		CCR4–NOT transcription complex, subunit 6-like
			angel	CG12273		ANGEL1	angel homolog 1 (*Drosophila*)
						ANGEL2	angel homolog 2 (*Drosophila*)
			d3635	CG31759; *3635*		PDE12	phosphodiesterase 12
			noctunin	CG31299; *curled (cu)*		CCRN4L	CCR4 carbon catabolite repression 4-like (*S. cerevisiae*)

Continued

Table 9.1 Subunits of the CCR4–NOT complex—cont'd

| *Saccharomyces cerevisiae* (baker's yeast) | | *Drosophila melanogaster* (fruit fly) | | | *Homo sapiens* (human) | | |
Protein names	Alternative names	Homolog protein	Related proteins	Annotation ID; gene and alternative names	Closest homolog (es)	Related proteins	Names
POP2	CAF1, YNR052C	*Pop2*		CG5684; *CAF1*	CNOT7		CCR4–NOT transcription complex, subunit 7, CAF1
					CNOT8		CCR4–NOT transcription complex, subunit 8, POP2, CALIF, CAF1, hCAF1
						TOE1	hCaf1z
CAF 40	YNL288W	*Caf40*		CG14213; *Required for cell differentiation 1 ortholog (Rcd-1), CAF40,*	RQCD1		RCD1 homolog (*S. pombe*), CNOT9, Rcd1
CAF 130	YGR134W	*n.e.*			n.e.		
NOT4	MOT2, SIG1, *NOT4* YER068W	*NOT4*		CG31716; *Cnot 4 homologue (Cnot4)*	CNOT4		CCR4–NOT transcription complex, subunit 4, NOT4

[a]http://yeastgenome.org
[b]http://flybase.org
[c]http://genenames.org

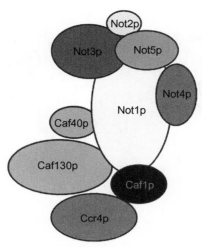

Figure 9.4 Composition of the yeast CCR4–NOT complex. The cartoon roughly corresponds to the L-shaped structure seen in electron micrographs [79]. The arrangement of the subunits reflects the known interactions and roughly indicates the three "modules" described in the text. The stoichiometry of one subunit each is not established and may in fact be wrong. In metazoans, NOT4 is not stably associated with the complex, and NOT3 and NOT5 are represented by a single type of polypeptide (see text). (See color plate section in the back of the book.)

(rA)-specific 3′ exonuclease [84,87]. Although cleaved-off nucleotides have, to our knowledge, not been analyzed directly, polynucleotides appear to be shortened in steps of single nucleotides, and the family membership suggests a hydrolytic reaction liberating 5′-AMP. X-ray crystallography of the nuclease domain of CNOT6L [87] reveals a monomeric structure closely corresponding to other members of the EEP class of enzymes (Fig. 9.5). Structures of the catalytic domain complexed with an AMP product or oligo(dA) (which is a poor substrate) demonstrate specific recognition of at least one adenosine base in the active site. Two Mg^{2+} ions are also found in the active site, and amino acid side chains contacting either are required for enzymatic activity, suggesting an involvement of both metal ions in catalysis. The active site of CNOT6L is extremely similar to that of the endonuclease APE1, which also contains two metal ions and has been proposed to use a two-metal ion mechanism for hydrolysis [88] (Fig. 9.6). Note, however, that the two metal ions in CNOT6L are seen contacting the "wrong" phosphate, predicting the release of a di- rather than the expected mononucleotide. Thus, the mechanism of CCR4 remains to be defined more precisely.

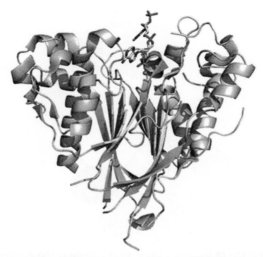

Figure 9.5 Structure of the catalytic domain of human CNOT6L. The oligo(dA) substrate shown at the top (light blue) marks the active site. Two catalytic metal ions are also visible (red spheres). Based on PDB 3NGO [87]. (See color plate section in the back of the book.)

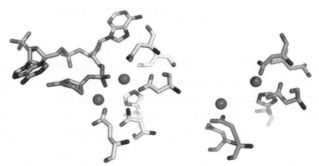

Figure 9.6 A comparison of the active sites of CNOT6L and APE1. CNOT6L (PDB 3NGO) [87] is shown on the left, and APE1 (PDB 1E9N) [88] is shown on the right. Amino acid carbon atoms are in lime and cyan, respectively, oligo(dA) carbon atoms are in light blue, and metal ions in magenta. (See color plate section in the back of the book.)

In addition to CCR4, three families of conserved related proteins have been identified: Nocturnin, Angel, and 3635 [81]. All these share the catalytic domain with CCR4, but lack the LRR thought to connect CCR4 with the rest of the complex. Nevertheless, Nocturnin has been found associated with other subunits of the CCR4–NOT complex [74,89]. Nocturnin also has poly (A) degrading activity [90,91] and the *Drosophila* orthologue, encoded by the *curled* gene [92], may be involved in mRNA degradation [74]. Nocturnin is thought to be involved in circadian rhythm [90] but probably has other functions as well [92]. One of two Angel proteins encoded in the human genome, originally described as hCCR4d [89] and now called Angel homolog 2, is associated with a variant CAF1-type protein (see below). The 3635 protein is identical with phosphodiesterase 12, which has recently been discovered to act as a mitochondrial deadenylase [93,94].

Like PAN2 and PARN, CAF1 (= POP2; 51 kDa in yeast) is a member of the DEDD family of nucleases. The protein seems to be more widely conserved than CCR4 [75]. In humans, there are two closely related versions of CAF1, CNOT7 (= hCAF1 = CAF1A) and CNOT8 (= hPOP2 = CAF1B = CALIF), which associate with the CCR4–NOT complex in a mutually exclusive manner [77]. The human genome also encodes a more distant relative dubbed hCAF1z (= TOE1), which contains a zinc finger motif in addition to the nuclease domain. This CAF1 variant associates with Angel homolog 2, as mentioned earlier. Interestingly, these two proteins are not associated with CNOT2, suggesting the existence of a complex distinct from the "canonical" CCR4–NOT complex. Also, whereas the latter is localized mostly in the cytoplasm, hCAF1z and hCCR4d were detected in Cajal bodies, nuclear structures in which maturation of snoRNAs takes place [89].

In vitro deadenylase activity has been demonstrated for several CAF1 proteins, including all three mammalian paralogs, CNOT7, CNOT8, and hCAF1z [46,89,95–98]. Most assays showed a preference for poly(A), but various members of the CAF1 family are able to digest nonpoly(A) sequences with different efficiencies, being inhibited by secondary structures (e.g., [97]). Release of nucleoside monophosphates by CNOT7 and CNOT8 confirms the expected hydrolytic activity [95]. In yeast Caf1p, one of the catalytic motifs is not conserved. Whether or not the protein has catalytic activity is controversial [70,85,86,99,100]. Crystal structures of Pop2p/Caf1p from *S. cerevisiae* and *S. pombe* and of hCAF1 (Fig. 9.7) show a monomeric protein with overall structural similarity to the founding members of the DEDD class of enzymes, the proofreading ε-subunit of *E. coli* DNA polymerase III and the proofreading exonuclease domain of DNA polymerase I [46,98,99]. Not surprisingly, the active site is also highly similar to that of other members of DEDD enzymes, allowing for the sequence divergence of the *S. cerevisiae* enzyme. The two metal sites can be occupied by different ions; the highest activity is observed with site A occupied by Mg^{2+} and site B by Mn^{2+}. However, as it is often observed with nucleic acid polymerizing or degrading enzymes, Mn^{2+} relaxes the specificity of the enzyme,

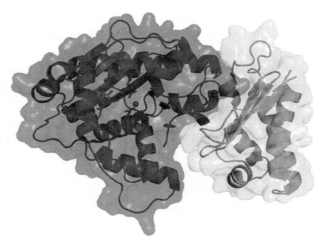

Figure 9.7 Structure of a CNOT7/TOB complex. Based on (PDB 2D5R) [98]. A ribbon diagram and transparent surface of CNOT7 is in magenta, and the equivalent structure of an N-terminal domain of TOB in green. Active site residues of CNOT7 are shown as sticks, and catalytic metal ions (spheres) were placed into the active site by alignment with spPOP2 (PDB 1UOC). (See color plate section in the back of the book.)

permitting efficient degradation of nonpoly(A) sequences [46,96]. Which ions occupy the two metal sites inside a living cell is obviously hard to tell. Although a complex of CAF1 with either substrate or product has not been crystallized so far, models can be derived from comparison with the corresponding structures of other DEDD enzymes, including PARN (Fig. 9.2). These predict specific recognition of the 3' adenosine by hydrogen bonding of its N3 with a particular serine side chain. In fact, replacement of this serine causes a loss of poly(A) specificity [96]. Effects of substrate length and structure on the catalytic activities of both CCR4 and CAF1 have been reported, suggesting the possibility that binding sites exist outside the active sites [84,97].

The very large NOT1 protein (240 kDa in yeast) is considered the scaffold subunit of the complex, as it appears to be touched by most of the other subunits; CAF1 and, via this protein, CCR4 associate with an internal segment, and the NOT proteins bind to a C-terminal region [83,101]. No functional motifs are apparent in the NOT1 sequence. NOT2 is a 22 kDa protein in yeast. While the protein does not contain motifs of known function, its C-terminus harbors the so-called Not box, which is conserved not only between species, but also in Not3p and Not5p [102]. The yeast proteins Not3p (94 kDa) and Not5p (66 kDa) are homologous to each other. In addition to the Not box, a conserved N-terminal coiled-coil region is predicted for both [76]. Genetic data indicate a partial functional redundancy in yeast [103], and both are represented by a single protein in flies and humans, which behaves like a subunit of the CCR4–NOT complex [74,77,78]. The C-terminal Not box of human CNOT3 is required for association with the complex via interaction with CNOT2 [77]. Similarly, yeast Not5p associates with Not2p [83,101]. In agreement with these interactions, *not2* mutations lead to a reduced abundance of Not3p and impaired incorporation of Not5p into the yeast complex [104]. Similar effects have been reported in metazoan cells [74,105].

The Not4 protein is not stably associated with the complex in either *Drosophila* [74] or humans [72,77,78], and knockdown of Not4 had no effect on mRNA deadenylation in *Drosophila* S2 cells [74]. Likewise, deletion of yeast NOT4 had little if any effect on the decay of the MFA mRNA [86]. Therefore, NOT4, which is an active ubiquitin ligase [106], does not appear to be a conserved subunit of the CCR4–NOT complex and may not play a role in mRNA deadenylation. Nevertheless, the physical connection of yeast Not4p with the CCR4–NOT complex is in agreement with many other data supporting a functional connection (e.g., [101,107,108]). *Drosophila* NOT4 has been proposed to be required for deadenylation of the cyclin B mRNA by mediating an interaction between the CCR4–NOT complex and the regulatory protein Nanos [109].

Caf40p (41 kDa in yeast) and its orthologues in flies and human (CNOT9 = Rcd-1 = RQCD1) are considered stably associated subunits of the CCR4–NOT complex [72,74,77,78]. However, whereas all other subunits of the yeast CCR4–NOT complex are present in the cell in roughly comparable concentrations (~2000–4000 monomers per cell), Caf40p is reported to occur at a five- to tenfold higher concentration (19,000 copies per cell) (http://yeastgfp.yeastgenome.org), suggesting the possibility of functions unrelated to the complex. The isolated human protein is a homodimer, and X-ray crystallography [110] revealed each monomer to be composed of six *armadillo* repeats, structural motifs known to mediate protein–protein interactions. Unexpectedly, RQCD1 binds a number of single- and double-stranded RNA and DNA homopolymers *in vitro*, discriminating against poly(A) [110]. The Caf130 protein (129 kDa) appears to be unique to yeast. Both Caf40p and Caf130p can associate with Not1p independently of the other subunits [71].

Additional proteins have been found in affinity-purified CCR4–NOT complexes, but nothing is known about potential functions in the complex. These proteins include human CNOT10, listed as AAH02928 in Ref. [72,78], with its orthologues also present in the fly and, possibly, the trypanosome complex [74,75]; human TNKS1BP1 = TAB182 [77] = KIAA1741 [78], a binding protein for tankyrase, involved in telomere biology; and human C2ORF29 [77].

Studies of the CCR4–NOT complex in yeast initially led to the suggestion of a bipartite structure already alluded to above: a Ccr4p–Caf1p complex and a Not2p–Not5p complex are linked via independent association with Not1p [83,101]. However, interactions have also been reported between CAF1 and NOT2 and NOT3 [76,111]. Later, Caf4p and Caf130p were suggested to form a third "module" bound to Not1p [71]. An L-shaped structure of the complex has recently been found by electron microscopy at 31–33 Å resolution, in agreement with at least a bipartite complex [79]. The subunit stoichiometry of the complex is not known. Dimerization of CAF40 has already been mentioned. Also, the homology of Not3 to Not5 and their representation by a single protein in higher eukaryotes suggest the possibility that this protein oligomerizes. However, the total size of the "core" complex, estimated as ∼1 MDa in both yeast and humans [71,77,79,101], is not very different from the sum of the molecular weights of the subunits of the yeast complex (including Not4p), 800 kDa. Thus, it is likely that most subunits are present as single copies.

4.2. Biological function

The CCR4–NOT complex is the main deadenylase in *S. cerevisiae* [17], *Drosophila* [25,73], human cells [24], and trypanosomes [75]. Interestingly, though, biochemical purification of the main deadenylating activities from yeast or mammalian cells resulted in the identification of PAN or PARN, respectively. This may be due to the fact that nonspecific deadenylation assays were used to monitor the purifications. Presumably, the CCR4–NOT complex has a low basal activity, which makes its detection in such assays difficult, and relies on mRNA-specific activators. Due to its role in the decay of many different mRNAs (and possibly in other regulatory phenomena), it is not surprising that the CCR4–NOT complex affects many different biological processes. Because of the limited scope of this review, we will limit ourselves to discussions of mechanistic aspects and largely ignore the wider biological context.

In yeast, the catalytic activity of Ccr4p is involved in mRNA deadenylation, as both a deletion of CCR4 and point mutations in the active site lead to deadenylation defects. In contrast, point mutations affecting the vestigial active site of Caf1p have no detectable phenotype, although a deletion does have a defect in deadenylation [17,70,85,86,97], presumably because Caf1p connects Ccr4p with the rest of the complex (see above).

A comparison to the situation in other organisms reveals that the division of catalytic labor between CCR4 and CAF1 has shifted during evolution: in *Drosophila* S2 cells, neither knockdown of CCR4 nor overexpression of an inactive variant had detectable effects on mRNA deadenylation, whereas the equivalent experiments with CAF1 resulted in the expected deadenylation phenotypes, suggesting that CAF1 carries the main burden of catalysis [73,74]. *Drosophila* CCR4 is encoded by the *twin* gene. In contrast to CCR4 knockdown in S2 cells, *twin* mutants, which have a severe defect in female germline development, display elongated poly(A) tails and a reduced rate of deadenylation [73,112,113]. However, it is currently unclear whether this reflects a structural role of CCR4 in the CCR4–NOT complex or a catalytic function in deadenylation. In *C. elegans*, *ccf-1* encodes the only CAF1 orthologue. Animals homozygous for an allele that deletes part of the conserved nuclease active site arrest development at larval stage 4. RNAi against *ccf-1* caused sterility in both males and hermaphrodites, in contrast with RNAi against the orthologues of CCR4, PAN, and PARN, which had no obvious phenotypes [114] (and references cited therein). Trypanosomes possess no CCR4 orthologue, and deadenylation relies on CAF1 [75].

In human cells, both CCR4 and CAF1 catalyze deadenylation, as shown both by the phenotypes of knockdowns and by dominant-negative effects of active site mutants, and this applies to stable RNAs as well as to RNAs destabilized either by an ARE (see below) or a premature stop codon [21,24,75,115,116]. Depletion of hCCR4b (CNOT6L) causes a growth defect by stabilization of the mRNA encoding $p27^{Kip1}$, an inhibitor of cyclin-dependent kinases [78]. The data indicated that none of the other catalytic subunits can replace the function of Ccr4b in the deadenylation of $p27^{Kip1}$. Knockdown of CNOT7 and CNOT8 also interferes with cell-cycle progression; the proteins play largely redundant roles [117]. Supporting redundancy of CNOT7 and CNOT8 in mammals, $Cnot7^{-/-}$ mice are viable and healthy, except for male sterility due to defects in spermatogenesis; thus, CNOT7 has a specific function in this process [118,119].

Roles of the noncatalytic subunits of the CCR4–NOT complex in mRNA deadenylation have been more difficult to investigate. Knockdown of NOT1 in *Drosophila* S2 cells affects mRNA deadenylation, as does depletion of CAF1, NOT2, and NOT3, suggesting, at first sight, that all are required for this process. However, knockdown of most individual subunits also affects the abundance of other subunits, conceivably due to a reduced stability of partial assemblies; thus, defects observed in RNAi experiments do not necessarily reflect a direct role of the depleted subunit in mRNA deadenylation [74]. The same caveat may apply to the phenotype of yeast deletion mutants. For example, the *not2-1* and *not5Δ* alleles, which show synthetic lethality with *ccr4Δ* and *caf1Δ* mutations, also reduce the amount of the essential Not1p subunit [101], and the importance of Not2p for maintaining normal levels of Not3p and the integrity of the complex has been discussed above.

In a systematic study carried out in yeast, consequences of individual deletions of most subunits of the CCR4–NOT complex were investigated by microarray experiments. The results supported the model of three "modules" in the complex derived from protein–protein interaction studies: gene expression profiles were very similar in the *ccr4* and *caf1* mutants, the *not* mutants were also similar to each other but distinct from *ccr4* and *caf1*, and *caf40* and *caf130* formed another, although less distinct group [107]. However, it is unclear to which extent the different expression profiles depended on changes in deadenylation rates.

Among the genes encoding subunits of the CCR4–NOT complex in yeast, only NOT1 is essential. In contrast, even a combined deletion of CCR4 and CAF1 is not lethal [101]. That the deletion of NOT1 causes a more severe phenotype than the deletion of the catalytic subunit(s) is perhaps the most definitive evidence that the CCR4–NOT complex has functions in addition to mRNA deadenylation (for a specific example, see [120]). Additional functions in posttranscriptional regulation that have been reported are deadenylation-independent inhibition of translation [121,122] and stimulation of mRNA decapping [108].

Whereas, to our knowledge, mRNA deadenylation in conditional *not1* alleles in yeast has not been examined, a knockdown of human CNOT1 affected the deadenylation of unstable reporters [123], in agreement with the *Drosophila* data cited above. Yeast *not2Δ* and *not5Δ* mutants display slower deadenylation of the MFA2 RNA [86]. NOT2 is encoded by the essential *Regena* gene in flies [124], and mutant larvae have modestly elongated poly(A) tails [73]. Yeast *caf40Δ* mutants share some phenotypes with *ccr4* mutants, but synthetic interactions with mutations in other subunits have not been found [71]. To our knowledge, mRNA deadenylation has not been examined in the mutant. Knockdown of CAF40 had no detectable effect on mRNA decay in *Drosophila* S2 cells [74]. A yeast *caf130Δ* mutant does not share phenotypes with other mutants affecting the CCR4–NOT complex but, as mentioned, has similarities in its gene expression profile with *caf40Δ* [71,107].

Numerous experiments show that the CCR4–NOT complex catalyzes rapid deadenylation of specific unstable mRNAs. This includes a role in nonsense-mediated decay [24,115]. In the case of "normal" unstable messages, the complex is recruited by mRNA-specific activators binding to specific sites, usually in the 3′ UTR. Among such mRNA sequence elements directing rapid deadenylation and decay, AREs are prominent [2]. One particularly well-studied protein mediating ARE-dependent mRNA instability is tristetraprolin (TTP), which modulates the inflammatory response by controlling, among others, the mRNA encoding tumor necrosis factor-α [125,126]. TTP accelerates mRNA deadenylation by recruiting the CCR4–NOT complex. Interactions with several subunits of the complex have been reported, but it is not clear which are direct [65,123,127]. Recruitment of the deadenylase complex is prevented by MAPKAP kinase 2-catalyzed phosphorylation of TTP [65].

PUF proteins are 3′-UTR-binding proteins that induce deadenylation of their mRNA ligands, sometimes in cooperation with partner proteins [128]. *Saccharomyces cerevisiae* Puf5p (= Mpt5p) mediates deadenylation of the mRNA encoding the HO endonuclease required for mating type switching. The reaction can be reconstituted *in vitro* from recombinant Mpt5p and affinity-purified CCR4–NOT complex. Although Puf5p interacts with Caf1p/Pop2p, catalysis of poly(A) degradation relies on Ccr4p [100,129]. The data suggest the possibility that no other proteins are essential for the reaction, although the presence of such factors in the CCR4–NOT complex preparation cannot be excluded. PUF proteins have also been found to repress translation independently of deadenylation and to promote decapping [128]. As the CCR4–NOT complex has also been suggested to have these two functions (see above), it is conceivable that all three functions of PUF proteins can be traced to a single mechanism, recruitment of the CCR4–NOT complex.

Additional proteins recruiting the CCR4–NOT complex include *Drosophila* Bicaudal-C [130], *Drosophila* Nanos, which cooperates with the PUF protein Pumilio [109], mammalian NANOS2 [131], and *Drosophila* Smaug and it yeast orthologue Vts1p [113,132–134].

MicroRNAs exert their regulatory functions not only by inhibiting translation but predominantly by accelerating the 5′-3′ pathway of mRNA decay, in particular the first step, deadenylation [135]. MiRNA-dependent deadenylation is carried out mostly by the CCR4–NOT complex with a minor role played by PAN [31,32,67,136,137]. GW182 proteins, which, together with Argonaute proteins, are essential components of the miRNA-induced silencing complex, recruit both deadenylases through interactions with PAN3 and NOT1, respectively [32–34,122].

"Tethering" yeast Caf1p (= Pop2p) to mRNA by fusion to an RNA-binding protein and insertion of the corresponding binding site into the 3′ UTR of a reporter RNA is sufficient to induce rapid deadenylation [138]. Similar effects have been seen upon tethering of *Xenopus* CCR4 and CAF1 orthologues [121]. These experiments suggest that mere recruitment of the CCR4–NOT complex to a specific RNA by association with an RNA-bound factor may suffice to bring about deadenylation. However, cell-free systems faithfully reproducing deadenylation depending on regulatory RNA sequences are ATP-dependent [133,139]. The energy-requiring step has not been identified, but the data indicate that deadenylation governed by mRNA-specific factors may be more complex than mere recruitment of the deadenylase.

In contrast to the mRNA-specific factors, the BTG/TOB family of proteins comprises general activators of deadenylation: overexpression enhances the deadenylation of various reporter RNAs, including NMD substrates, and of bulk mRNA [116,140] (reviewed in [141]). The BTG/TOB proteins, which have antiproliferative and tumor-suppressive functions, are conserved mostly in metazoans, with six family members in humans (BTG1-4, TOB1, and Tob2). The proteins, which are regulated by posttranslational modifications [141,142], also have interesting functions in differentiation; for example, the single *C. elegans* TOB protein, FOG-3, is required for germline cells to develop into sperm rather than oocytes [142]. All family members tested associate with CAF1 [141], and the BTG2–CAF1 interaction is essential for accelerated deadenylation [116]. A cocrystal structure [98] (Fig. 9.7) reveals that TOB uses two conserved motifs, termed box A and box B, to bind CAF1 in a manner not affecting the active site. The interaction surface of TOB seen in the crystal structure agrees with mutagenesis results and is conserved among all family members. Likewise, the CAF1 surface interacting with TOB is also conserved in those organisms that have TOB. Thus, the TOB–CAF1 interaction is likely to be shared among all members of the respective families [141]. In addition,

TOB proteins may also interact with other subunits of the CCR4–NOT complex [143]. TOB1 and TOB2 each contain two PAM2 motifs interacting with the C-terminal domain of PABPC, and it has been proposed that TOB may enhance deadenylation by recruiting the CCR4–NOT complex via a CAF1–PABPC interaction [21,140]. However, the PAM2 motifs are not conserved in BTG1-4 [141]; nevertheless, family members lacking the PAM2 motifs also stimulate deadenylation [116]. It is possible that BTG1-4 bind PABPC in some other manner; it is also possible that deadenylation is enhanced by a different mechanism. Stimulation of deadenylation of a specific mRNA through an interaction of TOB with the RNA-binding protein CPEB3 has also been reported [144]. Interestingly, the sperm–oocyte decision in *C. elegans*, which requires the TOB orthologue *fog-3*, also involves a CPEB orthologue, *fog-1*; whether or not they participate in the regulation of the same mRNAs is not known [142].

Finally, the role of Pab1p/PABPC in deadenylation is not entirely clear. Inefficient deadenylation in yeast pab1Δ mutants [9,26] is not explained by Pabp1-dependence of PAN, as PAN is not the dominant nuclease (see above). Possibly, poly(A) tails devoid of Pab1p associate with other proteins that inhibit deadenylation efficiently. *In vitro*, Pab1p inhibits deadenylation by the CCR4–NOT complex [86,145]. A deletion of the linker region connecting the RRMs with the globular C-terminal domain of Pab1p enhances the inhibitory effect *in vitro* and leads to a reduced rate of deadenylation *in vivo*, presumably through altered packing of the RNP [145,146]. Thus, a—presumably indirect—interplay between the CCR4–NOT complex and the cytoplasmic poly(A) binding protein contributes to maintaining the desired rates of mRNA deadenylation.

ACKNOWLEDGMENTS

We are grateful to Haiwei Song for supplying Fig. 9.1. We apologize to all colleagues whose work could not be adequately cited due to space limitations.

REFERENCES

[1] Houseley J, Tollervey D. The many pathways of RNA degradation. Cell 2009;136:763–76.
[2] Schoenberg DR, Maquat LE. Regulation of cytoplasmic mRNA decay. Nat Rev Genet 2012;13:246–59.
[3] Goldstrohm AC, Wickens M. Multifunctional deadenylase complexes diversify mRNA control. Nat Rev Mol Cell Biol 2008;9:337–44.

[4]   Benoit P, Papin C, Kwak JE, Wickens M, Simonelig M. PAP- and GLD2-type poly
      (A) polymerases are required sequentially in cytoplasmic polyadenylation and
      oogenesis in Drosophila. Development 2008;135:1969–79.
[5]   Barnard DC, Ryan K, Manley JL, Richter JD. Symplekin and xGLD-2 are required
      for CPEB-mediated cytoplasmic polyadenylation. Cell 2004;119:641–51.
[6]   Grzechnik P, Kufel J. Polyadenylation linked to transcription termination directs the
      processing of snoRNA precursors in yeast. Mol Cell 2008;32:247–58.
[7]   Berndt H, et al. Maturation of mammalian H/ACA box snoRNAs:
      PAPD5-dependent adenylation and PARN-dependent trimming. RNA
      2012;18:958–72.
[8]   Sachs AB, Deardorff JA. Translation initiation requires the PAB-dependent poly(A)
      ribonuclease in yeast. Cell 1992;70:961–73.
[9]   Sachs AB, Davis RW. The poly(A) binding protein is required for poly(A) shortening
      and 60S ribosomal subunit-dependent translation initiation. Cell 1989;58:857–67.
[10] Boeck R, Tarun S, Rieger M, Deardorff JA, Müller-Auer S, Sachs AB. The yeast Pan2
      protein is required for poly(A)-binding protein-stimulated poly(A)-nuclease activity.
      J Biol Chem 1996;271:432–8.
[11] Brown CE, Tarun SZ, Boeck R, Sachs AB. PAN3 encodes a subunit of the Pab1p-
      dependent poly(A) nuclease in Saccharomyces cerevisiae. Mol Cell Biol
      1996;16:5744–53.
[12] Mangus DA, Evans MC, Agrin NS, Smith M, Gongidi P, Jacobson A. Positive and
      negative regulation of poly(A) nuclease. Mol Cell Biol 2004;24:5521–33.
[13] Schwede A, et al. The role of deadenylation in the degradation of unstable mRNAs in
      trypanosomes. Nucleic Acids Res 2009;37:5511–28.
[14] Zuo Y, Deutscher MP. Exoribonuclease superfamilies: structural analysis and
      phylogenetic distribution. Nucleic Acids Res 2001;29:1017–26.
[15] Uchida N, Hoshino S, Katada T. Identification of a human cytoplasmic poly(A)
      nuclease complex stimulated by poly(A)-binding protein. J Biol Chem
      2004;279:1383–91.
[16] Lowell JE, Rudner DZ, Sachs AB. 3′-UTR-dependent deadenylation by the yeast
      poly(A) nuclease. Genes Dev 1992;6:2088–99.
[17] Tucker M, Valencia-Sanchez MA, Staples RR, Chen J, Denis CL, Parker R. The
      transcription factor associated Ccr4 and Caf1 proteins are components of the major
      cytoplasmic mRNA deadenylase in Saccharomyces cerevisiae. Cell 2001;104:377–86.
[18] Kozlov G, Trempe J-F, Khaleghpour K, Kahvejian A, Ekiel I, Gehring K. Structure
      and function of the C-terminal PABC domain of human poly(A)-binding protein.
      Proc Natl Acad Sci USA 2001;98:4409–13.
[19] Kozlov G, et al. Solution structure of the orphan PABC domain from Saccharomyces
      cerevisiae poly(A)-binding protein. J Biol Chem 2002;277:22822–8.
[20] Siddiqui N, Mangus DA, Chang T-C, Palermino J-M, Shyu A-B, Gehring K. Poly(A)
      nuclease interacts with the C-terminal domain of polyadenylate-binding protein do-
      main from poly(A) -binding protein. J Biol Chem 2007;282:25067–75.
[21] Funakoshi Y, et al. Mechanism of mRNA deadenylation: evidence for a molecular
      interplay between translation termination factor eRF3 and mRNA deadenylases.
      Genes Dev 2007;21:3135–48.
[22] Mangus DA, Smith MM, McSweeney JM, Jacobson A. Identification of factors reg-
      ulating poly(A) tail synthesis and maturation. Mol Cell Biol 2004;24:4196–206.
[23] Brown CE, Sachs AB. Poly(A) tail length control in Saccharomaces cerevisiae occurs
      by message-specific deadenylation. Mol Cell Biol 1998;18:6548–59.
[24] Yamashita A, et al. Concerted action of poly(A) nucleases and decapping enzyme in
      mammalian mRNA turnover. Nat Struct Mol Biol 2005;12:1054–63.

[25] Bönisch C, Temme C, Moritz B, Wahle E. Degradation of hsp70 and other mRNAs in Drosophila via the 5'-3' pathway and its regulation by heat shock. J Biol Chem 2007;282:21818–28.

[26] Caponigro G, Parker R. Multiple functions for the poly(A)-binding protein in mRNA decapping and deadenylation in yeast. Genes Dev 1995;9:2421–32.

[27] Dunn EF, Hammell CM, Hodge CA, Cole CN. Yeast poly(A)-binding protein, Pab1, and PAN, a poly(A) nuclease recruited by Pab1, connect mRNA biogenesis to export. Genes Dev 2005;19:90–103.

[28] Chekanova JA, Shaw RJ, Belostotsky DA. Analysis of an essential requirement for the poly(A) binding protein function using cross-species complementation. Curr Biol 2001;11:1207–14.

[29] Chekanova JA, Belostotsky DA. Evidence that poly(A) binding protein has an evolutionarily conserved function in facilitating mRNA biogenesis and export. RNA 2003;9:1476–90.

[30] Huh W-K, et al. Global analysis of protein localization in budding yeast. Nature 2003;425:686–91.

[31] Chen C-YA, Zheng D, Xia Z, Shyu A-B. Ag-TNRC6 triggers microRNA-mediated decay by promoting two deadenylation steps. Nat Struct Mol Biol 2009;16:1160–6.

[32] Braun JE, Huntzinger E, Fauser M, Izaurralde E. GW182 proteins directly recruit cytoplasmic deadenylase complexes to miRNA targets. Mol Cell 2011;44:120–33.

[33] Fabian MR, et al. miRNA-mediated deadenylation is orchestrated by GW182 through two conserved motifs that interact with CCR4-NOT. Nat Struct Mol Biol 2011;18:1211–7.

[34] Kuzuoglu-Öztürk D, Huntzinger E, Schmidt S, Izaurralde E. The caenorhabditis elegans GW182 protein AIN-1 interacts with PAB-1 and subunits of the PAN2-PAN3 and CCR4-NOT deadenylase complexes. Nucleic Acids Res 2012;40: 5651–65.

[35] Astrom J, Astrom A, Virtanen A. In vitro deadenylation of mammalian mRNA by a HeLa cell 3' exonuclease. EMBO J 1991;10:3067–71.

[36] Astrom J, Astrom A, Virtanen A. Properties of a HeLa cell 3' exonuclease specific for degrading poly(A) tails of mammalian mRNA. J Biol Chem 1992;267:18154–9.

[37] Körner C, Wahle E. Poly(A) tail shortening by a mammalian poly(A)-specific 3'-exoribonuclease. J Biol Chem 1997;272:10448–56.

[38] Martinez J, Ren Y-G, Thuresson A-C, Hellman U, Astrom J, Virtanen A. A 54-kDa fragment of the poly(A)-specific ribonuclease is an oligomeric, processive and cap-interacting poly(A)-specific 3' exonuclease. J Biol Chem 2000;275:4222–30.

[39] Copeland PR, Wormington M. The mechanism and regulation of deadenylation: identification and characterization of Xenopus PARN. RNA 2001;7:875–86.

[40] Körner CG, Wormington M, Muckenthaler M, Schneider S, Dehlin E, Wahle E. The deadenylating nuclease (DAN) is involved in poly(A) tail removal during the meiotic maturation of Xenopus oocytes. EMBO J 1998;17:5427–37.

[41] Wu M, Reuter M, Lilie Y, Liu Y, Wahle E, Song H. Structural insight into poly(A) binding and catalytic mechanism of human PARN. EMBO J 2005;24:4082–93.

[42] Wu M, et al. Structural basis of m7G pppG binding to poly(A)-specific ribonuclease. Structure 2009;17:276–86.

[43] Niedzwiecka A, Lekka M, Nilsson P, Virtanen A. Global architecture of human poly (A)-specific ribonuclease by atomic force microscopy in liquid and dynamic light scattering. Biophys Chem 2011;158:141–9.

[44] Ren Y-G, Martinez J, Virtanen A. Identification of the active site of poly(A)-specific ribonuclease by site-directed mutagenesis and Fe^{2+}-mediated cleavage. J Biol Chem 2002;277:5982–7.

[45] Ren Y-G, Kirsebom LA, Virtanen A. Coordination of divalent metal ions in the active site of poly(A)-specific ribonuclease. J Biol Chem 2004;279:48702–6.

[46] Jonstrup AT, Andersen KR, Van LB, Brodersen DE. The 1.4-A crystal structure of the S. pombe Pop2p deadenylase subunit unveils the configuration of an active enzyme. Nucleic Acids Res 2007;33:3153–64.

[47] Nilsson P, et al. A multifunctional RNA recognition motif in poly(A)-specific ribonuclease with cap and poly(A) binding properties. J Biol Chem 2007;282:32902–11.

[48] Henriksson N, Nilsson P, Wu M, Song H, Virtanen A. Recognition of adenosine residues by the active site of poly(A)-specific ribonuclease. J Biol Chem 2010;285:163–70.

[49] Dehlin E, Wormington M, Körner CG, Wahle E. Cap-dependent deadenylation of mRNA. EMBO J 2000;19:1079–86.

[50] Gao M, Fritz DT, Ford LP, Wilusz J. Interaction between a poly(A)-specific ribonuclease and the 5′ cap influences mRNA deadenylation rates in vitro. Mol Cell 2000;5:479–88.

[51] Martinez J, Ren Y-G, Nilsson P, Ehrenberg M, Virtanen A. The mRNA cap structure stimulates rate of poly(A) removal and amplifies processivity of degradation. J Biol Chem 2001;276:27923–9.

[52] Monecke T, Schell S, Dickmanns A, Ficner R. Crystal structure of the RRM domain of poly(A)-specific ribonuclease reveals a novel m^{7}G-cap-binding mode. J Mol Biol 2008;382:827–34.

[53] Nagata T, et al. The RRM domain of poly(A)-specific ribonuclease has a non-canonical binding site for mRNA cap analog recognition. Nucleic Acids Res 2008;36:4754–67.

[54] Reverdatto SV, Dutko JA, Chekanova JA, Hamilton DA, Belostotsky DA. mRNA deadenylation by PARN is essential for embryogenesis in higher plants. RNA 2004;10:1200–14.

[55] Chiba Y, Johnson MA, Lidder P, Vogel JT, van Erp H, Green PJ. AtPARN is an essential poly(A) ribonuclease in Arabidopsis. Gene 2004;328:95–102.

[56] Nishimura N, et al. ABA *Hypersensitive Germination2-1* causes the activation of both abscisic acid and salicylic acid responses in Arabidopsis. Plant Cell Physiol 2009;50:2112–22.

[57] Nishimura N, et al. Analysis of ABA hypersensitivie germination 2 revealed the pivotal functions of PARN in stress response in Arabidopsis. Plant J 2005;44:972–84.

[58] Radford HE, Meijer HA, De Moor CH. Translational control by cytoplasmic poly-adenylation in *Xenopus* oocytes. Biochim Biophys Acta 2008;1779:217–29.

[59] Kim JH, Richter JD. Opposing polymerase-deadenylase activities regulate cytoplas-mic polyadenylation. Mol Cell 2006;24:173–83.

[60] Minshall N, Reiter MH, Weil D, Standart N. CPEB interacts with an ovary-specific eIF4E and 4E-T in early Xenopus oocytes. J Biol Chem 2007;282:37389–401.

[61] Cevher MA, et al. Nuclear deadenylation/polyadenylation factors regulate 3′ processing in response to DNA damage. EMBO J 2010;29:1674–87.

[62] Chou C-F, et al. Tethering KSRP, a decay-promoting AU-rich element-binding pro-tein, to mRNAs elicits mRNA decay. Mol Cell Biol 2006;26:3695–706.

[63] Lin W-J, Duffy A, Chen C-Y. Localization of AU-rich element-containing mRNA in cytoplasmic granules containing exosome subunits. J Biol Chem 2007;282:19958–68.

[64] Lai WS, Kennington EA, Blackshear PJ. Tristetraprolin and its family members can promote the cell-free deadenylation of AU-rich element-containing mRNAs by poly(A) ribonuclease. Mol Cell Biol 2003;23:3798–812.

[65] Marchese FP, Aubareda A, Tudor C, Saklatvala J, Clark AR, Dean JLE. MAPKAP kinase 2 blocks tristetraprolin-directed mRNA decay by inhibiting CAF1 deadenylase recruitment. J Biol Chem 2010;285:27590–600.

[66] Chang T-C, et al. UNR, a new partner of poly(A)-binding protein, plays a key role in translationally coupled mRNA turnover mediated by the *c-fos* major coding-region determinant. Genes Dev 2004;18:2010–23.

[67] Piao X, Zhang X, Wu L, Belasco JG. CCR4-NOT deadenylates mRNA associated with RNA-induced silencing complexes in human cells. Mol Cell Biol 2010;30:1486–94.

[68] Zhu Y, et al. Zinc-finger antiviral protein inhibits HIV-1 infection by selectively targeting mutiply spliced viral mRNAs for degradation. Proc Natl Acad Sci USA 2011;108:15834–9.

[69] Collart MA, Panasenko OO. The Ccr4-Not complex. Gene 2012;492:42–53.

[70] Daugeron M-C, Mauxion F, Seraphin B. The yeast POP2 gene encodes a nuclease involved in mRNA deadenylation. Nucleic Acids Res 2001;29:2448–55.

[71] Chen J, Rappsilber J, Chiang Y-C, Russell P, Mann M, Denis CL. Purification and characterization of the 1.0 MDa CCR4-NOT complex identifies two novel components of the complex. J Mol Biol 2001;314:683–94.

[72] Gavin AC. Functional organization of the yeast proteome by systematic analysis of protein complexes. Nature 2002;415:141–7.

[73] Temme C, Zaessinger S, Simonelig M, Wahle E. A complex containing the CCR4 and CAF1 proteins is involved in mRNA deadenylation in Drosophila. EMBO J 2004;23:2862–71.

[74] Temme C, et al. Subunits of the Drosophila CCR4-NOT complex and their roles in mRNA deadenylation. RNA 2010;16:1356–70.

[75] Schwede A, Ellis L, Luther J, Carrington M, Stoecklin G, Clayton C. A role for Caf1 in mRNA deadenylation and decay in trypanosomes and human cells. Nucleic Acids Res 2008;36:3374–88.

[76] Albert TK, Lemaire M, van Berkum NL, Gentz R, Collart M, Timmers HTM. Isolation and characterization of human orthologs of yeast CCR4-NOT complex subunits. Nucleic Acids Res 2000;28:809–17.

[77] Lau N-C, et al. Human Ccr4-Not complexes contain variable deadenylase subunits. Biochem J 2009;422:443–53.

[78] Morita M, Suzuki T, Nakamura T, Yokoyama K, Miysaka T, Yamamoto T. Depletion of mammalian CCR4b deadenylase triggers elevation of the $p27^{Kip1}$ mRNA levels and impairs cell growth. Mol Cell Biol 2007;27:4980–90.

[79] Nasertorabi F, Batisse C, Diepholz M, Suck D, Böttcher B. Insights into the structure of the CCR4-NOT complex by electron microscopy. FEBS Lett 2011;585:2182–6.

[80] Bartlam M, Yamamoto T. The structural basis for deadenylation by the CCR4-NOT complex. Protein Cell 2010;1:443–52.

[81] Dupressoir A, Morel A-P, Barbot W, Loireau M-P, Corbo L, Heidmann T. Identification of four families of yCCR4- and Mg-dependent endonuclease-related proteins in higher eukaryotes, and characterization of orthologs of yCCR4 with a conserved leucine-rich repeat essential for hCAF1/hPOP2 binding. BMC Genomics 2001;2:9–22.

[82] Clark LB, et al. Systematic mutagenesis of the leucine-rich repeat (LRR) domain of CCR4 reveals specific sites for binding to CAF1 and a separate critical role for the LRR in CCR4 deadenylase activity. J Biol Chem 2004;279:13516–623.

[83] Bai Y, Salvadore C, Chiang Y-C, Collart M, Liu H-Y, Denis CL. The CCR4 and CAF1 proteins of the CCR4-NOT complex are physically and functionally separated from NOT2, NOT4 and NOT5. Mol Cell Biol 1999;19:6642–51.

[84] Viswanathan P, Chen J, Chiang Y-C, Denis CL. Identification of multiple RNA features that influence CCR4 deadenylation activity. J Biol Chem 2003;278:14949–55.

[85] Chen J, Chiang Y, Denis CL. CCR4, a 3′-5′ poly(A) RNA and ssDNA exonuclease, is the catalytic component of the cytoplasmic deadenylase. EMBO J 2002;21:1414–26.

[86]   Tucker M, Staples RR, Valencia-Sanchez MA, Muhlrad D, Parker R. Ccr4p is the catalytic subunit of a Ccr4p/Pop2/Notp mRNA deadenylase complex in Saccharomyces cerevisiae. EMBO J 2002;21:1427–36.

[87]   Wang H, et al. Crystal structure of the human CNOT6L nuclease domain reveals strict poly(A) substrate specificity. EMBO J 2010;29:2566–76.

[88]   Beernink PT, Segelke BW, Hadi MZ, Erzberger JP, Wilson III DM, Rupp B. Two divalent metal ions in the active site of a new crystal form of human apurinic/apyrimidinic endonuclease, Ape1: implications for the catalytic mechanism. J Mol Biol 2001;307:1023–34.

[89]   Wagner E, Clement SL, Lykke-Andersen J. An unconventional human Ccr4-Caf1 deadenylase complex in nuclear Cajal bodies. Mol Cell Biol 2007;27:1686–95.

[90]   Baggs JE, Green CB. Nocturnin, a deadenylase in Xenopus laevis retina: a mechanism for posttranscriptional control of circadian-related mRNA. Curr Biol 2003;13:189–98.

[91]   Garbarino-Pico E, Niu S, Rollag MD, Strayer CA, Besharse JC, Green CB. Immediate early response of the circadian poly(A) ribonuclease nocturnin to two extracellular stimuli. RNA 2007;13:745–55.

[92]   Grönke S, Bickmeyer I, Wunderlich R, Jäckle H, Kühnlein RP. curled encodes the Drosophila homolog of the vertebrate circadian deadenylase Nocturnin. Genetics 2009;183:219–32.

[93]   Poulsen JB, et al. Human 2′-phosphodiesterase localizes to the mitochondrial matrix with a putative function in mitochondrial RNA turnover. Nucleic Acids Res 2011;39:3754–70.

[94]   Rorbach J, Nicholls TJ, Minczuk M. PDE12 removes mitochondrial RNA poly(A) tails and controls translation in human mitochondria. Nucleic Acids Res 2011;39:7750–63.

[95]   Bianchin C, Mauxion F, Sentis S, Seraphin B, Corbo L. Conservation of the deadenylase activity of proteins of the Caf1 family in human. RNA 2005;11:487–94.

[96]   Andersen KR, Jonstrup AT, Van LB, Brodersen DE. The activity and selectivity of fission yeast Pop2p are affected by a high affinity for Zn^{2+} and Mn^{2+} in the active site. RNA 2009;15:850–61.

[97]   Viswanathan P, Ohn T, Chiang Y-C, Chen J, Denis CL. Mouse CAF1 can function as a processive deadenylase/3′-5′ exonuclease in vitro but in yeast the deadenylase function of CAF1 is not required for mRNA poly(A) removal. J Biol Chem 2004;279:23988–95.

[98]   Horiuchi M, et al. Structural basis for the antiproliferative activity of the Tab-hCaf1 complex. J Biol Chem 2009;284:13244–55.

[99]   Thore S, Mauxion F, Seraphin B, Suck D. X-ray structure and activity of the yeast Pop2 protein: a nuclease subunit of the mRNA deadenylase complex. EMBO Rep 2003;4:1150–5.

[100] Goldstrohm AC, Seay DJ, Hook BA, Wickens M. PUF protein-mediated deadenylation is catalyzed by Ccr4p. J Biol Chem 2007;282:109–14.

[101] Maillet L, Tu C, Hong YK, Shuster EO, Collart M. The essential function of Not1 lies within the Ccr4-Not complex. J Mol Biol 2000;303:131–43.

[102] Zwartjes CGM, Jayne S, van den Berg DLC, Timmers HTM. Repression of promoter activity by CNOT2, a subunit of the transcription regulatory Ccr4-Not complex. J Biol Chem 2004;279:10848–54.

[103] Oberholzer U, Collart M. Characterization of NOT5 that encodes a new component of the Not protein complex. Gene 1998;207:61–9.

[104] Russell P, Benson JD, Denis CL. Characterization of mutations in NOT2 indicates that it plays an important role in maintaining the integrity of the CCR4-NTO complex. J Mol Biol 2002;322:27–39.

[105] Ito K, Inoue T, Yokoyama K, Morita M, Suzuki T, Yamamoto T. CNOT2 depletion disrupts and inhibits the CCR4-NOT deadenylase complex and induces apoptotic cell death. Genes Cells 2011;16:368–79.

[106] Albert TK, et al. Identification of a ubiquitin-protein ligase subunit within the CCR4-NOT transcription repressor complex. EMBO J 2002;21:355–64.

[107] Cui Y, Ramnarain DB, Chiang Y-C, Ding L-H, McMahon JS, Denis CL. Genome wide expression analysis of the CCR4-NOT complex indicates that it consists of three modules with the NOT module controlling SAGA-responsive genes. Mol Genet Genom 2008;279:323–37.

[108] Muhlrad D, Parker R. The yeast EDC1 mRNA undergoes deadenylation-independent decapping stimulated by Not2p, Not4p, and Not5p. EMBO J 2005;24:1033–45.

[109] Kadyrova LY, Habara Y, Lee TH, Wharton RP. Translational control of maternal Cyclin B mRNA by Nanos in the Drosophila germline. Development 2007;134:1519–27.

[110] Garces RG, Gillon W, Pai EF. Atomic model of human Rcd-1 reveals an armadillo-like repeat protein with in vitro nucleic acid binding properties. Protein Sci 2007;16:176–88.

[111] Liu H-Y, Badarinarayana V, Audino DC, Rappsilber J, Mann M, Denis CL. The NOT proteins are part of the CCR4 transcriptional complex and affect gene expression both positively and negatively. EMBO J 1998;17:1096–998.

[112] Morris JZ, Hong A, Lilly MA, Lehmann R. twin, a CCR4 homolog, regulates cyclin poly(A) tail length to permit Drosophila oogenesis. Development 2005;132:1165–74.

[113] Zaessinger S, Busseau I, Simonelig M. Oskar allows nanos mRNA translation in Drosophila embryos by preventing its deadenylation by Smaug/CCR4. Development 2006;133:4573–83.

[114] Molin L, Puisieux A. C. elegans homologue of the Caf1 gene, which encodes a subunit of the CCR4-NOT complex, is essential for embryonic and larval development and for meiotic progression. Gene 2005;358:73–81.

[115] Zheng D, Ezzedine N, Chen C-Y, Zhu W, He X, Shyu A-B. Deadenylation is prerequisite for P-body formation and mRNA decay in mammalian cells. J Cell Biol 2008;182:89–101.

[116] Mauxion F, Faux C, Seraphin B. The BTG2 protein is a general activator of mRNA deadenylation. EMBO J 2008;27:1039–48.

[117] Aslam A, Mittal S, Koch F, Andrau J-C, Winkler GS. The Ccr4-Not deadenylase subunits CNOT7 and CNOT8 have overlapping roles and modulate cell proliferation. Mol Biol Cell 2009;20:3840–50.

[118] Berthet C, et al. CCR4-associated factor CAF1 is an essential factor for spermatogenesis. Mol Cell Biol 2004;24:5808–20.

[119] Nakamura T, et al. Oligo-astheno-teratozoospermia in mice lacking Cnot7, a regulator of retinoid X receptor beta. Nat Genet 2004;36:528–33.

[120] Traven A, Beilharz TH, Lo TL, Lueder F, Preiss T, Heierhorst J. The Ccr4-Pop2-NOT mRNA deadenylase contributes to septin organization in Saccharomyces cerevisiae. Genetics 2009;182:955–66.

[121] Cooke A, Prigge A, Wickens M. Translational repression by deadenylases. J Biol Chem 2010;285:28506–13.

[122] Chekulaeva M, et al. miRNA repression involves GW182-mediated recruitment of CCR4-NOT through conserved W-containing motifs. Nat Struct Mol Biol 2011;18:1218–26.

[123] Sandler H, Kreth J, Timmers HTM, Stoecklin G. Not1 mediates recruitment of the deadenylase Caf1 to mRNAs targeted for degradation by tristetraprolin. Nucleic Acids Res 2011;39:4373–86.

[124] Frolov MV, Benevolenskaya EV, Birchler JA. Regena (Rga), a Drosophila homolog of the global negative transcriptional regulator CDC36 (NOT2) from yeast, modifies gene expression and suppresses position effect variegation. Genetics 1998;148:317–29.

[125] Carballo E, Lai WS, Blackshear PJ. Feedback inhibition of macrophage tumor necrosis factor-alpha production by tristetraprolin. Science 1998;281:1001–5.

[126] Lai WS, Carballo E, Strum JR, Kennington EA, Philipps RS, Blackshear PJ. Evidence that tristetraprolin binds to AU-rich elements and promotes the deadenylation and destabilization of tumor necrosis factor alpha mRNA. Mol Cell Biol 1999;19:4311–23.

[127] Lykke-Andersen J, Wagner E. Recruitment and activation of mRNA decay enzymes by two ARE-mediated decay activation domains in the proteins TTP and BRF-1. Genes Dev 2005;19:351–61.

[128] Wickens M, Bernstein DS, Kimble J, Parker R. A PUF family portrait: 3′ UTR regulation as a way of life. Trends Genet 2002;18:150–7.

[129] Goldstrohm AC, Hook BA, Seay DJ, Wickens M. PUF proteins bind Pop2p to regulate messenger RNAs. Nat Struct Mol Biol 2006;13:533–9.

[130] Chicoine J, Benoit P, Gamberi C, Paliouras M, Simonelig M, Lasko P. Bicaudal-C recruits CCR4-NOT deadenylase to target mRNAs and regulates oogenesis, cytoskeletal organization, and its own expression. Dev Cell 2007;13:691–704.

[131] Suzuki A, Igarashi K, Aisaki K, Kanno J, Saga Y. NANOS2 ineracts with the CCR4-NOT deadenylation complex and leads to suppression of specific RNAs. Proc Natl Acad Sci USA 2010;107:3594–9.

[132] Aviv T, Lin Z, Lau S, Rendl LM, Sicheri F, Smibert CA. The RNA binding SAM domain of Smaug defines a new family of post-transcriptional regulators. Nat Struct Biol 2003;10:614–21.

[133] Jeske M, Meyer S, Temme C, Freudenreich D, Wahle E. Rapid ATP-dependent deadenylation of nanos mRNA in a cell-free system from Drosophila embryos. J Biol Chem 2006;281:25124–33.

[134] Semotok JL, Cooperstock RL, Pinder BD, Vari HK, Lipshitz HD, Smibert CA. Smaug recruits the CCR4/POP2/NOT deadenylase complex to trigger maternal transcript localization in the early Drosophila embryo. Curr Biol 2005;15:284–94.

[135] Huntzinger E, Izaurralde E. Gene silencing by microRNAs: contributions of translational repression and mRNA decay. Nat Rev Genet 2011;12:99–110.

[136] Behm-Ansmant I, Rehwinkel J, Doerks T, Stark A, Bork P, Izaurralde E. mRNA degradation by miRNAs and GW182 requires both CCR4:NOT deadenylase and DCP1:DCP2 decapping complexes. Genes Dev 2006;20:1885–98.

[137] Fabian MR, et al. Mammalian miRNA RISC recruits CAF1 and PABP to affect PABP-dependent deadenylation. Mol Cell 2009;35:868–80.

[138] Finoux A-L, Séraphin B. In vivo targeting of the yeast Pop2 deadenylase Subunit to reproter transcripts induces their rapid degradation and generates new decay intermediates. J Biol Chem 2006;281:25940–7.

[139] Iwasaki S, Kawamata T, Tomari Y. Drosophila argonaute 1 and argonaute 2 employ distinct mechanisms for translational repression. Mol Cell 2009;34:58–67.

[140] Ezzedine N, et al. Human TOB, an antiproliferative trasncription factor, is a poly(A)-binding protein-dependent positive regulator of cytoplasmic mRNA deadenylation. Mol Cell Biol 2007;27:7791–801.

[141] Mauxion F, Chen C-AA, Seraphin B, Shyu A-B. BTG/TOB factors impact deadenylases. Trends Biochem Sci 2009;34:640–7.

[142] Lee M-H, Kim KW, Morgan CT, Morgan DE, Kimble J. Phosphorylation state of a Tob/BTG protein, FOG-3, regulates initiation and maintenance of the Caenorhabditis elegans sperm fate program. Proc Natl Acad Sci USA 2011;108:9125–30.

[143] Miyasaka T, et al. Interaction of antiproliferative protein Tob with the CCR4-NOT deadenylase complex. Cancer Sci 2008;99:755–61.

[144] Hosoda N, et al. Anti-proliferative protein Tob negatively regulates CPEB3 target by recruiting Caf1 deadenylase. EMBO J 2011;30:1311–23.

[145] Simón E, Séraphin B. A specific role for the C-terminla region of the poly(A)-binding protein in mRNA decay. Nucleic Acids Res 2007;35:6017–28.

[146] Yao G, Chian Y-C, Zhang C, Lee DJ, Laue TM, Denis CL. PAB1 self-association precludes its binding to poly(A), thereby accelerating CCR4 deadenylation in vivo. Mol Cell Biol 2007;27:6243–53.

CHAPTER TEN

The Diverse Functions of Fungal RNase III Enzymes in RNA Metabolism

Kevin Roy, Guillaume F. Chanfreau[1]

Department of Chemistry and Biochemistry and the Molecular Biology Institute, University of California, Los Angeles, California, USA
[1]Corresponding author: e-mail address: guillom@chem.ucla.edu

Contents

1. Introduction   214
2. Phylogenetic Distribution and Conservation of RNase III Enzymes in Fungi   214
3. Ribosomal RNA Processing and RNA Polymerase I Transcriptional Termination   217
4. Small Nuclear RNAs Processing   219
5. Functions in Small Nucleolar RNAs Processing   221
   5.1 Processing of independently transcribed snoRNA precursors   222
   5.2 Cleavage of polycistronic transcription units   224
   5.3 Processing of intron-encoded snoRNAs   224
6. (Pre)-mRNA Surveillance, Degradation, and Regulation   226
7. RNA Polymerase II Termination   228
8. Conclusions and Perspectives   230
Acknowledgements   231
References   231

Abstract

Enzymes from the ribonuclease III family bind and cleave double-stranded RNA to initiate RNA processing and degradation of a large number of transcripts in bacteria and eukaryotes. This chapter focuses on the description of the diverse functions of fungal RNase III members in the processing and degradation of cellular RNAs, with a particular emphasis on the well-characterized representative in *Saccharomyces cerevisiae*, Rnt1p. RNase III enzymes fulfill important functions in the processing of the precursors of various stable noncoding RNAs such as ribosomal RNAs and small nuclear and nucleolar RNAs. In addition, they cleave and promote the degradation of specific mRNAs or improperly processed forms of certain mRNAs. The cleavage of these mRNAs serves both surveillance and regulatory functions. Finally, recent advances have shown that RNase III enzymes are involved in mediating fail-safe transcription termination by RNA polymerase II (Pol II), by cleaving intergenic stem-loop structures present downstream from Pol II transcription units. Many of these processing functions appear to be conserved in fungal species close to the *Saccharomyces* genus, and even in more distant eukaryotic species.

The Enzymes, Volume 31
ISSN 1874-6047
http://dx.doi.org/10.1016/B978-0-12-404740-2.00010-0

1. INTRODUCTION

RNase III enzymes were discovered over 45 years ago as enzymatic activities specific for cleaving double-stranded RNA in prokaryotes [1,2]. Representatives of this class of enzymes were characterized only more recently in eukaryotic cells. This chapter focuses on fungal representatives of this class of enzymes, with a special emphasis on describing their diverse functions in RNA metabolism. We mostly exclude from this discussion RNase III from the Dicer and Drosha families and their involvement in the RNA interference and microRNA biogenesis pathways, as these topics will be extensively treated in chapters from the next volume of this series. Instead, we focus on the function of RNase III enzymes in the processing of noncoding RNAs involved in splicing and ribosomal RNA processing and in the cleavage of specific mRNAs and read-through transcripts. Many of these enzymes are not essential in fungi, facilitating the steady-state analysis of potential RNA substrates of these enzymes. However, the fact that strains carrying deletions of the gene encoding these enzymes exhibit substantial growth defects reflects their functional importance in RNA metabolism. In the next sections, we will review the phylogenetic distribution and conservation of RNase III enzymes in fungi and describe the various RNA processing and degradation pathways that are known to involve cleavage by these enzymes.

2. PHYLOGENETIC DISTRIBUTION AND CONSERVATION OF RNase III ENZYMES IN FUNGI

RNase III enzymes are present in all kingdom and species except in *Archaea*, where the bulge-helix-bulge nucleases cleave endogenous dsRNA substrates [3]. RNase III proteins are identified by a characteristic catalytic domain exhibiting the signature motif NERLEFLGD (Fig. 10.1A and B). In addition, nearly all contain a C-terminal dsRNA-binding domain (dsRBD; Fig. 10.1A) with the conserved $\alpha\beta\beta\beta\alpha$ fold, with the one known exception of *Giardia* Dicer [5]. In addition to the catalytic and dsRBD domains, eukaryotic RNases III can contain additional accessory domains important for substrate specificity and proper positioning on the substrate (Fig. 10.1A; [6]). Based on domain architecture and evolutionary relationships, RNase III proteins can be divided into four classes [7]: class I, bacterial RNase III; class II, Drosha; class IIIa, canonical Dicer; and class IIIb, fungal

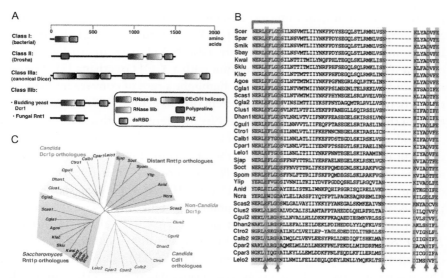

Figure 10.1 Architecture and evolutionary relationships of fungal RNase III representatives. (A) Classification scheme for RNase III enzymes, which range in size from ~200 (class I) to ~2000 amino acids (class IIIa). Class I includes bacterial RNase III (homologues of *E. coli rnc*). Class II contains Drosha, which is involved in pre-rRNA processing and in the conversion of primary microRNA (pri-miRNA) transcripts to precursor microRNA (pre-miRNA) hairpins. Class IIIa contains the canonical Dicer enzymes involved in processing pre-miRNA to mature miRNA. Class IIIb includes the fungal RNase III enzymes, which are further divided into two types depending on the presence (Dcr1) or absence (Rnt1) of a second dsRBD at the C-terminus. (C) Alignment of the class IIIb RNase III domains around the highly conserved signature motif NERLEFLGDS indicated by the bracket. The six catalytic residues found in all active eukaryotic RNase III domains are highlighted in red; for *S. cerevisiae* Rnt1, these correspond to E241, D245, N278, K313, D317, and E320 (class I prokaryotic RNase III contains only four catalytic residues, lacking residues corresponding to N278 and K313). In the *Candia dicer*-like (Cdl1/2) class of enzymes, these residues have been altered to become catalytically inactive. The function of these catalytically "dead" RNase III orthologues remains unknown. The multiple sequence alignment was performed using sequences from the Fungal Orthogroups Repository (http://www.broadinstitute.org/regev/orthogroups/) with the exception that *Candida parapsilosis* orthologues Cpar1/2/3 were identified by BLAST. ClustalX 2.1 program produced the alignment with default settings. Species abbreviations: *Scer, Saccharomyces cerevisiae; Spar, Saccharomyces paradoxus; Smik, Saccharomyces mikatae; Sbay, Saccharomyces bayanus; Kwal, Kluyveromyces waltii; Sklu, Saccharomyces kluyveri; Klac, Kluyveromyces lactis; Agos, Ashbya gossypii; Cgla, Candida glabrata; Scas, Saccharomyces castelli; Clus, Clavispora lusitaniae; Dhan, Debaryomyces hansenii; Cgui, Candida guilliermondi; Ctro, Candida tropicalis; Calb, Candida albicans; Cpar, Candida parapsilosis; Lelo, Lodderomyces elongisporus; Sjap, Schizosaccharomyces japonicus; Soct, Schizosaccharomyces octosporus; Spom, Schizosaccharomyces pombe; Ylip, Yarrowia lipolytica; Anid, Aspergillus nidulans; Ncra, Neurospora crassa.* (C) Phylogenetic tree of class IIIb fungal RNase III enzymes using the multiple sequence alignment from (B) visualized in the DrawTree 3.66 program on the Phylogeny.fr server [4]. (See color plate section in the back of the book.)

RNases III, including the noncanonical Dicers (Dcr1) and the orthologues of *Saccharomyces cerevisiae* Rnt1p (Fig. 10.1A). In addition to lacking helicase and PAZ domains, noncanonical Dicers direct dsRNA cleavage internally, while canonical Dicers require dsRNA termini with a 3′ OH overhang [7]. These mechanistic and sequence differences highlight a case of convergent evolution where RNase III enzymes of different origins evolved independently to take on the function of Dicer in the RNA interference pathway.

Modern-day budding yeasts outside of the *Candida* clade exhibit two different types of class IIIb enzymes, the first typified by *S. cerevisiae* Rnt1p and the second by *Saccharomyces castelli* noncanonical Dicer (Dcr1), which contains a second C-terminal dsRBD (Fig. 10.1A). The *Candida* lineage presents an exceptional case, as it contains two types of RNase III similar in domain architecture to Dcr1, but not syntenic with either *DCR1* or *RNT1* [8]. Class IIIb orthologues are found in all fungi, indicating that they evolved from an ancient fungal ancestor (Fig. 10.1B). The noncanonical Dicers are exclusive to the budding yeast clade, whereas class IIIa canonical dicers are exclusive to nonbudding yeasts and higher eukaryotes. The sequence similarities of noncanonical Dicers to Rnt1 suggest that they evolved from the *RNT1* gene duplication and acquisition of a second dsRBD in an early budding yeast ancestor, perhaps with concomitant loss of the canonical Dicer [8]. Interestingly, the *Candida* lineage appears to have lost Rnt1p, while members of the *Saccharomyces* lineage, including *S. cerevisiae*, have lost Dcr1 and thus RNAi as well. In *Candida albicans*, a single noncanonical Dicer has evolved to perform both RNAi- and Rnt1-like processing functions, highlighting the functional plasticity of RNase III enzymes [8]. Rnt1p orthologue functions in non-RNAi dsRNA processing related to the maturation of noncoding RNA and surveillance of mRNAs are discussed in further detail in the next sections.

Biochemical and structural analyses have shown that *S. cerevisiae* Rnt1p specifically binds and cleaves dsRNAs capped by tetraloops exhibiting an NGNN sequence [9–12], as long as the sequence of the loop adopts the AGNN-type fold and does not form a GNRA- or UNCG-type fold [9,12,13]. An exception in the NGNN sequence was described for one substrate [14,15], but this loop sequence was found to adopt an AGNN-type fold upon binding of the Rnt1p dsRBD [11]. Bioinformatics analyses have shown that some of the budding yeast Rnt1p enzymes have evolved to conserve that specificity [16]. It is currently unknown how the dual-function Dcr1 enzymes present in the *Candida* clade recognize both specific stem-loop substrates, some of which contain the NGNN tetraloops and others which do not, such as the generic substrates of the RNAi pathway. Interestingly, the *Candida* clade also contains catalytically inactive Dicer-like proteins termed Cdl (*Candida* Dicer-like). This protein family contains the same domain architecture as the Dcr1 protein family, but all six active site residues have been mutated to inactive amino acids (Fig. 10.1B). Other conserved residues within the catalytic domain have been retained in Cdl proteins, suggesting that the catalytic domain retains the overall domain fold and may serve a noncatalytic function [8]. As this protein family retains both dsRBDs, it is possible that it has a role in binding dsRNA *in vivo*. Notably, RNase III cleavage requires two catalytic domains assembled on the dsRNA substrate, so class I and class IIIb enzymes must form intermolecular dimers for cleavage to occur. It is possible that heterodimers may form in species with more than one RNase III enzyme and that the catalytically inactive Cdl1 proteins may interact with the active Dcr1 enzymes and regulate binding and/or cleavage activity.

3. RIBOSOMAL RNA PROCESSING AND RNA POLYMERASE I TRANSCRIPTIONAL TERMINATION

One of the first discovered physiological functions of fungal RNases III was the cleavage of the precursor of ribosomal RNA to generate the 35S pre-rRNA [17,18]. This cleavage occurs at a stem–loop structure located downstream from the 25S rRNA in the 3′ external transcribed spacer (ETS) region (Fig. 10.2), and this stem–loop is found in the rDNA of many Hemiascomycetes [8,16]. This processing function is conserved throughout fungi [16], but in the case of fungal species such as *Candida*, which lack a classical Rnt1p-like enzyme, this function is fulfilled by Dcr1, which is the only active RNase III enzyme in this organism [8]. A function for RNase III in the processing of the pre-rRNA seems highly conserved across eukaryotes, as RNase III homologues have been shown to be involved in pre-rRNA processing in human cells [19] and in plants [20].

This processing of the pre-rRNA occurs cotranscriptionally, as *S. cerevisiae* Rnt1p was shown to localize at the site of transcription of the ribosomal DNA and can be crosslinked to the ribosomal DNA [21]. In addition, Rnt1p can interact with several RNA polymerase I (Pol I) subunits [22], suggesting that a physical interaction with individual subunits of the polymerase in the context of elongation might help bring Rnt1p to its processing site once the polymerase has passed the region of the 25S rRNA.

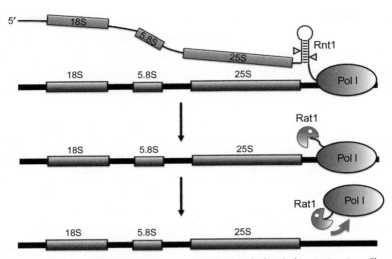

Figure 10.2 Cotranscriptional rRNA processing is coupled to Pol I termination. Cleavage of nascent pre-rRNA precursor by Rnt1p generates the 35S precursor and provides an entry site for the Rat1p exonuclease to degrade the downstream nascent RNA and simultaneously release Pol I from the template. (For color version of this figure, the reader is referred to the online version of this chapter.)

In the case of *Schizosaccharomyces pombe*, processing at the 3′-end of the pre-rRNA by the RNase III Pac1 occurs within a large complex called the RAC complex [23], which influences the cleavage efficiency by Pac1.

RNase III not only initiates processing of the pre-rRNA [24], but it also triggers termination of transcription by RNA Pol I, as evidenced by the fact that Rnt1p-deficient *S. cerevisiae* strains are defective in transcription termination [25–27]. The current model, as depicted in Fig. 10.2, proposes that Rnt1p cleavage in the 3′-ETS introduces an entry site for the 5′-3′ exonuclease Rat1p, which would then catch up with the elongating RNA Pol I and act as a "torpedo" to trigger dissociation of the polymerase from the rDNA, resulting in termination [27,28]. Additional evidence for the fact that the processing activity of RNase III influences rDNA transcription was provided by the observation that the absence of RNase III leads to an increase in the number of transcriptionally active ribosomal DNA repeats and to an opening of repressed rDNA chromatin [22]. While this appears to be an attempt by the cell to compensate for the severe defects in pre-rRNA processing in the absence of Rnt1p, the precise mechanism that results in the change in rDNA chromatin architecture remains unknown.

While important to promote efficient production of ribosomal RNAs, the function of RNases III in pre-rRNA processing is not absolutely essential, as many fungal species are viable in the absence of RNase III activities. While this might be due to the duplication of genes encoding RNase III enzymes for some of these species, species such as *S. cerevisiae*, express only one member of this class of enzymes. In the absence of Rnt1p cleavage, a second fail-safe termination event is known to take place, which results in the production of an elongated form of the 25S ribosomal RNA transcript [26]. Thus, the production of functional rRNAs in the absence of RNase III activity reflects the ability of other ribonucleases to fulfill its 3′-processing functions, albeit less efficiently than if 3′-end trimming is preceded by Rnt1p endonucleolytic cleavage.

4. SMALL NUCLEAR RNAs PROCESSING

Small nuclear RNAs comprise five transcripts (U1, U2, U4, U5, and U6), one of which is transcribed by RNA Pol III (U6) and the others by RNA Pol II. Most of these RNAs, with the exception of U6, are processed at their 3′-ends by multiple redundant 3′-processing pathways, some of which involve cleavage of the 3′-extended small nuclear RNA (snRNA) precursor by RNase III (Fig. 10.3). The first hint of a function for fungal RNases III in snRNA processing was suggested by genetic suppression data showing that overexpression of the *S. pombe* RNase III Pac1 was sufficient to suppress defects in snRNA biogenesis in this species [29]. Direct evidence for a function for RNase III in the processing of snRNAs was first shown by the reconstitution of the U5 snRNA processing pathway *in vitro* [30]. U5 snRNA exists in *S. cerevisiae* in two forms, which differ at their 3′-end. The longer form, U5L, is processed in an Rnt1p-dependent pathway, while the short form, U5S, is processed in an Rnt1p-independent manner [30]. Cleavage of a stem–loop structure located close to the mature 3′-end of U5L generates an entry site for 3′–5′ exonucleases to generate the mature 3′-end. This final trimming is carried out by the nuclear exosome [31] and/or the Rex exonucleases [32]. The 3′-to-5′ exonuclease activity of the exosome is likely stimulated by polyadenylation of the Rnt1p cleavage product [33]. Exonuclease progression is blocked at the mature 3′-end by the binding of the Sm proteins [34], which ensure the production of the correct 3′-end. This pathway has proven to be similar for the U1, U2, and U4 snRNAs in *S. cerevisiae* [31,34,35], except that these snRNAs exhibit only one major 3′-end. For each of these snRNAs, cleavage by Rnt1p initiates one

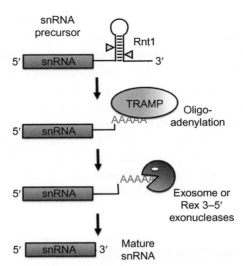

Figure 10.3 Small nuclear RNA processing by RNase III Rnt1p. Pol II-transcribed U1, U2, U4, and U5 snRNA precursors contain Rnt1p target stem-loops downstream of the mature 3'-end. Cleavage by Rnt1 provides an entry site for Rrp6 and the nuclear exosome, and/or the Rex 3'-to-5' exonucleases, to produce the mature 3'-end. The exonuclease activity of the exosome is stimulated by TRAMP-mediated oligoadenylation of the Rnt1 cleavage product. This pathway is similar for the U3 snoRNA. (For color version of this figure, the reader is referred to the online version of this chapter.)

possible processing pathway, but alternative 3'-processing pathways ensure the production of functional snRNAs in the absence of Rnt1p cleavage. For instance, the cleavage and polyadenylation machinery can generate polyadenylated but functional U2 snRNA when Rnt1p cleavage activity is inhibited [35], and Rnt1p-independent pathways have also been described for the processing of the U1 snRNA [34].

The function of RNase III in snRNA 3'-end processing is probably conserved in many fungi, at least in species evolutionarily close to *S. cerevisiae*. The presence of potential stem-loop structures that obey the cleavage specificity of Rnt1p was predicted based on the analysis of sequences downstream from snRNA genes in some Hemiascomycetes genomes [16]. Furthermore, the involvement of RNase III in snRNA processing was demonstrated experimentally for *S. pombe* Pac1 for U2 [36] and for U4 in *Candida* [8]. In *S. pombe*, cleavage of the precursor by Pac1p also promotes termination of transcription by RNA Pol II [37]. However, the quantitative contribution of RNase III to snRNA 3'-end processing might vary depending upon the species. For instance, in *Candida*, the major form of the U5 snRNA is the U5S form [38], which, assuming functional conservation with *S. cerevisiae*, does not involve RNase III cleavage. Only a small amount of U5L can be detected in *Candida* [38], suggesting that the contribution of RNase III to U5 processing might be minor in this species in comparison to *S. cerevisiae*. In addition, in several species close to *S. cerevisiae* such as *S. dairensis* and *S. kluyveri*, only one form of U5 can be detected by Northern blot [39], which corresponds in size to the short form of U5. Thus, it appears that in these yeast species, the stem-loop structure that forms the Rnt1p recognition site might have been lost or is no longer efficiently recognized by the RNase III homologue, indicative of rapid evolution of RNA processing pathways in these species. Alternatively, it is possible that RNase III cleavage intermediates can be more readily trimmed to the shorter form in species other than *S. cerevisiae*.

5. FUNCTIONS IN SMALL NUCLEOLAR RNAs PROCESSING

Small nucleolar RNAs (snoRNAs) are noncoding RNAs involved in the processing and modification of ribosomal RNAs. They are grouped in two distinct families, the box C/D family, which catalyzes methylation of $2'$-hydroxyls of the pre-rRNA precursor, and the box H/ACA family, which catalyzes the modification of uridines into pseudouridines in various RNAs (reviewed in Refs. [24] and [40]). Both families of snoRNAs are bound *in vivo* by four core proteins, different in each family, to form complexes termed small nucleolar ribonucleoproteins (snoRNPs), where one protein provides the enzymatic activity and the snoRNA provides target specificity via the guide sequence (reviewed in Refs. [24] and [40]). Expression of these snoRNAs is relatively diverse in fungi, as some of them are expressed from independent transcription units, others from polycistronic transcripts containing multiple snoRNAs in the same precursor, while a few are produced from the introns of host mRNA genes (which is how most snoRNAs are expressed in metazoans; reviewed in Ref. [41]). Because of this diversity in their genomic organization, it is not surprising that RNase III cleavage can have very diverse functions in the processing of snoRNAs. Depending on the location of the target stem-loop structures, cleavage of these precursors can promote $5'$- and/or $3'$-end processing, and in some cases, releases individual snoRNAs from polycistronic precursors or from introns of pre-mRNA transcripts.

5.1. Processing of independently transcribed snoRNA precursors

For most snoRNAs produced by independent transcription units, the main function of RNase III cleavage is to initiate the 5'-end processing. Cleavage of a stem–loop structure present in the 5'-extension of the precursors typically provides an entry site for 5'–3' exonucleolytic processing, which is fulfilled by the exonucleases Xrn1p and Rat1p (Fig. 10.4A; [42]). These exonucleases stop at the mature 5'-end, probably because the binding of the snoRNP core proteins to the snoRNA block progression into the mature sequence. This 5'-end processing pathway is frequently found for *S. cerevisiae* box C/D snoRNAs [14,42,43] and less frequent for box H/ACA snoRNAs [43,44]. The conservation of stem–loop structures in other Hemiascomycetes suggests that at least some snoRNAs undergo a similar 5'-end processing pathway in these other fungal species [16]. As shown for other processing pathways, the contribution of Rnt1p cleavage to the production of mature snoRNAs depends on snoRNA species; some of these snoRNAs can be processed relatively efficiently in the absence of Rnt1p [42], showing that the 5'–3' exonucleases Xrn1p and Rat1p can process the snoRNAs directly from decapped precursors.

For most of these independently transcribed snoRNAs, RNase III cleavage is not a major determinant of 3'-end processing. There are a few exceptions to this rule. One is the *snR40* transcript, for which the target stem–loop that promotes processing loops out the snoRNA sequence [43]. In this case, Rnt1p cleavage initiates both the 5' and 3' processing of the snoRNA. Another exception is the U3 snoRNA, which is involved in ribosomal RNA processing [24]. In *S. cerevisiae*, U3 is produced from two independent transcription units in the genome. The U3 snoRNAs undergo capping, splicing [45], and processing at the 3'-end. The U3 3'-processing pathway strongly resembles that of the U1,U2, U4, and U5 snRNAs (Fig. 10.3), as a stem–loop is present downstream from the mature 3'-end, and Rnt1p cleavage introduces an entry site for 3'–5' processing by the exosome [46], which is stimulated by TRAMP-mediated polyadenylation of the Rnt1-cleaved U3 intermediate [33]. The involvement of RNase III in the 3'-end processing of U3 is conserved in *S. pombe* [47], and in that organism, Pac1 cleavage also dictates the efficiency of termination of U3 transcription. It is unknown why U3 undergoes a processing pathway that differs dramatically from that of other snoRNAs. However, this might be linked to the unique protein composition of the U3 snoRNP compared to the canonical box C/D snoRNPs and to the essential function of U3 in rRNA processing [48].

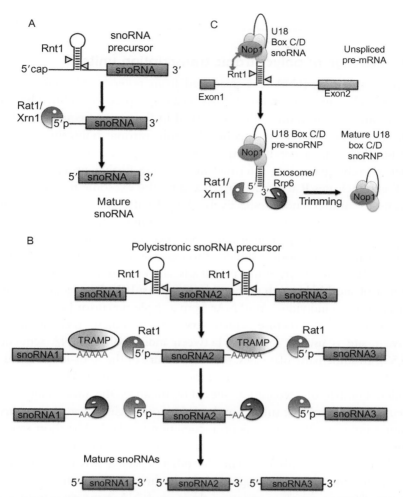

Figure 10.4 Small nucleolar RNA processing by RNase III Rnt1p. (A) Monocistronic snoRNA nascent transcripts typically contain Rnt1p target stem-loops upstream of the mature 5′-end. Rnt1 cleavage provides an entry site for the Rat1 and/or Xrn1p 5′-to-3′ exonucleases to bypass the 5′ cap structure and produce the proper 5′-end. (B) Polycistronic small nucleolar RNAs processing is initiated by the RNase III Rnt1p. Individual snoRNAs within a polycistronic array are released from one another by Rnt1 cleavage of the precursor RNA at target stem-loops in between the mature snoRNA sequences. The released snoRNAs are then trimmed from the 5′- to the 3′-end by Rat1 and/or Xrn1p, and from the 3′- to the 5′-end by Rrp6 and the nuclear exosome, which is likely stimulated by the TRAMP oligoadenylation activity. (C) RNase III Rnt1 processes some intron-embedded snoRNAs. In a pathway alternative to debranching of the lariat splicing product, Rnt1p acts to release some of these box C/D snoRNAs from the intron with concerted cleavage of a stem composed of sequences upstream and downstream of the mature snoRNA. The U18 precursor lacks an Rnt1p tetraloop, and assembly of the Box C/D core protein Nop1p on the intronic snoRNA precursor instead recruits Rnt1p for dsRNA cleavage. After Rnt1p cleavage, exonucleolytic trimming produces both termini of the mature transcript. (For color version of this figure, the reader is referred to the online version of this chapter.)

5.2. Cleavage of polycistronic transcription units

In yeast, some snoRNAs are produced from polycistronic transcription units, which express several snoRNAs on a single precursor [49]. This type of genomic organization is also conserved in fungi and frequent in plants [50]. For many snoRNAs encoded by polycistronic transcription units, RNase III cleavage serves the purpose of separating the snoRNAs from each other, thereby generating intermediates that are further processed by the 5′–3′ exonucleases Rat1p and Xrn1p and by the exosome to generate the mature 5′- and 3′-ends, respectively (Fig. 10.4B; [14,43,51–53]). Many snoRNAs are encoded as precursors containing only two or three snoRNAs; for these precursors, the presence of one or two stem–loop structures in the precursor is sufficient to separate snoRNA from each other (Fig. 10.4B). However, some precursors are known to contain a larger number of snoRNAs, such as the polycistronic snR72-78 precursor, which generates seven individual snoRNAs from a single precursor [43,52]. In this case, a stem–loop structure which serves as an Rnt1p recognition and cleavage site is usually found in between individual snoRNA sequences [43,52]. In some particular cases, the architecture of the stem–loops seems to have been adapted to maximize the processing function of Rnt1p using the smallest number of recognition sites. For instance, the processing of the three snoRNAs snR55, snR57, and snR61 from their tricistronic precursor requires only one Rnt1p tetraloop recognition site [14]. The short stem in which this tetraloop is located can stack onto two distinct longer stems, one of which loops out snR55, while the other separates snR57 from snR61 [14]. Thus, one tetraloop recognition site in the precursor can mediate two distinct cleavage events, separating three snoRNA transcripts from each other. As observed for other snRNA and snoRNA processing events, the presence of tetraloop signals in polycistronic snoRNA precursors is conserved in many Hemiascomycetes [16], suggesting that RNase III processing of polycistronic snoRNA precursors is a common feature in fungi.

5.3. Processing of intron-encoded snoRNAs

In mammalian cells, almost all snoRNAs are contained within introns, and the processing of most of these snoRNAs is splicing dependent and occurs by debranching of the lariat structures of the introns by the debranching enzyme and subsequent exonucleolytic trimming from both directions [41]. In *S. cerevisiae*, a few snoRNAs are also found within introns, and their processing pathway was thought to occur mostly through debranching as well [53,54], with speculations that random cleavage in the lariat structure could also contribute to processing [54]. However, *S. cerevisiae* RNase III can provide an alternative or complementary mechanism of initiating the processing of snoRNAs from introns (Fig. 10.4C). The first example described was the U18 snoRNA, for which Rnt1p cleavage occurs in an intronic stem-loop structure that loops out the snoRNA sequence [55]. The target stem-loop is a noncanonical structure that lacks a *bona fide* AGNN tetraloop, and the binding of Rnt1p to this noncanonical site is thought to be assisted by an interaction with the snoRNP protein Nop1p, which binds the snoRNA and recruits Rnt1p to this cleavage site [55]. The observation that Rnt1p can cleave and initiate the processing of U18 provides an explanation for the observation that a substantial amount of mature U18 snoRNA can be produced when the debranching enzyme Dbr1p is inactivated [53]. Thus, although the level of U18 snoRNA is not affected in a strain lacking Rnt1p, this enzyme can clearly provide an alternative processing pathway to splicing and debranching.

Two other examples of processing of intron-encoded snoRNAs by Rnt1p cleavage were described for the snR39 and snR59 snoRNAs [14]. Interestingly, the quantitative contribution and cleavage mechanism of Rnt1p for the processing of these intron-encoded snoRNAs seem to vary depending on the individual snoRNA. For snR39, which is encoded in the second intron of the ribosomal protein gene *RPL7A*, Rnt1p cleavage plays only a minor role in the maturation of this snoRNA, as the production of most of the mature *snR39* is strongly decreased in a strain lacking the debranching enzyme Dbr1p. However, the production of mature snR39 is completely blocked in a *dbr1Δ rnt1Δ* double mutant, showing that Rnt1p cleavage contributes to some extent to the production of the mature snoRNA [14]. Interestingly, cleavage by Rnt1p occurs only on the lariat form of the excised intron, but not on the unspliced precursor form [14]. Thus, as described above for U18, Rnt1p function in the processing of snR39 seems to act as an alternative to Dbr1p in initiating the processing of the snoRNA from the excised lariat introns. By contrast, the cleavage by Rnt1p to initiate processing of snR59 occurs *only* on the unspliced precursor form of the host transcript *RPL7B* and this cleavage initiates degradation of the entire pre-mRNA [14]. Thus, as shown for cleavage in other unspliced precursors such as *RPS22B* (see below), cleavage by Rnt1p is mutually exclusive to splicing of the pre-mRNA, suggesting a competition between production of mature mRNA and production of mature snoRNA from the same precursor by Rnt1p. It is interesting to note that, while snR39 and snR59 are encoded in introns of duplicated genes coding for the same ribosomal protein (the Rpl7 paralogues), differences in processing pathways have emerged.

6. (PRE)-mRNA SURVEILLANCE, DEGRADATION, AND REGULATION

Because most of the *S. cerevisiae* RNase III localizes in the nucleus [21,56], and in particular, at the site of transcription of the rRNA, it was initially thought that the function of fungal RNase III was limited to the processing of nuclear noncoding RNAs. However, multiple studies have shown that fungal RNases III can also cleave a number of mRNAs. This cleavage serves two main purposes. First, it can provide a way to discard unprocessed or improperly processed mRNAs. Additionally, cleavage can provide an alternative mechanism to degrade mRNAs or limit their accumulation in conditions in which they should be repressed.

A role for fungal RNase III in (pre)-mRNA degradation was first discovered by the identification of target stem-loop structures in the intronic sequences of the ribosomal protein genes *RPL18A* and *RPS22B* [57]. For these transcripts, Rnt1p cleavage can serve two purposes (Fig. 10.5). The first function is to cleave the lariat introns produced by the splicing reaction, providing a mechanism of degradation of these lariats complementary to that provided by the debranching enzyme [57]. The second function is to directly compete with splicing or degrade unspliced pre-mRNAs that have escaped the splicing pathway [57]. As described above, cleavage by Rnt1p provides an entry site for exonucleolytic degradation by the exosome and the 5'–3' exonucleases Xrn1p and Rat1p. Interestingly, the relative dependence upon the nuclear Rat1p or the cytoplasmic Xrn1p exonucleases for degradation of the 3' cleavage products seem to vary between *RPS22B* and *RPL18A* [57], suggesting that Rnt1p cleavage intermediates can be exported to the cytoplasm or retained in the nucleus with varying efficiencies. This function in the surveillance of unspliced pre-mRNA species was also demonstrated for the *MATa1* transcript [58]. Inhibition of Rnt1p cleavage was found to result in an increase in mature mRNA levels [57,58]. This observation is consistent with the idea that cleavage of the pre-mRNAs by Rnt1p occurs very early in the biogenesis pathway and competes with splicing, thus titrating a fraction of the precursors away from the splicing pathway. This could potentially constitute a means of regulating the

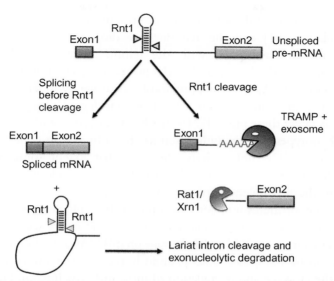

Figure 10.5 RNase III regulation of unspliced pre-mRNAs and lariat degradation. Various intron-containing protein coding genes in *S. cerevisiae* contain intronic Rnt1p target stem-loops. Rnt1p cleavage serves to limit the accumulation of the unspliced precursor and also acts redundantly with the debranching enzyme Dbr1p to degrade the lariat released after splicing of these precursors. Competition between splicing and Rnt1p cleavage regulates levels of the mature mRNA in at least one case, the *RPS22B* ribosomal protein mRNA. (For color version of this figure, the reader is referred to the online version of this chapter.)

expression of these transcripts; however, it is unknown if the cleavage of these unspliced pre-mRNAs by Rnt1p can be regulated during specific growth conditions. The surveillance function for Rnt1p in the degradation of unspliced pre-mRNAs can be partially redundant with those of other degradation systems, such as degradation by Rat1p or by the nonsense-mediated decay (NMD) pathway. For instance, degradation of *MATa*1 pre-mRNAs seems to be the most effective only when both Rat1p and Rnt1p are active [58], and the full accumulation of unspliced *RPS22B* pre-mRNAs is observed only when both the Rnt1p stem-loop structure and the NMD system are inactivated [59]. The existence of multiple degradation pathways that localize to different compartments highlights the importance of limiting the accumulation and aberrant translation of unspliced pre-mRNAs.

In addition to these functions in the surveillance of unprocessed mRNAs, a large number of mature mRNAs appear to be cleaved by RNase III in *S. cerevisiae*. These include the mRNA encoding the transcriptional repressor Mig2p [60], mRNAs coding for proteins involved in iron uptake and assimilation [61] and the acireductone dioxygenase [62], mRNAs encoding proteins associated with the telomerase complex [63], and mRNAs encoding the Swi4p and Hsl1p proteins involved in the cell wall stress response [64]. As described previously, cleavage by RNase III generates entry sites for exonucleases to complete their degradation [61,65]. The lack of cleavage of these pre-mRNAs has been associated with a number of cellular phenotypes [64], including sensitivity to high iron concentrations [61]. However, it is not entirely clear to what extent some of these cellular phenotypes are directly due to the lack of cleavage of these mRNAs in the *rnt1* deletion strain or whether the general growth defect of this strain contributes indirectly to some of these phenotypes.

It does not appear that the genomic context or accessory sequences contribute in a major way to RNase III cleavage on mRNAs, as various stem–loop structures can be transposed onto plasmid-borne reporter mRNAs, which results in the cleavage and degradation of these mRNAs [65,66]. However, cleavage efficiency can vary depending on the stem–loop/reporter combination [65], suggesting that for some of these combinations, the presence of competing secondary structures may impede proper cleavage activity. The ability of Rnt1p to cleave these structures independently from their genomic context has been used to develop synthetic gene expression reporters. Stem–loop structures with various cleavage efficiencies can be inserted into reporter mRNAs to generate a range of reporters to fine-tune gene expression [67] as well as to design gene reporters that are responsive to the binding of metabolites to the reporter mRNA [66].

It is unclear whether other fungal homologues of Rnt1p also cleave specific mRNAs. Overexpression of *S. pombe* Pac1p inhibits mating and sporulation [68], quite possibly by cleaving mRNA(s) that are required for the proper completion of the mating and sporulation processes. However, despite this initial genetic observation, the identity of these mRNAs has yet to be determined. Besides this observation, it is unknown whether RNase III contributes to mRNA degradation independently from of the RNA interference pathway in other fungal species outside *S. cerevisiae*.

7. RNA POLYMERASE II TERMINATION

Transcription termination of RNA Pol II has emerged as an important step in gene expression that can directly affect the fate of the transcribed RNA. Furthermore, proper partitioning of transcription units via efficient termination is crucial to avoid transcriptional interference and aberrant production of overlapping sense–antisense transcripts [69]. Initial studies had shown that cleavage by fungal RNase III downstream from small RNA genes could trigger termination of the RNA Pol II for these ncRNA transcription units, suggesting that RNase III cleavage might serve a general role in triggering termination of RNA Pol II [37,47]. Recent work has highlighted the role of Rnt1p in providing a more widespread fail-safe mode of RNA Pol II termination, in situations where a failure to cleave at the poly(A) site results in transcription of a downstream Rnt1-target stem–loop [70,71]. Cotranscriptional cleavage by Rnt1p is followed by the binding of the 5′-to-3′ exonuclease Rat1p on the now unprotected 5′-end of the nascent transcript (Fig. 10.6). Upon reaching the elongating RNA Polymerase, Rat1p promotes the release of Pol II from the template [71], in a mechanism similar to the one proposed for the "torpedo" model of transcription termination of RNA Pol I by Rat1p [27,28]. Lack of cleavage at the poly(A) site can be the result of weak

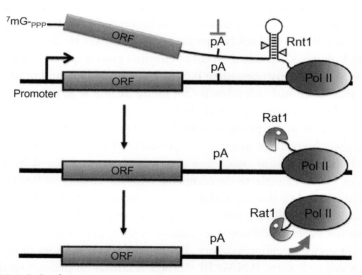

Figure 10.6 Fail-safe transcription termination of RNA Polymerase II by RNase III. Cleavage and polyadenylation of Pol II transcripts can fail to occur due to *cis*-acting elements (e.g., a weak polyadenylation (pA) signal) or *trans*-acting factors (e.g., RNA-binding proteins). In many cases, Rnt1p target stem-loops exist downstream of these pA signals and elicit cotranscriptional cleavage by Rnt1. This is followed by recruitment of Rat1 5′-to-3′ exonuclease, resulting in degradation of the nascent transcript and the release of Pol II from the template. (For color version of this figure, the reader is referred to the online version of this chapter.)

cis termination or polyadenylation signals, or the presence of *trans*-acting factors inhibiting recognition of the poly(A) site [70]. This effect on termination could also explain the presence of extended species of mRNAs showing target stem-loop structures when Rnt1p is inactivated [61,62]. It has been suggested that this function for RNase III in transcription termination may be a general feature of eukaryotic gene expression [72]. Indeed, the RNase III Dicer-like 4 of *Arabidopsis thaliana* has been shown to be required for transcription termination of the FCA gene involved in chromatin silencing [73]. This highlights the ability of Dicer enzymes to have RNAi-independent functions and suggests that different classes of RNase III may have independently evolved to provide fail-safe termination throughout eukaryotes.

8. CONCLUSIONS AND PERSPECTIVES

In this chapter, we have reviewed the diverse functions of fungal RNase III enzymes in the processing of noncoding RNAs and in the degradation of mRNAs. We have excluded from this discussion the role of this class of enzymes in the RNAi and siRNA pathways, as these will be described extensively in Volume 32. While this class of enzymes plays important functions in the processing of many noncoding RNAs (rRNA, snRNA, and snoRNA), it is clear that functional redundancies are found in almost all processing pathways described, such that inactivation of these enzymes usually results in only a partial inhibition of the production of the mature species. Similar redundancies are also observed for many of the pre-mRNA surveillance functions associated with Rnt1p, as other degradation pathways mediated by Dbr1p, Rat1p, or NMD can effectively compensate for the absence of Rnt1p cleavage. While it is clear that the redundancy is rampant in RNase III-mediated processing and surveillance functions, it is increasingly evident that RNase III cleavage has evolved additional specialized roles in gene expression that are important under specific conditions. For example, in conditions of high iron levels or cell wall stress, RNAse III cleavage is critical for preventing the excessive accumulation of target mRNAs, where high levels of protein expression from these mRNAs would be detrimental.

Most of the examples of RNase III-mediated RNA processing and degradation derive from the work done with the RNase III Rnt1p from the model Hemiascomycete *S. cerevisiae*. While it is clear that some of the processing pathways are conserved in other Hemiascomycetes, the available evidence suggests that there is also significant variation in the quantitative contributions of RNase III to these various processing pathways, even in closely related species. This variation probably reflects the existence of multiple processing and degradative pathways for individual transcripts, thus allowing for a rapid evolutionary loss of a particular processing pathway. Future work using other model yeast and fungi should further reveal the extent to which RNA processing pathways have been conserved or altered throughout fungal evolution.

Some cellular phenotypes linked to the absence of these enzymes still lack a clear molecular explanation. For example, inactivation of *S. cerevisiae* Rnt1p leads to cell-cycle defects, but these defects can be rescued by the expression of a catalytically inactive mutant [56]. This result suggests the existence of cleavage-independent functions of RNase III, at least for the *S. cerevisiae* enzyme, but it is unknown what molecular functions of the enzyme are directly responsible for these defects. The cell-cycle rescue was shown to be dependent on trafficking of Rnt1p from the nucleolus to the nucleoplasm at the G2/M transition of the cell cycle, suggesting that Rnt1p function may be regulated by the subcellular localization pattern of this enzyme. It has been speculated that this phenomenon may also regulate the Rnt1p cleavage activity on mRNAs, which would likely require the recruitment of Rnt1p from the nucleolus to the nucleoplasm for their cleavage [64]. Future work will elucidate the ways in which fungal RNAse III activity is regulated and whether this is conserved across species. It is clear that RNase III enzymes have evolved to provide cells with an additional tool to fine-tune gene expression and perform quality control. The presence of simple dsRNA structures embedded throughout the transcriptome provides a rapid and potent means of regulating gene expression by multiple mechanisms.

ACKNOWLEDGEMENTS

This work was supported by NIH Grant GM61518 to G. F. C.

K.R. was supported by a Whitcome Fellowship and Ruth L. Kirschtein National Research Service Award GM007185.

REFERENCES

[1] Robertson HD, Webster RE, Zinder ND. A nuclease specific for double-stranded RNA. Virology 1967;32(4):718–9.
[2] Robertson HD, Webster RE, Zinder ND. Purification and properties of ribonuclease III from Escherichia coli. J Biol Chem 1968;243(1):82–91.
[3] Dennis PP. Ancient ciphers: translation in Archaea. Cell 1997;89(7):1007–10.

[4] Dereeper A, Guignon V, Blanc G, Audic S, Buffet S, Chevenet F, et al. Phylogeny.fr: robust phylogenetic analysis for the non-specialist. Nucleic Acids Res 2008;36 (Web Server issue):W465–9.

[5] Macrae IJ, Zhou K, Li F, Repic A, Brooks AN, Cande WZ, et al. Structural basis for double-stranded RNA processing by Dicer. Science 2006;311(5758):195–8.

[6] MacRae IJ, Doudna JA. Ribonuclease revisited: structural insights into ribonuclease III family enzymes. Curr Opin Struct Biol 2007;17(1):138–45.

[7] Weinberg DE, Nakanishi K, Patel DJ, Bartel DP. The inside-out mechanism of Dicers from budding yeasts. Cell 2011;146(2):262–76.

[8] Bernstein DA, Vyas VK, Weinberg DE, Drinnenberg IA, Bartel DP, Fink GR. Candida albicans Dicer (CaDcr1) is required for efficient ribosomal and spliceosomal RNA maturation. Proc Natl Acad Sci USA 2012;109(2):523–8.

[9] Chanfreau G, Buckle M, Jacquier A. Recognition of a conserved class of RNA tetraloops by Saccharomyces cerevisiae RNase III. Proc Natl Acad Sci USA 2000;97 (7):3142–7.

[10] Wu H, Henras A, Chanfreau G, Feigon J. Structural basis for recognition of the AGNN tetraloop RNA fold by the double-stranded RNA-binding domain of Rnt1p RNase III. Proc Natl Acad Sci USA 2004;101(22):8307–12.

[11] Wang Z, Hartman E, Roy K, Chanfreau G, Feigon J. Structure of a yeast RNase III dsRBD complex with a noncanonical RNA substrate provides new insights into binding specificity of dsRBDs. Structure 2011;19(7):999–1010.

[12] Nagel R, Ares Jr. M. Substrate recognition by a eukaryotic RNase III: the double-stranded RNA-binding domain of Rnt1p selectively binds RNA containing a 5'-AGNN-3' tetraloop. RNA 2000;6(8):1142–56.

[13] Wu H, Yang PK, Butcher SE, Kang S, Chanfreau G, Feigon J. A novel family of RNA tetraloop structure forms the recognition site for Saccharomyces cerevisiae RNase III. EMBO J 2001;20(24):7240–9.

[14] Ghazal G, Ge D, Gervais-Bird J, Gagnon J, Abou Elela S. Genome-wide prediction and analysis of yeast RNase III-dependent snoRNA processing signals. Mol Cell Biol 2005;25(8):2981–94.

[15] Ghazal G, Elela SA. Characterization of the reactivity determinants of a novel hairpin substrate of yeast RNase III. J Mol Biol 2006;363(2):332–44.

[16] Chanfreau G. Conservation of RNase III processing pathways and specificity in hemiascomycetes. Eukaryot Cell 2003;2(5):901–9.

[17] Elela SA, Igel H, Ares Jr. M. RNase III cleaves eukaryotic preribosomal RNA at a U3 snoRNP-dependent site. Cell 1996;85(1):115–24.

[18] Kufel J, Dichtl B, Tollervey D. Yeast Rnt1p is required for cleavage of the pre-ribosomal RNA in the 3' ETS but not the 5' ETS. RNA 1999;5(7):909–17.

[19] Wu H, Xu H, Miraglia LJ, Crooke ST. Human RNase III is a 160-kDa protein involved in preribosomal RNA processing. J Biol Chem 2000;275(47):36957–65.

[20] Comella P, Pontvianne F, Lahmy S, Vignols F, Barbezier N, Debures A, et al. Characterization of a ribonuclease III-like protein required for cleavage of the pre-rRNA in the 3'ETS in Arabidopsis. Nucleic Acids Res 2008;36(4):1163–75.

[21] Henras AK, Bertrand E, Chanfreau G. A cotranscriptional model for 3'-end processing of the Saccharomyces cerevisiae pre-ribosomal RNA precursor. RNA 2004;10 (10):1572–85.

[22] Catala M, Tremblay M, Samson E, Conconi A, Abou Elela S. Deletion of Rnt1p alters the proportion of open versus closed rRNA gene repeats in yeast. Mol Cell Biol 2008;28(2):619–29.

[23] Spasov K, Perdomo LI, Evakine E, Nazar RN. RAC protein directs the complete removal of the 3' external transcribed spacer by the Pac1 nuclease. Mol Cell 2002;9 (2):433–7.

[24] Venema J, Tollervey D. Ribosome synthesis in Saccharomyces cerevisiae. Annu Rev Genet 1999;33:261–311.

[25] Prescott EM, Osheim YN, Jones HS, Alen CM, Roan JG, Reeder RH, et al. Transcriptional termination by RNA polymerase I requires the small subunit Rpa12p. Proc Natl Acad Sci USA 2004;101:6068–73.

[26] Braglia P, Kawauchi J, Proudfoot NJ. Co-transcriptional RNA cleavage provides a failsafe termination mechanism for yeast RNA polymerase I. Nucleic Acids Res 2011;39(4):1439–48.

[27] Kawauchi J, Mischo H, Braglia P, Rondon A, Proudfoot NJ. Budding yeast RNA polymerases I and II employ parallel mechanisms of transcriptional termination. Genes Dev 2008;22(8):1082–92.

[28] El Hage A, Koper M, Kufel J, Tollervey D. Efficient termination of transcription by RNA polymerase I requires the 5′ exonuclease Rat1 in yeast. Genes Dev 2008;22 (8):1069–81.

[29] Rotondo G, Gillespie M, Frendewey D. Rescue of the fission yeast snRNA synthesis mutant snm1 by overexpression of the double-strand-specific Pac1 ribonuclease. Mol Gen Genet 1995;247(6):698–708.

[30] Chanfreau G, Elela SA, Ares Jr. M, Guthrie C. Alternative 3′-end processing of U5 snRNA by RNase III. Genes Dev 1997;11(20):2741–51.

[31] Allmang C, Kufel J, Chanfreau G, Mitchell P, Petfalski E, Tollervey D. Functions of the exosome in rRNA, snoRNA and snRNA synthesis. EMBO J 1999;18(19):5399–410.

[32] van Hoof A, Lennertz P, Parker R. Three conserved members of the RNase D family have unique and overlapping functions in the processing of 5S, 5.8S, U4, U5, RNase MRP and RNase P RNAs in yeast. EMBO J 2000;19(6):1357–65.

[33] Egecioglu DE, Henras AK, Chanfreau GF. Contributions of Trf4p- and Trf5p-dependent polyadenylation to the processing and degradative functions of the yeast nuclear exosome. RNA 2006;12(1):26–32.

[34] Seipelt RL, Zheng B, Asuru A, Rymond BC. U1 snRNA is cleaved by RNase III and processed through an Sm site-dependent pathway. Nucleic Acids Res 1999;27 (2):587–95.

[35] Abou Elela S, Ares Jr. M. Depletion of yeast RNase III blocks correct U2 3' end formation and results in polyadenylated but functional U2 snRNA. EMBO J 1998;17(13):3738–46.

[36] Zhou D, Frendewey D, Lobo Ruppert SM. Pac1p, an RNase III homolog, is required for formation of the 3′ end of U2 snRNA in Schizosaccharomyces pombe. RNA 1999;5(8):1083–98.

[37] Nabavi S, Nazar RN. Cleavage-induced termination in U2 snRNA gene expression. Biochem Biophys Res Commun 2010;393(3):461–5.

[38] Mitrovich QM, Guthrie C. Evolution of small nuclear RNAs in S. cerevisiae, C. albicans, and other hemiascomycetous yeasts. RNA 2007;13(12):2066–80.

[39] Roiha H, Shuster EO, Brow DA, Guthrie C. Small nuclear RNAs from budding yeasts: phylogenetic comparisons reveal extensive size variation. Gene 1989;82(1):137–44.

[40] Henras AK, Soudet J, Gerus M, Lebaron S, Caizergues-Ferrer M, Mougin A, et al. The post-transcriptional steps of eukaryotic ribosome biogenesis. Cell Mol Life Sci 2008;65 (15):2334–59.

[41] Filipowicz W, Pogacic V. Biogenesis of small nucleolar ribonucleoproteins. Curr Opin Cell Biol 2002;14(3):319–27.

[42] Lee CY, Lee A, Chanfreau G. The roles of endonucleolytic cleavage and exonucleolytic digestion in the 5′-end processing of S. cerevisiae box C/D snoRNAs. RNA 2003;9 (11):1362–70.

[43] Chanfreau G, Legrain P, Jacquier A. Yeast RNase III as a key processing enzyme in small nucleolar RNAs metabolism. J Mol Biol 1998;284(4):975–88.

[44] Hiley SL, Babak T, Hughes TR. Global analysis of yeast RNA processing identifies new targets of RNase III and uncovers a link between tRNA 5' end processing and tRNA splicing. Nucleic Acids Res 2005;33(9):3048–56.

[45] Myslinski E, Segault V, Branlant C. An intron in the genes for U3 small nucleolar RNAs of the yeast Saccharomyces cerevisiae. Science 1990;247(4947):1213–6.

[46] Kufel J, Allmang C, Chanfreau G, Petfalski E, Lafontaine DL, Tollervey D. Precursors to the U3 small nucleolar RNA lack small nucleolar RNP proteins but are stabilized by La binding. Mol Cell Biol 2000;20(15):5415–24.

[47] Nabavi S, Nazar RN. Pac1 endonuclease and Dhp1p 5' → 3' exonuclease are required for U3 snoRNA termination in Schizosaccharomyces pombe. FEBS Lett 2010;584 (15):3436–41.

[48] Venema J, Tollervey D. Processing of pre-ribosomal RNA in Saccharomyces cerevisiae. Yeast 1995;11(16):1629–50.

[49] Samarsky DA, Fournier MJ. A comprehensive database for the small nucleolar RNAs from Saccharomyces cerevisiae. Nucleic Acids Res 1999;27(1):161–4.

[50] Brown JW, Clark GP, Leader DJ, Simpson CG, Lowe T. Multiple snoRNA gene clusters from Arabidopsis. RNA 2001;7(12):1817–32.

[51] Chanfreau G, Rotondo G, Legrain P, Jacquier A. Processing of a dicistronic small nucleolar RNA precursor by the RNA endonuclease Rnt1. EMBO J 1998;17 (13):3726–37.

[52] Qu LH, Henras A, Lu YJ, Zhou H, Zhou WX, Zhu YQ, et al. Seven novel methylation guide small nucleolar RNAs are processed from a common polycistronic transcript by Rat1p and RNase III in yeast. Mol Cell Biol 1999;19(2):1144–58.

[53] Petfalski E, Dandekar T, Henry Y, Tollervey D. Processing of the precursors to small nucleolar RNAs and rRNAs requires common components. Mol Cell Biol 1998;18 (3):1181–9.

[54] Ooi SL, Samarsky DA, Fournier MJ, Boeke JD. Intronic snoRNA biosynthesis in Saccharomyces cerevisiae depends on the lariat-debranching enzyme: intron length effects and activity of a precursor snoRNA. RNA 1998;4(9):1096–110.

[55] Giorgi C, Fatica A, Nagel R, Bozzoni I. Release of U18 snoRNA from its host intron requires interaction of Nop1p with the Rnt1p endonuclease. EMBO J 2001;20 (23):6856–65.

[56] Catala M, Lamontagne B, Larose S, Ghazal G, Elela SA. Cell cycle-dependent nuclear localization of yeast RNase III is required for efficient cell division. Mol Biol Cell 2004;15(7):3015–30.

[57] Danin-Kreiselman M, Lee CY, Chanfreau G. RNAse III-mediated degradation of unspliced pre-mRNAs and lariat introns. Mol Cell 2003;11(5):1279–89.

[58] Egecioglu DE, Kawashima TR, Chanfreau GF. Quality control of MATa1 splicing and exon skipping by nuclear RNA degradation. Nucleic Acids Res 2012;40(4):1787–96.

[59] Sayani S, Janis M, Lee CY, Toesca I, Chanfreau GF. Widespread impact of nonsense-mediated mRNA decay on the yeast intronome. Mol Cell 2008;31(3):360–70.

[60] Ge D, Lamontagne B, Elela SA. RNase III-mediated silencing of a glucose-dependent repressor in yeast. Curr Biol 2005;15(2):140–5.

[61] Lee A, Henras AK, Chanfreau G. Multiple RNA surveillance pathways limit aberrant expression of iron uptake mRNAs and prevent iron toxicity in S. cerevisiae. Mol Cell 2005;19(1):39–51.

[62] Zer C, Chanfreau G. Regulation and surveillance of normal and 3'-extended forms of the yeast aci-reductone dioxygenase mRNA by RNase III cleavage and exonucleolytic degradation. J Biol Chem 2005;280(32):28997–9003.

[63] Larose S, Laterreur N, Ghazal G, Gagnon J, Wellinger RJ, Elela SA. RNase III-dependent regulation of yeast telomerase. J Biol Chem 2007;282(7):4373–81.

[64] Catala M, Aksouh L, Abou Elela S. RNA-dependent regulation of the cell wall stress response. Nucleic Acids Res 2012 May 10. [Epub ahead of print].

[65] Meaux S, Lavoie M, Gagnon J, Abou Elela S, van Hoof A. Reporter mRNAs cleaved by Rnt1p are exported and degraded in the cytoplasm. Nucleic Acids Res 2011;39 (21):9357–67.

[66] Babiskin AH, Smolke CD. Engineering ligand-responsive RNA controllers in yeast through the assembly of RNase III tuning modules. Nucleic Acids Res 2011;39 (12):5299–311.

[67] Babiskin AH, Smolke CD. Synthetic RNA modules for fine-tuning gene expression levels in yeast by modulating RNase III activity. Nucleic Acids Res 2011;39 (19):8651–64.

[68] Iino Y, Sugimoto A, Yamamoto M. S. pombe pac1+, whose overexpression inhibits sexual development, encodes a ribonuclease III-like RNase. EMBO J 1991;10 (1):221–6.

[69] Kuehner JN, Pearson EL, Moore C. Unravelling the means to an end: RNA polymerase II transcription termination. Nat Rev Mol Cell Biol 2011;12(5):283–94.

[70] Ghazal G, Gagnon J, Jacques PE, Landry JR, Robert F, Elela SA. Yeast RNase III triggers polyadenylation-independent transcription termination. Mol Cell 2009;36 (1):99–109.

[71] Rondon AG, Mischo HE, Kawauchi J, Proudfoot NJ. Fail-safe transcriptional termination for protein-coding genes in S. cerevisiae. Mol Cell 2009;36(1):88–98.

[72] Nabavi S, Nazar RN. Fail-safe termination elements: a common feature of the eukaryotic genome? FASEB J 2010;24(3):684–8.

[73] Liu F, Bakht S, Dean C. Cotranscriptional role for Arabidopsis DICER-LIKE 4 in transcription termination. Science 2012;335(6076):1621–3.

AUTHOR INDEX

Numbers in regular font are reference numbers and indicate that an author's work is referred to although the name is not cited in the text. Numbers in italics refer to the page numbers on which the complete reference appears.

A

Abbasi, N., 43, *51*
Abou Elela, S., 81, 83, *92*, 216–220, 222, 224, 225, 228, 231, *232–235*
Adams, C., 82, *93*
Adelman, K., 148, *162*
Agrin, N. S., 183, 184, *204*
Aguilar-Henonin, L., 46, *52*
Ahn, S. H., 144, 146, *161*
Aiba, H., 5, *24*
Aisaki, K., 201, *210*
Aksouh, L., 228, 231, *234*
Albert, T. K., 190, 197, 198, *207, 209*
Aldrich, T. L., 133, *157*
Alen, C. M., 218, *233*
Alexandrov, A., 45, *51*, 152, *163*
Allen, M. A., 144, *161*
Allmang, C., 2, 3, 5, 8, 11, 14, 15, 17, 19, 23, *24*, 34, 40, *49*, 54–56, 67, 70, *72–74*, 82, *93*, 141, 145, *160, 161*, 219, 222, *233, 234*
Amberg, D. C., 133, *157*
Amblar, M., 63, 64, *73*
Ambros, V., 102, *111*
Ames, J. M., 19, *28, 29*, 80, 81, 83, *92, 93*
Amit, I., 106, *113*
Anantharaman, V., 136, *158*
Andersen, K. R., 13–15, *27*, 68, *75*, 187, 195, 196, *206, 208*
Anderson, J., 17, *28*, 78, 80–82, 87, *91*
Anderson, J. S., 5, 20, *24*, 32, 41, 42, *48*
Anderson, J. T., 15, 17, *27, 28*, 45, *51*, 54, 72, 78, 80–83, 87, 88, *91, 93, 94*, 152, *163*
Anderson, P., 173, *179*
Andrau, J.-C., 199, *209*
Andrulis, E. D., 11, *26*, 38, *49*, 54, 68, *72*
Ansari, A. Z., 144, 145, *161*
Araki, Y., 41, *50*
Aravind, L., 87, *94*, 136, *158*

Arcus, V. L., 64, *74*
Ares, M. Jr., 3, *23*, 80, 81, 84, *92*, 216, 217, 219, 220, *232, 233*
Arigo, J. T., 3, 19, *23, 29*, 80, 81, 83, 84, *92*, 145, *161*
Armon, A., 126, *129*
Arraiano, C. M., 63, 64, *73*
Artymiuk, P. J., 121, *128*
Aslam, A., 199, *209*
Assenholt, J., 13–15, *27*, 68, 69, *75*
Astrom, A., 185, *205*
Astrom, J., 185, 186, *205*
Astuti, D., 13, *27*, 157, *163*
Asuru, A., 219, 220, *233*
Atack, J. M., 121, 122, *129*
Aubareda, A., 189, 201, *206*
Audic, S., 215, *231*
Audino, D. C., 198, *209*
Aviv, T., 201, *210*
Azzalin, C. M., 89, *95*
Azzouz, N., 79, *91*

B

Babajko, S., 104, *112*
Babak, T., 222, *234*
Babiskin, A. H., 228, *235*
Bäckbro, K., 64, *74*
Badarinarayana, V., 198, *209*
Badis, G., 32, 40, 41, 45, *48*, 67, *74*
Baek, K., 121, *128*
Bagga, S., 102, *112*
Baggs, J. E., 194, 195, *208*
Bah, A., 89, *95*
Bahler, J., 6, *24*, 89, *95*
Bai, X., 20, *29*, 41, *50*
Bai, Y., 190, 196–198, *207*
Baker, E. N., 64, *74*
Baker, K. E., 101, 104, 105, 107, *111–113*, 154, *163*
Baker, R. E., 20, 21, *29*

Bakht, S., 230, *235*
Ballarino, M., 148, 152, *162*
Ballin, J. D., 79, 87, 88, *91*
Barbas, A., 36, 39, *49*, 61, 63–65, *73*
Barbezier, N., 217, *232*
Barbot, W., 190, 194, *207*
Bar-Nahum, G., 124, *129*
Barnard, D. C., 182, *204*
Barrass, D., 38, *49*
Bartel, D. P., 6, *25*, 102, *111*, *112*, 214, 216, 217, 220, *232*
Bartlam, M., 190, *207*
Bartunik, H. D., 59, *73*
Basavappa, R., 83, *93*
Baserga, S. J., 15, *27*, 138, *159*
Bashkirov, V. I., 104, *112*
Basquin, J., 7, 11, 12, 14, *25*, 38–40, *49*, 59, 61, 63–65, 67, *73*
Batisse, C., 193, 198, *207*
Bayne, E. H., 83, 89, *93*
Bazzini, A. A., 102, *112*
Beelman, C. A., 100, 104, 105, *111*, *112*
Beernink, P. T., 193, 194, *208*
Behm-Ansmant, I., 102, 104, 105, *111–113*, 121, *129*, 136, *158*, 202, *210*
Beilharz, T. H., 200, *209*
Belasco, J. G., 102, *112*, 189, 202, *207*
Belgard, T. G., 106, *113*
Belostotsky, D. A., 11, *26*, 38, 39, 46, *49*, *52*, 184, 188, 189, *205*, *206*
Benard, L., 21, *29*, 125, *129*
Bénard, L., 136, *159*
Benevolenskaya, E. V., 200, *210*
Benoit, P., 182, 201, *204*, *210*
Benson, J. D., 197, *208*
Ben-Tal, N., 126, *129*
Bentley, D. L., 141, 143, 144, 146, 154, 155, *160*
Berndt, H., 90, *95*, 182, 188, 190, *204*
Bernstein, D. A., 216, 217, 220, *232*
Bernstein, D. S., 201, *210*
Bernstein, J., 15, *27*, 79, 87, 88, *91*
Bernstein, K. A., 15, *27*
Berretta, J., 3, *23*, 105, 106, *113*
Berthet, C., 199, *209*
Bertrand, E., 217, 226, *232*
Besharse, J. C., 194, *208*
Bianchin, C., 195, *208*
Bickmeyer, I., 194, 195, *208*

Bilen, B., 87, 90, *94*
Birchler, J. A., 200, *210*
Birney, E., 106, *113*
Birren, B. W., 21, *30*
Blackshear, P. J., 189, 201, *206*, *210*
Blanc, A., 32, 42, *48*
Blanc, G., 215, *231*
Blank, D., 32, 41, 43, 45, *48*
Bliss, T., 18, *28*
Blumenthal, T., 144, *161*
Boeck, R., 183, 184, *204*
Boeke, J. D., 141, *160*, 225, *234*
Bohnsack, M. T., 79–81, *92*, 151, *163*
Bönisch, C., 184, 198, *205*
Bonneau, F., 7, 11, 12, 14, 15, 17, 21, *25*, 28, 38–40, 42, *49*, *50*, 59, 61, 64, 65, 67, *73*, 85–87, 89, *94*
Bork, P., 202, *210*
Borovjagin, A. V., 138, *159*
Bottcher, B., 11, *26*
Böttcher, B., 70, *75*, 193, 198, *207*
Boulay, J., 5, 19, *23*, *29*, 32, 40, 41, 45, *48*, 67, *74*, 80, 81, 83, 84, *92*
Bourenkov, G. P., 59, *73*
Bousquet-Antonelli, C., 5, *23*, *24*, 149, 150, 154, *162*
Boyer, B., 105, *113*
Bozzoni, I., 141, 148, 152, 153, *160*, *162*, *163*, 225, *234*
Braglia, P., 144–147, *161*, 218, 219, 229, *233*
Branlant, C., 222, *234*
Brannan, K., 147, 148, *162*
Bratkowski, M. A., 14, *27*, 65, *74*
Brauer, M. J., 173, *179*
Braun, J. E., 185, 202, *205*
Brick, P., 68, *75*
Briggs, M. W., 13, 14, *27*
Bringmann, G., 37, *49*
Brodersen, D. E., 7, 13–15, *25*, *27*, 35, 41, *49*, 68, 69, *75*, 187, 195, 196, *206*, *208*
Bronkhorst, A. W., 63, *73*
Brooks, A. N., 214, *232*
Brouwer, R., 56, *73*, 157, *163*
Brow, D. A., 19, *29*, 80, *93*, 144, 145, *161*, 221, *233*
Brown, C. E., 183, 184, *204*
Brown, J., 39, 40, *50*, 82, *93*
Brown, J. T., 12, 15, 20, *26*, *29*, 41, *50*

Brown, J. W., 224, *234*
Brown, P. O., 105, *113*
Brown, S. E., 176, *180*
Brunkard, J. O., 42, *51*
Brzoska, P. M., 82, *93*
Buchan, J. R., 173, *179*
Buckle, M., 216, *232*
Buerstedde, J. M., 104, *112*
Buffet, S., 215, *231*
Buhler, D., 119, *128*
Buhler, M., 6, *24*, *25*, 79, 86, 89, *91*, *95*
Bühler, M., 67, *74*
Buratowski, S., 6, 19, *25*, *28*, *29*, 78, 80, 81,
    83, *91*, *93*, 107, *114*, 144–146, *161*
Burch-Smith, T. M., 42, *51*
Burge, C. B., 102, *111*
Burkard, K. T., 13, 14, *27*
Busseau, I., 199, 201, *209*
Bussing, I., 155, *163*
Butcher, S. E., 216, *232*
Butler, J. S., 13–15, *27*, 43, *51*, 65, 68, *74*,
    75, 79, 81, 82, 88, 89, *92*, *93*, *95*, 135,
    151, 153, 154, *158*, 176, *180*
Büttner, K., 55, 61, *72*
Byam, J., 79, 81, *92*
Byrne, M. E., 43, *51*

C

Cabili, M. N., 106, *113*
Caffarelli, E., 141, *160*
Caizergues-Ferrer, M., 221, *233*
Calabrese, J. M., 105, *112*
Caldo, R. A., 36, 37, *49*
Callaghan, A. J., 55, *72*
Callahan, K. P., 13, 15, *27*, 65, 68, *74*, 88,
    89, *95*
Camblong, J., 3, *23*, 45, *52*, 81, 84, *92*, 106,
    113
Campbell, J. L., 81, 82, *92*, *93*
Campbell, R. D., 176, *180*
Canaday, J., 32, 39–41, 43, 45, 47, *48*, *50*
Cande, W. Z., 214, *232*
Caponigro, G., 100, *111*, 184, 203, *205*
Carballo, E., 201, *210*
Carpousis, A. J., 55, *72*
Carrington, M., 190, 195, 198, 199, *207*
Carroll, K., 32, 42, *48*
Carroll, K. L., 3, 19, 21, *23*, *28*, *29*, 80, 81,
    83, 84, *92*, *93*, 145, *161*

Carr-Schmid, A., 166, 168, *178*
Carson, D. R., 82, 87, *93*
Carson, J. P., 82, 87, *93*
Castano, I. B., 80, 82, *92*, *93*
Catala, M., 217, 218, 226, 228, 231,
    232, *234*
Causton, H., 6, *25*
Causton, H. C., 173, *179*
Cavalli, G., 119, *128*
Ceradini, F., 141, 153, *160*, *163*
Cerutti, H., 47, *52*
Ceska, T. A., 121, *128*
Ceulemans, H., 43, *51*
Cevher, M. A., 189, *206*
Chak, K. F., 6, 7, *25*, 56, *72*
Chamnongpol, S., 104, 105, *112*
Chan, E. K. L., 104, *112*
Chan, S., 143, *160*
Chanfreau, G., 3, 15, 17, 19, *23*, 67, *74*, 82,
    93, 141, 145, 146, 151, *160–162*, 216,
    217, 219, 220, 222, 224, 226, 228, 230,
    232–234
Chanfreau, G. F., 80, 81, 83, *92*, 107, *114*,
    151, *162*, 219, 222, 226, 227, *233*, *234*
Chang, H. Y., 106, *114*
Chang, J. H., 116–121, 124, *127*, 136, *159*
Chang, S. K., 6, *25*
Chang, T.-C., 184, 189, *204*, *207*
Chang, Y. F., 149, *162*
Chapados, B. R., 121, *128*
Charette, J. M., 138, *159*
Charroux, B., 42, *50*
Chatterjee, S., 135, 155, *158*, *163*
Chekanova, J. A., 3, 11, *23*, *26*, 32, 35,
    37–39, 44–47, *48*, *49*, 141, *160*, 184, 188,
    189, *205*, *206*
Chekulaeva, M., 200, 202, *209*
Chen, C. L., 105, 107, *113*
Chen, C. Y., 5, *24*, 176, *180*
Chen, C.-A. A., 202, *210*
Chen, C.-Y., 189, 199, 201, *206*, *209*
Chen, C.-Y. A., 185, 202, *205*
Chen, E., 67, *74*
Chen, H., 32, 35, 37–39, 44–47, *48*, *52*, 82,
    93
Chen, H. W., 172, *179*
Chen, J., 184, 190, 193, 195–199, 201, *204*,
    207, *208*
Chen, L., 21, *30*

Chen, S., 133, *157*
Chen, X., 44, *51*
Cheng, Y., 44, *51*
Cheng, Z., 20, *29*
Cheng, Z. F., 63, *73*, 78, *91*
Chernukhin, I. V., 21, *29*
Chernyakov, I., 45, *51*, 108, *114*, 133, 147, 152–154, *157*, *163*
Chevenet, F., 215, *231*
Chian, Y.-C., 203, *211*
Chiang, Y., 190, 195, 199, *207*
Chiang, Y.-C., 190, 193, 196–201, *207–209*
Chiba, Y., 188, 189, *206*
Chicoine, J., 201, *210*
Chinchilla, K., 144, 147, *161*
Chiu, Y. L., 175, *179*
Chlebowski, A., 12, *26*, 32, 39, 40, *48*, *49*, 55, 63, 65, 67, 68, 70, *72*
Cho, H., 166, *177*
Cho, Y., 6, *25*, 121, *128*
Choi, E. A., 143, *160*
Choi, J. M., 6, *25*
Choi, Y. G., 42, *51*
Choi, Y. S., 90, *95*
Chou, C.-F., 189, *206*
Christensen, M. S., 32, 43, 45, *49*, *52*
Christman, M. F., 80, 82, *92*, *93*
Chu, C., 166, 175, *177*
Chu, C. Y., 105, *113*
Chua, N.-H., 40, *50*
Churchman, L. S., 84, *94*
Ciais, D., 79–81, *92*, 151, *163*
Clare, D. K., 61, 65, *73*
Clark, A., 46, *52*
Clark, A. R., 189, 201, *206*
Clark, G. P., 224, *234*
Clark, L. B., 190, *207*
Clarke, N. D., 80, *93*
Clauder-Münster, S., 45, *52*, 106, *113*
Clayton, C., 11, *26*, 35, *49*, 68, 70, *75*, 190, 195, 198, 199, *207*
Clement, S. L., 194, 195, *208*
Clissold, P. M., 121, *129*
Clurman, B. E., 90, *95*
Cognat, V., 39–41, 43, 45, 47, *50*
Cohen, L. S., 168, *178*
Cohen, S. N., 78, *91*
Colau, G., 79, *91*

Cole, C. N., 133, *157*, 184, *205*
Collart, M., 190, 196–198, 200, *207*, *208*
Collart, M. A., 79, *91*, 190, *207*
Coller, J., 104, 105, 107, *112*, *113*, 154, *163*, 167, *178*
Colmenares, S. U., 89, *95*
Comella, P., 141, *160*, 217, *232*
Conconi, A., 217, 218, *232*
Condon, C., 125, *129*, 136, *159*
Connelly, S., 141, *160*, 175, *180*
Conrad, N. K., 19, *29*, 80, *93*, 144, 145, *161*
Conti, E., 2, 5, 7, 8, 11, 12, 14, 15, 17, 21, 22, *24*, *25*, 28, *29*, 38–40, 42, *49*, *50*, 55, 56, 59, 61, 63–65, 67, 70, *72*, *73*, 85–87, 89, *94*, *95*, 121, *129*, 136, *158*
Cook, M. A., 59, *73*
Cooke, A., 200, 202, *209*
Cooper-Morgan, A., 116, 117, 122, 124–126, *127*, 135–137, 143, 146, *158*, 170, 172, 176, *179*
Cooperstock, R. L., 201, *210*
Copela, L. A., 81, 83, *92*
Copeland, P. R., 185, 189, *205*
Corbo, L., 190, 194, 195, *207*, *208*
Corden, J. L., 3, 19, *23*, 28, *29*, 80, 81, 83, 84, *92*, *93*, 144, 145, *161*
Core, L. J., 84, *94*
Corsini, L., 119, *128*
Costello, J. L., 13, 18, *27*, 68, *75*
Cougot, N., 104, *112*
Counsell, D., 176, *180*
Coyle, S. M., 117, 118, 122–124, *127*, 136, 153, *159*
Cramer, P., 143, *161*
Creamer, T. J., 3, 19, *23*
Cristodero, M., 11, *26*, 68, 70, *75*
Crooke, S. T., 217, *232*
Crouch, R. J., 64, *74*
Cudny, H., 6, *25*
Cui, M., 144, *161*
Cui, Y., 197, 200, 201, *209*
Culbertson, M. R., 144, 147, *161*
Curcio, M. J., 83, *94*

D
Dai, J., 169, *178*
Daiss, J. L., 121, *129*
Dal Peraro, M., 64, *74*

Dandekar, T., 108, *114*, 138, 141, 146, 153, 154, *159*, 224, 225, *234*
Daniel, E. L., 64, *74*
Danin-Kreiselman, M., 151, *162*, 226, *234*
Daou, R., 136, *159*
Darnell, J. E. J., 141, *160*
Das, B., 135, 151, *158*, *162*
Dastidar, E. G., 36, 39, *49*, 61, 64, 65, *73*
d'Aubenton-Carafa, Y., 45, *52*, 67, *74*, 84, *94*, 105, 107, *113*
Daugeron, M.-C., 190, 199, *207*
David, L., 84, *94*
Davidson, L., 137, 152, *159*
Davis, C. A., 3, *23*, 80, 81, 84, *92*
Davis, K., 116, 118, *127*, 136, *159*
Davis, R. E., 168, *178*
Davis, R. W., 79, *91*, 183, 203, *204*
Dawson, J., 18, *28*
De Gregorio, E., 141, *160*
de la Cruz, J., 15, *27*, 79, 81, 82, *91*
De Las Penas, A., 82, *93*
De Moor, C. H., 189, *206*
De Vivo, M., 64, *74*
Dean, C., 230, *235*
Dean, J. L. E., 189, 201, *206*
Deardorff, J. A., 183, 184, *204*
Debures, A., 217, *232*
Decker, C. J., 5, *24*, 100, 105, *111*, *113*, 155, *163*, 166, *178*
Decourty, L., 43, *51*
Dees, M., 154, *163*
Dehlin, E., 185–189, *205*
Delano, W. L., 57, *72*
Demple, B., 82, 87, *93*
Dengl, S., 143, *161*
Denis, C. L., 184, 190, 193, 195–201, 203, 204, *207–209*, *211*
Dennis, P. P., 214, *231*
Dereeper, A., 215, *231*
Deshmukh, M. V., 168, *178*
Dettwiler, S., 32, 41, 43, 45, *48*
Deutscher, M. P., 6, 13, *25*, *27*, 63, *73*, 78, *91*, 183, *204*
Deutscher, P., 55, *72*
Devaux, F., 38, *49*
Devos, J. M., 121, *128*
Dez, C., 80–82, *92*
Di Marco, S., 151, *163*
Di Segni, G., 133, *157*

Dichtl, B., 109, *114*, 134, 153, *157*, 217, *232*
Dickmanns, A., 187, *206*
Diepholz, M., 11, *26*, 70, *75*, 193, 198, *207*
Dieppois, G., 3, *23*, 81, 84, *92*
Dietz, H. C., 5, 21, *24*, 41, 42, *50*
Ding, F., 14, *27*, 61, 65, *73*, *74*
Ding, L.-H., 197, 200, 201, *209*
Djuranovic, S., 102, *112*
Dlakic, M., 80, *92*
Dmochowska, A., 6, *25*
Doerks, T., 202, *210*
Doma, M. K., 5, 21, *24*, 101, *111*
Domanski, M., 20, *29*, 32, 43, *49*
Dominski, Z., 136, *158*
Dorcey, E., 42, 46, *50*
Dorner, S., 103, *112*
Doudna, J. A., 117, 118, 122–124, *127*, 136, 153, *159*, 214, *232*
Douglas, M. G., 133, 134, *157*, 175, *180*
Dower, K., 5, *23*
Draper, S., 89, *95*
Drazkowska, K., 11–13, *26*, 39, *50*
Dreyfuss, G., 59, *73*
Drinnenberg, I. A., 216, 217, 220, *232*
D'Silva, S., 108, *114*, 153, 154, *163*
D'Souza, V., 86, *94*
Duffy, A., 189, *206*
Dufour, M. E., 67, *74*
Dufour, M.-E., 32, 40, 41, 45, *48*
Dunckley, T., 100, *111*, 166, *178*
Dunn, E. F., 184, *205*
Dupressoir, A., 190, 194, *207*
Dutko, J. A., 11, *26*, 188, 189, *206*
Dutta, A., 106, *113*
Dutta, D., 145, *161*
Dziembowski, A., 2, 6–8, 11–15, 18, *22*, *25*, *26*, 32, 39, *48*, *50*, 55, 56, 59, 61, 63–65, 67, 70, *72*, *73*

E

Eberle, A. B., 64, *74*, 101, *111*
Ebert, J., 7, 11, 12, 14, *25*, 38–40, *49*, 59, 61, 64, 65, 67, *73*
Eckmann, C. R, 87, *94*
Edge, R. E., 59, *73*
Edwards, L., 59, *73*
Edwards, S., 82, *93*
Egberts, W. V., 56, *73*

Egecioglu, D. E., 80, 81, 83, *92*, 151, *162*, 219, 222, 226, 227, *233*, *234*
Ehrenberg, M., 187, 188, *206*
Ekiel, I., 184, *204*
El Hage, A., 6, 17, *25*, 79–82, 84, *91*, 135, 138, 144, 145, 147, 154, *158*, 218, 229, *233*
Elela, S. A., 216, 217, 219, 226, 228–231, *232–235*
Ellis, L., 190, 195, 198, 199, *207*
Eom, S. H., 121, *129*
Epshtein, V., 124, *129*, 145, *161*
Erdjument-Bromage, H., 11, *26*
Erickson, S. L., 173, *179*
Erzberger, J. P., 193, 194, *208*
Estevez, A. M., 11, *26*
Estévez, A. M., 35, *49*, 68, *75*
Eulalio, A., 101–105, *111–113*
Evakine, E., 218, *232*
Evans, M. C., 183, 184, *204*
Evguenieva-Hackenberg, E., 7, 12, *25*, 39, *50*, 55, 56, *72*
Eyler, D. E., 3, 19, *23*, 80, 81, 83, 84, *92*, 145, *161*
Eystathioy, T., 104, *112*
Ezzedine, N., 199, 201–203, *209*, *210*

F

Fabian, M. R., 185, 202, *205*, *210*
Fabre, A., 42, *50*
Falck-Pedersen, E., 141, *160*
Falkenby, L. G., 32, 43, *49*
Fan, J., 102, *112*
Fang, F., 79, 81, 82, *92*, *93*, 135, 153, 154, *158*, 176, *180*
Fasken, M. B., 18, *28*, 85–87, 90, *94*
Fasler, M., 155, *163*
Fatica, A., 38, *49*, 141, *160*, 225, *234*
Fauser, M., 64, *74*, 101, *111*, 185, 202, *205*
Faux, C., 199, 202, 203, *209*
Fazo, J., 33, *49*
Fei, Z., 135, *158*
Feigon, J., 216, *232*
Feldser, D., 106, *113*
Fernandez, C. F., 81, 83, *92*, 107, *114*
Feuerhahn, S., 153, *163*
Fickentscher, C., 3, *23*, 81, 84, *92*
Ficner, R., 187, *206*

Filipowicz, W., 101, *111*, 140, 141, *159*, *160*, 221, 224, *233*
Finger, L. D., 121, 122, *129*
Fink, G. R., 3, *23*, 106, *113*, 216, 217, 220, *232*
Finkel, J. S., 144, 147, *161*
Finoux, A.-L., 202, *210*
Fischer, P. M., 166, *177*
Fischer, U., 119, *128*
Fitzgerald-Bocarsly, P., 169, *178*
Fitzhugh, D., 82, *93*
Fitzhugh, D. J., 80, 82, *92*, *93*
Ford, L. P., 186, *206*
Fortner, D. M., 100, *111*
Fournier, M., 141, *160*
Fournier, M. J., 224, 225, *234*
Fragapane, P., 141, *160*
Franks, T. M., 104, *112*, 167, *178*
Frazao, C., 12, *27*
Frazão, C., 63, 64, *73*
French, C., 106, *113*
Frendewey, D., 219, 220, *233*
Freudenreich, D., 201, 202, *210*
Fribourg, S., 7, 12, *25*, 39, *50*, 55, 56, *72*
Friedlein, A., 32, 41, 43, 45, *48*
Frischmeyer, P. A., 5, 21, *24*, 41, 42, *50*
Fritz, D. T., 186, *206*
Fritzler, M. J., 104, *112*
Frolov, M. V., 200, *210*
Fromont-Racine, M., 5, *23*
Funakoshi, Y., 184, 199, 203, *204*
Furuichi, Y., 98, *110*, 166, *177*

G

Gadal, O., 5, *23*
Gagliardi, D., 11, *26*, 32, 39, 43–45, 47, *48*, *51*, 89, *95*
Gagneur, J., 45, *52*, 106, *113*
Gagnon, J., 216, 222, 224, 225, 228–230, *232*, *234*, *235*
Gaidamakov, S. A., 64, *74*
Gajiwala, K. S., 119, *128*
Galitski, T., 3, *23*, 106, *113*
Gallouzi, I. E., 151, *163*
Galy, V., 5, *23*
Gamberi, C., 201, *210*
Gao, L., 175, *179*
Gao, M., 186, *206*

Garbarino-Pico, E., 194, *208*
Garber, M., 106, *113*, *114*
Garces, R. G., 197, *209*
Garfinkel, D. J., 83, *94*
Garneau, N. L., 167, *178*
Gasch, A., 59, 61, *73*
Gasch, A. P., 173, *179*
Gasciolli, V., 47, *52*
Gavin, A. C., 18, *28*, 190, 197, 198, *207*
Ge, D., 216, 222, 224, 225, 228, *232*, *234*
Geerlings, T. H., 108, *114*, 138, 154, *159*
Gehring, K., 184, *204*
Geisler, S., 105, 107, *113*, 154, *163*
Gellon, L., 82, 87, *93*
Gentz, R., 190, 197, 198, *207*
Gerber, S. A., 89, *95*
Gerbi, S. A., 138, *159*
Gerstein, M., 106, *113*
Gerus, M., 221, *233*
Gervais-Bird, J., 216, 222, 224, 225, *232*
Ghazal, G., 216, 222, 224–226, 228–231, *232*, *234*, *235*
Gherzi, R., 5, *24*
Ghirlando, R., 19, *28*, 80, *93*
Ghosh, A., 166, *177*
Ghosh, S., 98, *110*, 132, *157*
Ghosh, T., 169, *178*
Gillespie, M., 219, *233*
Gillon, W., 197, *209*
Gilmour, D. S., 148, *162*
Gingeras, T. R., 106, *113*
Gingras, A. C., 166, *177*
Giorgi, C., 141, *160*, 225, *234*
Giraldez, A. J., 102, *111*, *112*
Glaser, F., 60, *73*
Glass, C. K., 18, *28*
Glavan, F., 121, *129*, 136, *158*
Goebl, M., 135, *158*
Goldstein, A. L., 133, *157*
Goldstrohm, A. C., 182, 195, 201, *203*, *208*, *210*
Gombert, J., 47, *52*
Gongidi, P., 183, 184, *204*
Goodfellow, I. G., 166, *177*
Goodman, H. M., 47, *52*
Goodstein, D. M., 33, *49*
Gourvennec, S., 81, 84, *92*, 105, 107, *113*
Graham, A. C., 11, *26*, 38, *49*, 54, 68, 72

Granek, J. A., 80, *93*
Granneman, S., 15, 17, 19, *27*, *28*, 79–81, 83, 84, 87, 88, 90, *91*, 140, 153, *159*
Grasby, J. A., 121, 122, *129*
Graur, D., 126, *129*
Grayhack, E. J., 45, *51*, 108, *114*, 133, 147, 152–154, *157*
Green, C. B., 194, 195, *208*
Green, P. J., 46, 47, *52*, 103, *112*, 116, *127*, 135, 136, 140, 141, 154, *158*, *159*, 188, 189, *206*
Green, R., 102, *112*
Greenblatt, J. F., 144, 146, *161*
Gregory, B. D., 32, 35, 37–39, 44–47, *48*, *52*
Greimann, J. C., 7, 8, 12, 14, *25*, 35, 39, *49*, 55, 56, 58–61, 65, *72*, 88, *94*
Grewal, S. I., 67, *74*
Griffith, K., 104, *112*
Grisafi, P. L., 3, *23*, 106, *113*
Gromak, N., 142–144, 148, 152, *160–162*, 175, *180*
Grönke, S., 194, 195, *208*
Großhans, H., 135, 155, *158*, *163*
Gross, D. S., 175, *179*
Gruber, J. J., 20, *29*
Gruissem, W., 33, *49*
Grzechnik, P., 6, 17, *25*, *28*, 80, 81, 83, *92*, 182, *204*
Gu, W., 45, *51*
Gudipati, R. K., 19, *29*
Guerrerio, A. L., 5, *24*
Guevara, P., 147, *162*
Guignon, V., 215, *231*
Guigó, R., 106, *113*
Gunderson, S. I., 168, *178*
Guo, H., 46, *52*, 102, *112*
Guo, Q., 124, *129*
Gustafson, M. P., 90, *95*
Guthrie, C., 219, 221, *233*
Guttman, M., 106, *113*, *114*
Gy, I., 47, *52*, 154, 156, *163*
Gygi, S. P., 67, *74*, 89, *95*

H
Ha, I., 102, *111*
Ha, T., 14, *27*
Haaning, L. L., 146, 149, *162*, 175, *179*

Haas, W., 67, *74*, 89, *95*
Habara, Y., 197, 201, *209*
Hadi, M. Z., 193, 194, *208*
Hagiwara, M., 80, *92*
Haile, S., 11, *26*, 68, *75*
Hajnsdorf, E., 78, *91*
Halbach, F., 21, *29*, 89, *95*
Hall, B. D., 133, *157*
Hamill, S., 85, 86, *94*
Hamilton, D. A., 188, 189, *206*
Hamlin, R., 68, *75*
Hammell, C. M., 133, *157*, 184, *205*
Han, M., 144, *161*
Handa, H., 166, 175, *177*
Haracska, L., 87, *94*
Harigaya, Y., 67, *74*
Harlow, L. S., 6, *25*, 56, *72*
Harshman, K., 42, 46, *50*
Hartman, E., 216, *232*
Hashimoto, J., 98, *110*
Hatfield, L., 100, *111*
Hausmann, S., 83, *93*
Hayes, R. D., 33, *49*
Hayles, J., 67, *74*
Hazelbaker, D. Z., 107, *114*
He, F., 155, *163*
He, X., 199, 201, *209*
Heath-Pagliuso, S., 80, 82, *92*
Heck, A. J., 63, *73*
Heck, A. J. R., 41, *50*
Heidmann, T., 190, 194, *207*
Heierhorst, J., 200, *209*
Hellman, U., 185, 186, *205*
Hennig, L., 33, *49*
Henras, A., 216, 224, *232, 234*
Henras, A. K., 80, 81, 83, *92*, 138, 151, *159, 162*, 217, 219, 221, 222, 226, 228, 230, *232–234*
Henriksson, N., 186, *206*
Henry, Y., 108, *114*, 138, 141, 146, 153, 154, *159*, 176, *180*, 224, 225, *234*
Hentschel, J., 15, 17, 21, *28*, 42, *50*, 85–87, 89, *94*
Henz, S. R., 45, *52*
Hernandez, H., 12, *26*
Herschlag, D., 105, *113*
Hess, D., 5, *24*, 79, 86, 89, *91*
Hesson, J., 47, *52*

Heyer, W. D., 104, *112*, 136, *159*
Heyer, W.-D., 116, *127*
Hieronymus, H., 82, *93*
Higgins, C. F., 6, *25*
Hildebrandt, U., 37, *49*
Hiley, S. L., 45, *51*, 222, *234*
Hilleren, P., 5, *24*
Hilleren, P. J., 150, *162*
Hinnebusch, A. G., 17, *28*, 78, 80–82, 87, *91*
Hintze, B. J., 15, 17, 21, *28*, 42, *51*, 79, 85–87, 89, *91*
Hirsch-Hoffmann, M., 33, *49*
Hirsh, M., 6, *25*
Hitomi, M., 15, *27*, 133, *157*
Ho, C. K., 166, 175, *177, 179*
Ho, E. S., 168, *178*
Hobor, F., 80, *93*
Hochleitner, E., 55, *72*
Hodge, C. A., 184, *205*
Holec, S., 39–41, 43, 45, 47, *50*
Holub, P., 15, *28*, 79, 85–87, 89, *91*
Hong, A., 199, *209*
Hong, S. W., 175, *179*
Hong, Y. K., 196–198, 200, *208*
Hongay, C. F., 3, *23*, 106, *113*
Hook, B. A., 195, 201, *208, 210*
Hooker, T. S., 32, 33, 35, 37–39, 44–47, *48, 49*
Hopfner, K. P., 55, 61, *72*
Horiuchi, M., 195, 196, 202, *208*
Hoshino, S., 41, *50*, 183–185, *204*
Hosoda, N., 203, *211*
Houalla, R., 18, 19, *28, 29*, 38, 43, *49, 51*, 84, *94*
Houdusse, A., 124, *129*
Houseley, J., 3, 5, 6, 17, 18, *23, 25, 28*, 32, 41, 43, 46, *48, 51*, 54, 56, *71*, 79–82, 84, 89, *91, 92, 94, 95*, 182, *203*
Howson, R., 33, *49*
Hsu, C. L., 5, *24*, 100, *111*, 166, 167, *178*
Hu, W., 104, 105, *112, 113*
Hu, Y. H., 15, *27*
Huarte, M., 106, *114*
Huber, R., 59, *73*
Hughes, T. R., 45, *51*, 222, *234*
Huh, W. K., 8, *26*
Huh, W.-K., 185, *205*

Huntzinger, E., 64, 74, 101, 103, *111*, *112*, 185, 202, *205*, *210*
Hurt, E., 146, *162*
Hwang, H. C., 90, *95*
Hwang, K. Y., 121, *128*
Hyde, C. C., 121, *128*

I

Ibrahim, F., 47, *52*
Igarashi, K., 201, *210*
Igel, H., 217, *232*
Iglesias, N., 3, *23*, 81, 84, *92*, 106, 107, *113*, 145–147, 153, 154, *162*, *163*
Iino, Y., 228, *235*
Ikeda, H., 116, *128*, 134, 135, *157*, *158*, 172, *179*
Imam, J. S., 149, *162*
Inada, T., 5, *24*
Ingolia, N. T., 102, *112*
Inoue, K., 80, *92*
Inoue, T., 197, *209*
Ishida, J., 45, *52*
Ishii, R., 6, *25*, 56, *72*
Isken, O., 100, *111*
Ito, K., 197, *209*
Iwasaki, S., 202, *210*
Izaurralde, E., 5, 21, *24*, *29*, 47, *52*, 64, 74, 101, 103–105, *111–113*, 121, *129*, 136, *158*, 185, 202, *205*, *210*

J

Jäckle, H., 194, 195, *208*
Jackson, R. N., 15, 17, 21, *28*, 42, *51*, 79, 85–87, 89, *91*
Jackson, S. P., 18, *28*
Jacobs, E., 133, *157*
Jacobson, A., 98, 104, *110*, *112*, 132, 155, *157*, *163*, 183, 184, *204*
Jacques, P. E., 229, 230, *235*
Jacquier, A., 3, 5, *23*, 32, 41, 43, 45, *48*, *51*, *52*, 67, 74, 84, *94*, 145, *161*, 216, 222, 224, *232–234*
Janis, M., 227, *234*
Jankowsky, E., 15, 17, *27*, *28*, 45, *51*, 88, *94*
Januszyk, K., 8, 13, *26*, *27*, 56, 63, 68–70, 72, 75
Jayne, S., 196, *208*
Jensen, K. F., 6, *25*, 56, *72*

Jensen, T. H., 3, 5, 7, 11, 13–15, *23–27*, 35, 41, *49*, 56, 63, 64, 68, 69, *72*, 74, *75*, 101, *111*, 146, 149, *162*, 175, *179*
Jeong, W.-J., 47, *52*
Jeske, M., 201, 202, *210*
Jia, H., 15, 17, *27*, *28*, 45, *51*, 79, 87, 88, *91*, *94*
Jiao, X., 116, 117, 122, 124–126, *127*, *128*, 135–137, 143, 146, 152, *158*, 166–168, 170–173, 176, *178*, *179*
Jimeno-Gonzalez, S., 175, *179*
Jimeno-González, S., 146, 149, *162*
Jinek, M., 117, 118, 122–124, *127*, 136, 153, *159*
Johnson, A. W., 5, 20, *24*, *29*, 41, *50*, 116, 118, *127*, 134, 136, 141, 143, 144, 146, 147, 154, 155, *157*, *159*, *160*
Johnson, M. A., 188, 189, *206*
Johnson, R. E., 87, *94*
Johnson, S. J., 15, 17, 21, *28*, 42, *51*, 79, 85–87, 89, *91*
Jones, C. E., 121, *128*
Jones, C. I., 98, 100, *110*
Jones, G. H., 55, *72*
Jones, H. S., 218, *233*
Jonstrup, A. T., 13, *27*, 68, 69, *75*, 187, 195, 196, *206*, *208*
Ju, J. Y., 82, *93*

K

Kaberdin, V. R., 6, *25*
Kadaba, S., 15, 17, *27*, *28*, 45, *51*, 78, 80–82, 87, *91*, 152, *163*
Kadowaki, T., 133, *157*
Kadyk, L. C., 87, *94*
Kadyrova, L. Y., 197, 201, *209*
Kadziola, A., 6, *25*, 56, *72*
Kahvejian, A., 184, *204*
Kajiho, H., 41, *50*
Kammler, S., 45, *52*
Kaneko, S., 141, 144, *160*, *161*
Kang, S., 216, *232*
Kanno, J., 201, *210*
Kashima, I., 64, 74, 101, *111*
Kashlev, M., 175, *179*
Kastenmayer, J. P., 46, 47, *52*, 103, *112*, 116, *127*, 135, 136, *158*
Katada, T., 41, *50*, 183–185, *204*

Kawamata, T., 202, *210*
Kawashima, T. R., 151, *162*, 226, 227, *234*
Kawauchi, J., 80, *93*, 143–147, *160*, *161*, 218, 219, 229, *233*, *235*
Ke, A., 14, *27*, 61, *73*
Kearsey, S., 108, *114*, 138, 154, *159*, 176, *180*
Kearsey, S. E., 6, *24*, 89, *95*
Kedersha, N., 173, *179*
Keene, J. D., 104, *112*
Keller, C., 79, 86, 89, *91*
Keller, W., 87, 89, 90, *94*, *95*
Kellis, M., 21, *30*
Kempf, T., 11, *26*, 35, *49*
Kenna, M., 133, 134, *157*, 175, *180*
Kennington, E. A., 189, 201, *206*, *210*
Kerr, A., 137, 152, *159*
Khaleghpour, K., 184, *204*
Khalil, A. M., 105–107, *113*, *114*, 154, *163*
Khanna, R., 168, *178*
Kiledjian, M., 109, *114*, 116, 117, 122, 124–126, *127*, *128*, 135–137, 143, 146, 152, *158*, 166–173, 176, *177–179*
Kim, D. U., 67, *74*
Kim, H., 145, 146, *161*
Kim, H. B., 43, *51*
Kim, H. J., 175, *180*
Kim, H.-S., 43, *51*
Kim, H.-Y., 121, *128*
Kim, J. H., 6, *25*, 189, *206*
Kim, J.-M., 45, *52*
Kim, K. W., 202, 203, *210*
Kim, M., 6, 19, *25*, *29*, 80, 83, *93*, 135, 142–146, *158*, *161*, 175, *180*
Kim, Y., 121, *129*
Kim, Y.-K., 43, *51*
Kimble, J., 87, *94*, 201–203, *210*
Kinoshita, N., 135, *158*
Kireeva, M. L., 175, *179*
Kirsebom, L. A., 185, *206*
Kiss, D. L., 11, *26*, 38, *49*, *50*, 54, 68, *72*
Kiss, T., 141, *160*
Kisseleva-Romanova, E., 19, *29*, 80, 81, 83, 84, *92*
Klauer, A. A., 15, 17, 21, *28*, 42, 46, *51*, *52*, 79, 85–87, 89, *91*
Klein, M. L., 64, *74*

Klug, G., 7, 12, *25*, 39, *50*, 55, 56, *72*
Kobayashi, K., 42, *51*
Kobayashi, T., 41, *50*
Koch, F., 199, *209*
Kodama, Y., 98, *110*
Kolodner, R. D., 5, 20, *24*, 116, 118, *127*, 136, *159*
Koonin, E. V., 87, *94*
Koper, M., 135, 138, 144, 145, 147, 154, *158*, 218, 229, *233*
Kornberg, R. D., 79, *91*
Körner, C., 185, 187, *205*
Körner, C. G., 185–189, *205*, *206*
Kotelawala, L., 108, *114*, 133, 147, 152–154, *157*
Kotovic, K., 6, 17, *25*, 79–82, 84, *91*
Kozlov, G., 184, *204*
Krecic, A. M., 17, *28*, 78, 80–82, 87, *91*
Kressler, D., 15, *27*, 79, 81, 82, *91*
Kreth, J., 200, 201, *209*
Krisch, H. M., 6, *25*
Krishnakumar, S., 42, *51*
Kristiansen, M. S., 32, 39, 40, 43, *49*, 55, 63, 65, 67, 68, 70, *72*
Krogan, N. J., 144, 146, *161*
Krueger, A., 17, *28*, 78, 80–82, 87, *91*
Kuai, L., 82, *93*, 151, *162*
Kudla, G., 17, 19, *28*, 79–81, 83, 84, 87, 88, 90, *91*
Kuehner, J. N., 141, 144, 145, *160*, *161*, 229, *235*
Kufel, J., 3, 5, 15, 17, 19, *23*, *24*, *28*, 38, 47, *49*, *52*, 67, *74*, 80–83, *92*, *93*, 135, 138, 140, 141, 144, 145, 147, 154, *158–161*, 182, *204*, 217–219, 222, 229, *232–234*
Kühnlein, R. P., 194, 195, *208*
Kulkarni, M., 40, *50*
Kumagai, C., 133, *157*
Kumar, R., 32, 35, 37–39, 44–47, *48*
Kunkel, G., 168, *178*
Kunst, L., 33, 37, 38, *49*
Kurihara, Y., 47, *52*, 143, *160*
Kuzuoglu-Öztürk, D., 185, 202, *205*
Kwak, J. E., 87, *94*, 182, *204*
Kwapisz, M., 105, 107, *113*
Kyrpides, N. C., 119, *128*

L

LaCava, J., 15, 17, *27*, 32, 41, 43, *48*, 54, 56, *71*, 78–82, 87–89, *90*, *95*
LaFiandra, A., 98, *110*
Lafontaine, D. L., 138, 140, 146, *159*, 222, *234*
Lafontaine, D. L. J., 145, 153, *161*, *163*
LaGrandeur, T. E., 100, *111*
Lahmy, S., 217, *232*
Lai, W. S., 189, 201, *206*, *210*
Lam, P., 33, 37, 38, *49*
Lamontagne, B., 226, 228, 231, *234*
Landau, M., 60, *73*
Lander, E. S., 21, *30*
Landry, J. R., 229, 230, *235*
Lane, E. A., 108, *114*, 153, 154, *163*
Lang, W. H., 147, *162*
Lange, H., 11, *26*, 32, 39–41, 43–45, 47, *48*, *50*, *51*, 89, *95*
Lapik, Y. R., 90, *95*
Larimer, F. W., 100, *111*
Lario, P., 59, *73*
Larose, S., 81, 83, *92*, 226, 228, 231, *234*
Larsen, A., 144, *161*
Larsen, K. M., 32, 39, 40, *49*, 55, 63, 65, 67, 68, 70, *72*
Larsen, S., 6, *25*, 56, *72*
Lasko, P., 119, *128*, 201, *210*
Laterreur, N., 228, *234*
Lau, N.-C., 190, 195, 197, 198, *207*
Lau, S., 201, *210*
Laue, T. M., 203, *211*
Lauressergues, D., 47, *52*
Lavinsky, R. M., 18, *28*
Lavoie, M., 228, *235*
Lazar, M. A., 18, *28*
le Maire, A., 119, *128*
Le Ret, M., 39–41, 43, 45, 47, *50*
Leader, D. J., 224, *234*
Leavitt, A. D., 90, *95*
Lebaron, S., 221, *233*
Lebreton, A., 12, 14, 15, 18, *26*, 39, 41, *50*, 61, 64, 70, *73*
Lee, A., 141, 151, *162*, 222, 228, 230, *233*, *234*
Lee, A. V., 15, *27*

Lee, C. Y., 141, 146, 151, *160*, *162*, 222, 226, 227, *233*, *234*
Lee, D. J., 203, *211*
Lee, D.-S., 121, *129*
Lee, G., 14, *27*
Lee, M. T., 102, *112*
Lee, M.-H., 202, 203, *210*
Lee, T. H., 197, 201, *209*
Lee, W.-H., 156, *163*
Legrain, P., 145, 150, *161*, *162*, 222, 224, *233*, *234*
Lehmann, R., 199, *209*
Lehner, B., 35, *49*, 70, *75*, 176, *180*
Lejeune, F., 11, *26*, 68, *75*, 156, *163*
Lekka, M., 185, *205*
Lemaire, M., 190, 197, 198, *207*
Lennertz, P., 3, 15, 19, *23*, 67, *74*, 219, *233*
Leszczyniecka, M., 6, *25*
Leung, A. K., 105, *112*
Leung, E., 12, 15, *26*, 39, 40, *50*
Levin, N. A., 82, *93*
Levy, D. L., 82, *93*
Lewis, B. P., 102, *111*
Lewis, M. S., 41, *50*
Lhomme, F., 138, 140, 146, 153, *159*, *163*
Li, C. M., 82, *93*
Li, F., 214, *232*
Li, H., 46, *52*
Li, J., 148, *162*
Li, X., 11, *26*, 68, *75*, 156, *163*
Li, Y., 166–169, *178*
Li, Z., 106, 107, *113*, 145–147, 153, *162*
Liang, S., 15, *27*, 133, *157*
Libri, D., 5, 13–15, 19, *23*, *27*, *29*, 68, *75*, 80, 81, 83, 84, *92*
Lidder, P., 188, 189, *206*
Lilie, H., 185–187, *205*
Lilly, M. A., 199, *209*
Lim, L. P., 102, *111*
Lima, C. D., 7, 8, 12–14, *25*–*27*, 35, 39, *49*, 55, 56, 58–61, 63, 65, 67–71, *72*, *73*, *75*, 88, *94*, 166, *177*
Lin, M. F., 106, *113*
Lin, W.-J., 189, *206*
Lin, Z., 201, *210*
Lin-Chao, S., 6, 7, *25*, 56, *72*
Linder, P., 15, *27*, 79, 81, 82, *91*
Lindner, D., 7, *25*, 55, 56, *72*

Lingner, J., 106, 107, *113*, 145–147, 153, 154, *162, 163*
Lipshitz, H. D., 201, *210*
Lis, J. T., 11, *26*, 84, *94*
Lister, R., 46, 47, *52*
Liu, C. L., 105, *113*
Liu, F., 230, *235*
Liu, H., 109, *114*, 166, 167, *177, 178*
Liu, H.-Y., 190, 196–198, *207, 209*
Liu, J., 44, *51*
Liu, Q., 7, 8, 12–14, *25, 27*, 35, 39, *49*, 55, 56, 58–61, 65, 68–70, *72, 75*, 88, *94*
Liu, Y., 15, 20, *27, 29*, 185–187, *205*
Livingstone, M. J., 13, 18, *27*, 68, *75*
Lo, T. L., 200, *209*
Lo, W.-T., 156, *163*
Lobo Ruppert, S. M., 220, *233*
Loewer, S., 106, *113*
Logan, J., 141, *160*
Loh, Y.-H., 106, *113*
Loireau, M.-P., 190, 194, *207*
Lojek, L., 105, 107, *113*, 154, *163*
Lopez, M. E., 12, *26*
López, M. E. G., 32, 39, *48*
Lorentzen, E., 2, 7, 8, 11, 12, 14, 15, *22, 25*, 38–40, *49, 50*, 55, 56, 59, 61, 63–65, 67, 70, *72, 73*
Lottspeich, F., 55, *72*
Lowe, T., 224, *234*
Lowell, J. E., 183, *204*
Lu, C., 61, *73*
Lu, G., 169, *179*
Lu, Y., 157, *163*
Lu, Y. J., 224, *234*
Lubas, M., 11, 17–20, *26*, 32, 43, *49*, 79, 90, *92*
Lubkowska, L., 175, *179*
Lueder, F., 200, *209*
Luisi, B. F., 55, *72*
Luke, B., 106, 107, *113*, 145–147, 153, 154, *162, 163*
Lunde, B. M., 144, *161*
Luo, M. J., 146, *162*
Luo, W., 141, 143, 144, 146, 154, 155, *160*
Luther, J., 190, 195, 198, 199, *207*
Lykke-Andersen, J., 104, *112*, 166, 167, 173, *178, 179*, 194, 195, 201, *208, 210*

Lykke-Andersen, S., 7, 11, *25, 26*, 32, 35, 39–41, 43, 45, *49, 52*, 55, 63–65, 67, 68, 70, *72, 74*, 101, *111*

M

Mackie, G. A., 12, 17, *27, 28*, 59, *73*
Macmorris, M., 144, *161*
MacRae, I. J., 214, *232*
Macrae, I. J., 214, *232*
Maillet, L., 196–198, 200, *208*
Malabat, C., 45, *52*, 67, *74*, 84, *94*
Malagon, F., 146, 149, *162*, 175, *179*
Maldonado, E., 166, *177*
Malet, H., 14, *27*, 61, 65, *73*
Malhotra, A., 68, *75*
Mandal, S. S., 166, 175, *177*
Mangus, D. A., 104, *112*, 183, 184, *204*
Manickam, S., 15, *27*
Manley, J. L., 116–120, 122, 124–126, *127*, 135–137, 141, 143, 144, 146, *158–161*, 170, 172, 175, 176, *179, 180*, 182, *204*
Mann, M., 2, 8, 11, 13, 14, 18, 19, *22, 23*, *28*, 34, 40, 43, *49, 51*, 54, 55, 70, *72*, 190, 197, 198, 201, *207, 209*
Mao, X., 172, 175, *179*
Mapendano, C. K., 45, *52*
Maquat, L. E., 5, 11, *24, 26*, 68, *75*, 98, 100, 110, *111*, 156, *163*, 182, 183, 201, *203*
Marchese, F. P., 189, 201, *206*
Marquardt, S., 107, *114*
Marshall, G. T., 6, *25*
Martens, J. A., 106, *114*
Martin, C. E., 116, 125, *128*, 135, 137, 152, *158*, 167, 170–173, *178*
Martin, G., 32, 41, 43, 45, *48*, 89, *95*
Martinez, J., 185–188, *205, 206*
Martinez-Vinson, C., 42, *50*
Martz, E., 60, *73*
Marzluff, W. F., 90, *95*, 136, *158*
Masison, D. C., 21, *29*, 32, 42, *48*
Mathy, N., 136, *159*
Matsui, A., 45, *52*
Mattick, J. S., 106, *113*
Matunis, M. J., 59, *73*
Maupin, M. K., 100, *111*
Mauxion, F., 190, 195, 199, 202, 203, *207–210*
Mayer, B. J., 119, *128*
Mayrose, I., 60, *73*

Mazroui, R., 151, *163*
McCammon, M., 133, 134, *157*, 175, *180*
McCarthy, T., 5, *24*
McConaughy, B. L., 133, *157*
McMahon, J. S., 197, 200, 201, *209*
McManus, M. T., 90, *95*
McSweeney, J. M., 184, *204*
McVey, C. E., 63, 64, *73*
Meaux, S., 228, *235*
Mehler, M. F., 106, *113*
Meijer, H. A., 189, *206*
Meinhart, A., 19, *29*, 80, 83, *93*, 145, *161*
Meng, Y., 36, 37, *49*
Merrick, W. C., 166, *177*
Merry, C. R., 106, *113*
Meyer, S., 166, *177*, 201, 202, *210*
Meyerson, M., 144, *161*
Mian, I. S., 2, 11, *23*, *26*, 55, *72*
Michael, W. M., 59, *73*
Miczak, A., 6, *25*
Midtgaard, S. F., 13, *27*, 68, 69, *75*
Mikhli, C., 168, *178*
Milligan, L., 5, 15, 18, 19, *24*, *28*, 43, *51*
Minczuk, M., 6, *25*, 195, *208*
Mindrinos, M., 42, *51*
Minshall, N., 189, *206*
Miraglia, L. J., 217, *232*
Mischo, H., 144–147, *161*
Mischo, H. E., 80, *93*, 143, 144, 146, *160*, *161*, 218, 229, *233*, *235*
Mitchell, P., 2, 3, 8, 11, 13–15, 17–19, *22*, *23*, *27*, *28*, 34, 40, 41, 43, *49–51*, 54–56, 67, 68, 70, *71–75*, 82, *93*, 141, *160*, 219, *233*
Mitrovich, Q. M., 221, *233*
Mittal, S., 199, *209*
Miura, K. I., 98, *110*
Miyasaka, T., 203, *210*
Miysaka, T., 190, 197–199, *207*
Mizuno, T., 80, *92*
Moazed, D., 6, *24*, *25*, 67, 74, 89, *95*
Molin, L., 199, *209*
Molineux, C., 116, 118, *127*, 136, *159*
Monecke, T., 187, *206*
Moore, C., 141, 145, *160*, *161*, 229, *235*
Morel, A.-P., 190, 194, *207*
Morel, J.-B., 47, *52*
Morgan, C. T., 202, 203, *210*
Morgan, D. E., 202, 203, *210*

Morillon, A., 3, *23*, 81, 84, *92*, 105, 106, *113*
Morita, M., 190, 197–199, *207*, *209*
Moritz, B., 184, 198, *205*
Morlando, M., 148, 152, *162*
Morosawa, T., 45, 47, *52*
Morris, J. Z., 199, *209*
Morrissey, J. P., 108, *114*, 138, 154, *159*, 176, *180*
Moscou, M. J., 36, 37, *49*
Motamedi, M. R., 89, *95*
Mouaikel, J., 13–15, *27*, 68, *75*
Mougin, A., 221, *233*
Muckenthaler, M., 185, 188, 189, *205*
Mudd, E. A., 6, *25*
Mueser, T. C., 121, *128*
Muhlemann, O., 5, *23*, 101, *111*
Mühlemann, O., 64, *74*
Muhlrad, D., 5, *24*, 100, 101, *111*, 155, *163*, 190, 195, 197, 199, 200, 203, *208*, *209*
Mullen, T. E., 90, *95*
Müller-Auer, S., 183, 184, *204*
Murali, R., 121, *129*
Murphy, C., 39, 40, *50*
Murthy, H. M. K., 121, *129*
Mustaev, A., 124, *129*
Mutschler, H., 19, *29*, 80, 83, *93*, 145, *161*
Myslinski, E., 222, *234*

N

Nabavi, S., 220, 222, 229, 230, *233–235*
Nag, A., 83, 90, *94*
Nagalakshmi, U., 106, *113*
Nagamine, Y., 5, *24*
Nagata, T., 187, *206*
Nagel, R., 141, *160*, 216, 225, *232*, *234*
Nahvi, A., 102, *112*
Nakamura, R., 89, *95*
Nakamura, T., 190, 197–199, *207*, *209*
Nakanishi, K., 214, 216, *232*
Nasertorabi, F., 193, 198, *207*
Nazar, R. N., 218, 220, 229, *232*, *233*
Nazarian, A., 11, *26*
Nechaev, S., 148, *162*
Nehls, P., 18, *28*
Nehrbass, U., 5, *23*
Neil, H., 45, *52*, 67, 74, 84, *94*
Nelson, T. M., 43, *51*
Nery, C. R., 107, *114*

Nery, J. R., 47, *52*
Neupane, R., 33, *49*
Newbury, S., 103, *112*, 116, *127*
Newbury, S. F., 21, *29*, 98, 100, *110*
Nicholls, T. J., 195, *208*
Nicholson, A., 98, *110*
Niedzwiecka, A., 185, *205*
Nielsen, J., 45, *52*
Nielsen, P. R., 119, *128*
Niland, C. N., 106, *113*
Nilsson, P., 185–188, *205, 206*
Nishimura, N., 188, 189, *206*
Niu, S., 194, *208*
Noel, J. F., 81, 83, *92*
Norbury, C. J., 89, *95*
Nossal, N. G., 121, *128*
Nourizadeh, S. D., 46, *52*
Nowotny, M., 64, *74*
Nudler, E., 124, *129*, 145, *161*
Nureki, O., 6, *25*, 56, *72*
Nurmohamed, S., 55, *72*

O

Oberholzer, U., 197, *208*
Oddone, A., 59, 61, *73*
O'Donnell, K., 5, *24*
Odul, E., 42, *50*
Oeffinger, M., 136, 138, 146, 155, *158*
Oh, C., 167, 170–173, *178*
Oh, C.-S., 116, 125, *128*, 135, 137, 152, *158*
Ohn, T., 195, 196, 199, *208*
Okamoto, M., 45, *52*
Okubo-Kurihara, E., 47, *52*
Oliver, P. L., 106, *113*
Ollis, D. L., 68, *75*
Olmedo, G., 46, *52*
Olsen, P. H., 102, *111*
O'Malley, R. C., 46, 47, *52*
Ooi, S. L., 141, *160*, 225, *234*
Orans, J., 121, 124, *128*
Orban, T. I., 21, *29*, 47, *52*, 103, *112*
Osheim, Y. N., 218, *233*
Otegui, M. S., 42, *51*
Ouzounis, C. A., 119, *128*
Ozgur, S., 40, *50*

P

Pagano, F., 148, 152, *162*
Page, A. M., 116, 118, *127*, 136, *159*

Pai, E. F., 197, *209*
Palermino, J.-M., 184, *204*
Paliouras, M., 201, *210*
Panasenko, O. O., 79, *91*, 190, *207*
Panbehi, B., 144, 145, *161*
Panza, A., 106, 107, *113*, 145–147, 153, *162, 163*
Papin, C., 182, *204*
Park, E. Y., 6, *25*
Park, H. O., 67, *74*
Park, I. H., 106, *113*
Park, N.-I., 43, *51*
Park, Y.-I., 43, *51*
Parker, R., 67, *74*, 100, 101, 104–106, *111–113*, 150, 155, *162, 163*, 166–168, 173, *178, 179*, 184, 190, 195, 197–201, 203, *204, 205, 210*, 219, *233*
Parker, R. P., 3, 5, 15, 19–21, *23, 24, 29*, 32, 34, 41, 42, *48–50*
Paro, R., 119, *128*
Pascolini, D., 116, *127*, 136, *159*
Patel, D. J., 214, 216, *232*
Patel, N., 121, *128*
Patena, W., 90, *95*
Patterson, D. N., 15, *27*, 79, 87, 88, *91*
Pearce, J., 89, *95*
Pearson, E. L., 141, *160*, 229, *235*
Peculis, B. A., 169, *178, 179*
Pei, Y., 175, *179*
Pellegrini, O., 125, *129*, 136, *159*
Perdomo, L. I., 218, *232*
Perocchi, F., 45, *52*, 106, *113*
Pestov, D. G., 90, *95*, 138, 153, *159*
Peterson, B., 169, *178*
Petfalski, E., 2, 3, 8, 11, 13–15, 17–19, *22, 23, 28*, 32, 34, 40, 43, *48, 49, 51*, 54–56, 67, 70, 71–74, 82, *93*, 108, *114*, 138, 140, 141, 145, 146, 153, 154, *159–161*, 176, 180, 219, 222, 224, 225, *233, 234*
Petricka, J. J., 43, *51*
Pfeiffer, V., 154, *163*
Philipps, R. S., 201, *210*
Phillips, S., 68, *75*, 135, 153, 154, *158*, 176, *180*
Phipps, K. R., 138, *159*
Phizicky, E. M., 108, *114*, 133, 147, 152–154, *157, 163*
Piao, X., 189, 202, *207*
Piccirillo, C., 168, *178*

Pieuchot, L., 39–41, 43, 45, 47, *50*
Pillutla, R., 166, *177*
Pinder, B. D., 201, *210*
Pinskaya, M., 81, 84, *92*, 106, *113*
Piwowarski, J., 6, *25*
Podtelejnikov, A., 2, 8, 11, 14, 18, 19, *23*, *28*, 34, 40, 43, *49*, *51*, 54, 55, 70, *72*
Pogacic, V., 140, *159*, 221, 224, *233*
Ponting, C. P., 106, *113*, 119, 121, *128*, *129*
Pontvianne, F., 217, *232*
Poole, T. L., 116, *127*, 133–135, 146, *157*, 167, 170, 172, 175, *178*
Porro, A., 153, *163*
Poulsen, J. B., 195, *208*
Pradervand, S., 42, 46, *50*
Pradhan, D. A., 80, *93*
Prakash, L., 87, *94*
Prakash, S., 87, *94*
Preiss, T., 200, *209*
Preker, P., 3, 20, *23*, *29*, 45, *52*, 84, 90, *94*, *95*
Prescott, E. M., 146, 147, *162*, 218, *233*
Presutti, C., 5, *23*, 141, 149, 150, 154, *160*, *162*
Prigge, A., 200, 202, *209*
Prislei, S., 141, *160*
Proudfoot, N. J., 80, *93*, 141–148, 152, *160–162*, 175, *180*, 218, 219, 229, *233*, *235*
Proux, F., 47, *52*
Pruijn, G. J., 11, 18, 19, *26*, *28*, 157, *163*
Pruijn, G. J. M., 43, *51*
Puisieux, A., 199, *209*
Pupko, T., 60, *73*
Py, B., 6, *25*

Q

Qu, L. H., 141, 146, 154, *159*, 224, *234*
Qureshi, I. A., 106, *113*

R

Raats, J. M., 18, 19, *28*
Raats, J. M. H., 43, *51*
Radford, H. E., 189, *206*
Rafikov, R., 124, *129*
Raha, D., 106, *113*
Raijmakers, R., 18, 19, *28*, 43, *51*, 56, *73*
Raj, A., 106, *114*
Rammelt, C., 87, 90, *94*

Ramnarain, D. B., 197, 200, 201, *209*
Rana, T. M., 105, *113*, 175, *179*
Rando, O. J., 145, *161*
Rappsilber, J., 43, *51*, 190, 197, 198, 201, *207*, *209*
Rätsch, G., 45, *52*
Raue, H. A., 138, 154, *159*
Raué, H. A., 108, *114*
Raught, B., 166, *177*
Read, R. L., 89, *95*
Redon, S., 106, 107, *113*, 145–147, 153, 154, *162*, *163*
Reed, R., 146, *162*
Reeder, R. H., 147, *162*, 218, *233*
Regan, L., 59, *73*
Regnier, P., 78, *91*
Rehwinkel, J., 202, *210*
Reik, W., 106, *113*
Reinberg, D., 166, 175, *177*
Reinisch, K. M., 85, 86, *94*
Reis, C. C., 81, *92*
Reis, F. P., 36, 39, *49*, 61, 64, 65, *73*
Reiter, A., 147, *162*
Reiter, M. H., 189, *206*
Ren, Y.-G., 185–188, *205*, *206*
Rendl, L. M., 201, *210*
Repic, A., 214, *232*
Reuter, M., 185–187, *205*
Reverdatto, S. V., 32, 35, 37–39, 44–47, *48*, 188, 189, *206*
Reyes-Turcu, F. E., 67, *74*
Ribas, J. C., 32, 42, *48*
Richard, P., 141, *160*
Richardson, J. P., 145, *161*
Richter, J. D., 182, 189, *204*
Ridley, S. P., 20, *29*
Riederer, M., 37, *49*
Rieger, M., 183, 184, *204*
Rivea Morales, D., 106, *114*
Roan, J. G., 147, *162*, 218, *233*
Robert, F., 229, 230, *235*
Roberts, L. O., 166, *177*
Robertson, H. D., 214, *231*
Robinson, C. V., 12, *26*
Robinson, H., 15, 17, 21, *28*, 42, *51*, 79, 85–87, 89, *91*
Roche, V., 105, 107, *113*
Rode, M., 21, *29*, 89, *95*
Rodgers, N. D., 167, *178*

Rodriguez-Villalon, A., 42, 46, *50*
Rogozin, I. B., 87, *94*
Rohr, J., 47, *52*
Roiha, H., 221, *233*
Rollag, M. D., 194, *208*
Romano, A., 5, *23*
Rondon, A., 144–147, *161*
Rondon, A. G., 80, *93*
Rondón, A. G., 143, 146, *160*
Roos, A., 64, *74*
Roquelaure, B., 42, *50*
Rorbach, J., 195, *208*
Rosbash, M., 5, *23*, *24*, 150, *162*
Rosenberg, Y., 60, *73*
Rosenfeld, M. G., 18, *28*
Rosonina, E., 141, *160*
Roth, K. M., 79, 81, *92*
Rother, S., 146, *162*
Rotondo, G., 219, 224, *233*, *234*
Rougemaille, M., 5, 19, *24*, *29*, 32, 40, 41, 45, *48*, 67, *74*, 79–81, 83, 84, *92*
Rousselle, J. C., 67, *74*
Rousselle, J.-C., 32, 40, 41, 45, *48*
Roy, K., 216, *232*
Rozenblatt-Rosen, O., 144, *161*
Rubin, C., 67, *74*
Ruckenstein, A. E., 124, *129*
Rudner, D. Z., 183, *204*
Rupp, B., 193, 194, *208*
Ruppert, T., 35, *49*
Russell, P., 190, 197, 198, 201, *207*, *208*
Ruvkun, G., 102, *111*
Ryan, K., 182, *204*
Rybin, V., 59, 61, *73*
Rymond, B. C., 219, 220, *233*

S

Sachs, A. B., 79, *91*, 183, 184, 203, *204*
Sachsenberg, T., 45, *52*
Sadoff, B. U., 80, 82, *92*, *93*
Saga, Y., 201, *210*
Saguez, C., 5, *24*
Saha, N., 172, 175, *179*
Saklatvala, J., 189, 201, *206*
Sakuno, T., 41, *50*
Sakurai, S., 121, *128*
Salinas, P., 42, 46, *50*
Salvadore, C., 190, 196–198, *207*

Samarsky, D., 141, *160*
Samarsky, D. A., 224, 225, *234*
Samson, E., 217, 218, *232*
San Paolo, S., 3, *23*, 80, 81, 83, 84, 88, *92*
Sanchez-Rotunno, M., 36, 39, *49*, 61, 64, 65, *73*
Sanderson, C. M., 35, *49*, 70, *75*, 176, *180*
Sandler, H., 200, 201, *209*
Santuari, L., 42, 46, *50*
Sattler, M., 119, *128*
Saulière, J., 64, *74*, 101, *111*
Saveanu, C., 32, 41, 43, *48*, *51*
Sayani, S., 107, *114*, 227, *234*
Sayar, E., 42, *50*
Sayers, J. R., 121, *128*
Schaeffer, D., 12, 15, *26*, 36, 39, 46, *49*, *52*, 61, 64, 65, *73*
Scheffzek, K., 11, *26*, 70, *75*
Schell, S., 187, *206*
Scherthan, H., 104, *112*
Schilders, G., 11, 18, 19, *26*, *28*, 43, *51*, 63, *73*
Schillewaert, S., 138, 140, 146, 153, *159*, *163*
Schilling, M., 5, *24*
Schmid, M., 3, 5, *23*
Schmidt, S., 185, 202, *205*
Schmitz, R. J., 47, *52*
Schneider, C., 12, 15, *26*, 39, 40, 45, *50*, *51*, 54, 72, 82, 83, 88, *93*
Schneider, S., 185, 188, 189, *205*
Schoenberg, D. R., 5, *24*, 98, *110*, 182, 183, 201, *203*
Schubert, M., 59, *73*
Schultz, M. D., 47, *52*
Schuster, G., 6, *25*, 63, *73*
Schwartz, D. C., 168, *178*
Schwede, A., 183, 190, 195, 198, 199, *204*, *207*
Schweizer, D., 105, *113*
Schwer, B., 172, 175, *179*
Seago, J. E., 21, *29*
Seay, D. J., 195, 201, *208*, *210*
Segault, V., 222, *234*
Segelke, B. W., 193, 194, *208*
Seila, A. C., 84, *94*
Seipelt, R. L., 219, 220, *233*
Selenko, P., 119, *128*

Sement, F. M., 32, 39, 43–45, 47, *48*, *51*, 89, *95*

Semotok, J. L., 201, *210*

Semple, J. I., 176, *180*

Sentis, S., 195, *208*

Seraphin, B., 2, 7, 8, 12, 14, 15, 18, *22*, *25*, *26*, 104, *112*, 190, 195, 199, 202, 203, *207–209*

Séraphin, B., 32, 39, 41, *48*, *50*, 55, 56, 59, 61, 63, 64, 67, 70, *72*, *73*, 202, 203, *210*, *211*

Shafer, B. K., 175, *179*

Shapiro, R. A., 133, *157*

Sharkey, D. J., 121, *129*

Sharp, P. A., 105, *112*

Shatkin, A. J., 98, *110*, 166, 175, *177*

Shaw, R. J., 38, 39, *49*, 184, *205*

Shchepachev, V., 89, *95*

Shcherbik, N., 90, *95*

Shen, B., 47, *52*

Shenk, T., 141, *160*

Sherman, F., 82, *93*, 135, 151, *158*, *162*

Sherrer, R. L., 81, 83, *92*

Sheth, U., 104, *112*

Shevchenko, A., 2, 8, 13, *22*, 34, *49*, 55, 72

Shew, J.-Y., 156, *163*

Shi, Y., 143, *160*

Shi, Z., 6, 7, *25*, 56, 72

Shimotohno, K., 98, *110*

Shipman, T., 5, 15, *24*

Shobuike, T., 116, *128*, 134, 135, *157*, *158*, 172, *179*

Shou, C., 106, *113*

Shu, S., 33, *49*

Shuman, S., 166, 172, 175, *177*, *179*

Shuster, E. O., 196–198, 200, *208*, 221, *233*

Shyu, A. B., 5, *24*, 176, *180*

Shyu, A.-B., 184, 185, 199, 201, 202, *204*, *205*, *209*, *210*

Sicheri, F., 201, *210*

Siddiqui, N., 184, *204*

Sieburth, L. E., 39, 40, *50*

Sigler, P. B., 83, *93*

Silver, P. A., 82, *93*

Simón, E., 203, *211*

Simonelig, M., 182, 190, 198–201, *204*, 207, *209*, *210*

Simpson, C. G., 224, *234*

Sinturel, F., 125, *129*

Siomi, H., 59, *73*

Skourti-Stathaki, K., 144, *161*

Slomovic, S., 63, *73*

Smibert, C. A., 201, *210*

Smith, G. C., 18, *28*

Smith, G. R., 136, *158*

Smith, M., 183, 184, *204*

Smith, M. M., 184, *204*

Smolke, C. D., 228, *235*

Snow, P. M., 82, *93*

Soares, L. M., 6, *25*

Sobczyk, G. J., 47, *52*, 140, 141, 154, *159*

Solinger, J. A., 104, *112*, 116, *127*, 136, *159*

Sommer, S. S., 20, *29*

Sonenberg, N., 32, 42, *48*, 166, *177*

Song, H., 20, *29*, 185–187, *205*, *206*

Song, M., 166, 169, *178*

Song, M. G., 166–169, *178*

Soudet, J., 221, *233*

Souret, F. F., 46, 47, *52*, 103, *112*, 140, 141, 154, *159*

Sousa, R., 124, *129*

Spasov, K., 218, *232*

Spickler, C., 17, *28*

Spies, N., 6, *25*

Spitale, R. C., 106, *114*

Sprangers, R., 119, *128*

Srivastava, L., 90, *95*

Staals, R. H., 11, 13, *26*, 63, *73*

Stamatoyannopoulos, J. A., 106, *113*

Standart, N., 189, *206*

Stangl, M., 37, *49*

Staples, R. R., 20, 21, *29*, 184, 190, 195, 197–200, 203, *204*, *208*

Stark, A., 202, *210*

Stead, J. A., 13, 18, *27*, 68, *75*

Stefl, R., 78, *91*

Steiger, M., 168, *178*

Steinmetz, E. J., 19, *29*, 80, *93*, 144, 145, *161*

Steinmetz, L. M., 45, *52*, 67, *74*, 84, *94*

Steitz, J. A., 64, *74*, 83, 90, *94*

Steitz, T. A., 64, 68, *74*, *75*, 121, *129*

Stepien, P. P., 6, *25*

Stern, D. B., 135, *158*

Stevens, A., 5, *24*, 98–100, 105, 109, *110*, 111, *114*, 116, *127*, 133–135, 146, 153, *157*, 166, 167, 170, 172, 175, *178*, *180*

Stevenson, A. L., 6, *24*, 89, *95*
Stier, G., 119, 121, *128*
Stoecklin, G., 40, *50*, 190, 195, 198–201, 207, *209*
Storey, J. D., 105, *113*
Strasser, K., 146, *162*
Strathern, J. N., 175, *179*
Strayer, C. A., 194, *208*
Strum, J. R., 201, *210*
Strynadka, N. C., 59, *73*
Stutz, F., 3, *23*, 81, 84, *92*
Su, M.-I., 156, *163*
Subramanian, A. R., 59, *73*
Suck, D., 121, *128*, 193, 195, 198, *207*, *208*
Sugano, S., 116, *128*, 134, 135, *157*, *158*, 172, *179*
Sugimoto, A., 228, *235*
Sugino, A., 134, *157*
Suh, S. W., 121, *129*
Summers, M. F., 86, *94*
Suzuki, A., 201, *210*
Suzuki, T., 190, 197–199, *207*, *209*
Sweeney, H. L., 124, *129*
Sweet, T. J., 104, 105, *112*, *113*
Symmons, M. F., 55, *72*
Synowsky, S. A., 41, *50*
Szankasi, P., 136, *158*
Szczesny, R. J., 32, 39, 40, *49*, 55, 63, 65, 67, 68, 70, *72*

T

Tainer, J. A., 121, 122, *129*
Takahashi, S., 41, *50*
Takeda, T., 134, *157*
Tanaka, H., 67, *74*
Tanaka, K., 67, *74*
Tani, T., 116, *128*, 135, *158*, 172, *179*
Tartakoff, A. M., 15, *27*
Tarun, S., 183, 184, *204*
Tarun, S. Z., 183, 184, *204*
Tatebayashi, K., 116, *128*, 135, *158*, 172, *179*
Taverner, T., 12, *26*
Taylor, M. J., 169, *179*
Teixeira, D., 105, *113*
Temme, C., 166, *177*, 184, 190, 194, 195, 197–202, *205*, *207*, *210*
Tempst, P., 11, *26*

Tenenbaum, S. A., 104, *112*
Terzi, N., 6, *25*
Thiebaut, M., 19, *29*, 80, 81, 83, 84, *92*
Thomas, K., 106, *113*
Thompson, D. M., 105, 106, *113*
Thompson, E., 5, *24*, 32, 41, 43, *48*
Thomsen, R., 5, *23*
Thomson, T., 119, *128*
Thore, S., 195, *208*
Thuresson, A.-C., 185, 186, *205*
Tibshirani, R. J., 105, *113*
Till, D. D., 116, *127*, 136, *158*
Timmers, H. T. M., 190, 196–198, 200, 201, *207–209*
Tishkoff, D. X., 136, *159*
Toesca, I., 107, *114*, 227, *234*
Toh-e, A., 20, *29*
Tollervey, D., 2, 3, 5, 6, 8, 11–15, 17–19, *22–28*, 32, 34, 39–41, 43, 45, 46, *48–51*, 54–56, 67, 70, *71–74*, 79–84, 87–90, *91–93*, *95*, 108, 109, *114*, 134, 135, 138, 140, 141, 144–147, 149–151, 153, 154, *157–163*, 176, *180*, 182, *203*, 217–219, 221, 222, 224, 225, *232–234*
Tomanicek, S. J., 121, *128*
Tomari, Y., 202, *210*
Tomasevic, N., 169, *178*
Tomecki, R., 7, 11–15, 18–21, *25*, *26*, 32, 39, 40, *48–50*, 55, 61, 63–65, 67, 68, 70, *72*, *73*
Tong, L., 116–122, 124–126, *127–129*, 135–137, 143, 146, 152, *158*, *159*, 167, 170–173, 176, *178*, *179*
Tonti-Filippini, J., 46, 47, *52*
Toone, W. M., 134, *157*
Topf, M., 61, 65, *73*
Torchet, C., 5, 15, *24*, 38, *49*
Toth, E. A., 15, *27*, 79, 87, 88, *91*
Tran, H., 5, *24*
Traven, A., 200, *209*
Tremblay, M., 217, 218, *232*
Trempe, J.-F., 184, *204*
Trice, T., 17, *28*, 78, 80–82, 87, *91*
Tritschler, F., 104, *112*
Tsai, M-C., 106, *114*
Tsanova, B., 36, 39, 46, *49*, *52*, 61, 64, 65, *73*
Tsuboi, T., 21, *30*

Tsutakawa, S. E., 121, 122, 124, *128*, *129*
Tsutsumi, C., 67, *74*
Tu, C., 196–198, 200, *208*
Tucker, M., 184, 190, 195, 197–200, 203, *204*, *208*
Tudor, C., 189, 201, *206*

U

Uchida, N., 183–185, *204*
Urich, M. A., 46, 47, *52*
Ursic, D., 144, 147, *161*

V

Vaidialingam, B., 55, *72*
Valen, E., 148, *162*
Valencia-Sanchez, M. A., 103, *112*, 184, 190, 195, 197–200, 203, *204*, *208*
Valle, R. C., 21, *29*
Valverde, R., 59, *73*
van Aarssen, Y., 56, *73*
van Berkum, N. L., 190, 197, 198, *207*
van den Berg, D. L. C., 196, *208*
van den Elzen, A. M., 21, *30*
van Dijk, E., 11, 18, *26*, 43, *51*
van Dijk, E. L., 105, 107, *113*
van Erp, H., 188, 189, *206*
van Hoof, A., 3, 5, 15, 17, 19–21, *23*, *24*, *28–30*, 34, 41, 42, 46, *49–52*, 67, *74*, 79, 85–87, 89, *91*, 219, 228, *233*, *235*
Van, L. B., 13, *27*, 68, 69, *75*, 187, 195, 196, *206*, *208*
van Venrooij, W. J., 157, *163*
Vanacova, S., 15, 17, *27*, 32, 41, 43, 45, *48*, 78–83, 87–89, *90*, *91*
Vari, H. K., 201, *210*
Vasiljeva, L., 6, 19, *25*, *28*, *29*, 78, 80, 81, 83, *91*, *93*, 145, *161*
Venema, J., 138, *159*, 218, 221, 222, *232*, *234*
Verdel, A., 89, *95*
Vignols, F., 217, *232*
Villa, T., 19, *29*, 141, 153, *160*, *163*
Vincent, H. A., 63, *73*
Virtanen, A., 185–188, *205*, *206*
Viswanathan, P., 190, 193, 195, 196, 199, *207*, *208*
Vogel, J. T., 188, 189, *206*
Vogg, G., 37, *49*
von Roretz, C., 151, *163*

Vonrhein, C., 63, 64, *73*
Vos, J. C., 108, *114*, 138, 154, *159*
Vu, C., 82, *93*
Vyas, V. K., 216, 217, 220, *232*

W

Wacheul, L., 138, 140, 146, 153, *159*, *163*
Wada, K., 80, *92*
Wada, T., 166, 175, *177*
Wade, J., 145, *161*
Waern, K., 106, *113*
Wagner, E., 194, 195, 201, *208*, *210*
Wahl, M. C., 59, *73*
Wahle, E., 166, *177*, 184–190, 198–202, *205–207*, *210*
Walowsky, C., 82, *93*
Walter, P., 7, 12, *25*, 39, *50*, 55, 56, *72*
Wan, J., 157, *163*
Wang, C., 20, *29*
Wang, H., 190, 193, 194, *208*
Wang, H. W., 65, *74*
Wang, J., 65, *74*, 121, *129*
Wang, L., 41, *50*, 87, *94*
Wang, M., 90, *95*, 138, 153, *159*
Wang, S. W., 6, *24*, 89, *95*
Wang, X., 15, 17, *27*, *28*, 45, *51*, 78, 80–82, 87, 88, *91*, *94*, 152, *163*
Wang, Y., 68, *75*, 105, *113*
Wang, Z., 82, *93*, 106, 109, *113*, *114*, 166–168, *178*, 216, *232*
Warren, C. L., 144, 145, *161*
Wasmuth, E. V., 59, 67, 70, 71, *73*
Watanabe, Y., 67, *74*
Waterfall, J. J., 84, *94*
Watt, S., 6, *24*, 89, *95*
Webster, R. E., 214, *231*
Wei, C. L., 6, *25*
Wei, P.-C., 156, *163*
Wei, W., 45, *52*, 106, *113*
Weigel, D., 45, *52*
Weil, D., 189, *206*
Weinberg, D. E., 214, 216, 217, 220, *232*
Weir, J. R., 15, 17, 21, *28*, 42, *50*, 85–87, 89, *94*
Weissman, J. S., 84, *94*, 102, *112*
Welcker, M., 90, *95*
Wellinger, R. J., 81, 83, *92*, 228, *234*
Wen, T., 136, *159*
Wenig, K., 55, 61, *72*

Werner, D., 18, *28*
Werner, J., 11, *26*
West, S., 137, 142–144, 152, *159–161*, 175, *180*
Western, T. L., 44, *51*
Wharton, R. P., 197, 201, *209*
Whipple, J. M., 108, *114*, 133, 147, 152–154, *157*, *163*
Wickens, M., 87, *94*, 182, 195, 200–202, *203*, *204*, *208–210*
Wickner, R. B., 20, 21, *29*, 32, 42, *48*
Widmer, C. K., 45, *52*
Wightman, B., 102, *111*
Wilkinson, M. F., 149, *162*
Wills, M. A., 38, 39, *49*
Wilson, D. M. III., 193, 194, *208*
Wilson, G. M., 15, *27*, 79, 87, 88, *91*
Wilusz, C. J., 167, *178*
Wilusz, J., 167, *178*, 186, *206*
Win, T. Z., 89, *95*
Winkler, G. S., 199, *209*
Winston, F., 106, *114*
Wirbelauer, C., 5, *24*
Wischnewski, H., 89, *95*
Wlotzka, W., 17, 19, *28*, 79–81, 83, 84, 87, 88, 90, *91*
Woese, C. R., 119, *128*
Wolf, J., 32, 41, 43, 45, *48*
Wolin, S. L., 81, 83, 85, 86, *92*, *94*
Won, M., 67, *74*
Wood, H., 108, *114*, 138, 154, *159*, 176, *180*
Wood, V., 67, *74*
Woolcock, K., 79, 86, 89, *91*
Woollard, A., 103, *112*, 116, *127*
Worbs, M., 59, *73*
Wormington, M., 185–189, *205*, *206*
Wu, H., 216, 217, *232*
Wu, L., 102, *112*, 189, 202, *207*
Wu, M., 185–188, *205*, *206*
Wu, P.-Y. J., 106, *114*
Wunderlich, R., 194, 195, *208*
Wyers, F., 3, 17, *23*, 32, 40, 41, 45, *48*, 67, *74*, 78–81, 84, 87, *91*
Wyman, S. K., 90, *95*

X

Xi, L., 36, 37, *49*
Xia, Z., 185, 202, *205*

Xiang, K., 117–120, 124, *127*, 136, *159*
Xiang, S., 116–122, 124–126, *127–129*, 135–137, 143, 146, 152, *158*, *159*, 167, 170–173, 176, *178*, *179*
Xu, F., 78, *91*
Xu, H., 217, *232*
Xu, J., 40, *50*
Xu, W., 36, 37, *49*
Xu, Z., 17, *28*, 45, *52*, 67, *74*, 84, *94*, 106, *113*
Xue, Y., 116, 125, *128*, 135, 137, 138, 146, *158*, 170, 172, 176, *179*
Xue, Z., 45, *51*
Xuong, N. G., 68, *75*

Y

Yamamoto, M. S., 228, *235*
Yamamoto, T., 190, 197–199, *207*, *209*
Yamanaka, S., 67, *74*
Yamashita, A., 184, 185, 188–190, 198, 199, 201, *204*
Yamashita, T., 135, *158*
Yanagida, M., 135, *158*
Yang, P. K., 216, *232*
Yang W., 116, 121, 122, *127*
Yang, W. Z., 6, 7, *25*, 56, 64, 72, *74*
Yang, X. C., 136, *158*
Yao, G., 203, *211*
Yavuzer, U., 18, *28*
Yehudai-Resheff, S., 6, *25*
Yokoyama, K., 190, 197–199, *207*, *209*
Yokoyama, S., 6, *25*, 56, 72
Yu, M. C., 82, *93*
Yuan, H. S., 6, 7, *25*, 56, 72
Yue, Z., 166, *177*

Z

Zabka, V., 37, *49*
Zabolotskaya, M. V., 98, 100, *110*
Zaessinger, S., 190, 198–201, *207*, *209*
Zakrzewska-Placzek, M., 47, *52*, 140, 141, 154, *159*
Zambryski, P. C., 42, *51*
Zamir, I., 18, *28*
Zavolan, M., 87, 90, *94*
Zeller, G., 45, *52*
Zer, C., 228, 230, *234*
Zhang, C., 203, *211*

Zhang, K., 67, *74*
Zhang, M., 135, *158*
Zhang, W., 39, 40, *50*
Zhang, X., 189, 202, *207*
Zhelkovsky, A., 145, *161*
Zheng, B., 219, 220, *233*
Zheng, D., 176, *180*, 185, 199, 201, 202, *205*, *209*
Zheng, H., 33, 37, 38, *49*
Zhou, D., 220, *233*
Zhou, H., 224, *234*
Zhou, K., 214, *232*

Zhou, W. X., 224, *234*
Zhu, B., 21, *29*
Zhu, W., 199, 201, *209*
Zhu, Y., 189, *207*
Zhu, Y. Q., 224, *234*
Zimmer, S. L., 135, *158*
Zimmermann, P., 33, *49*
Zinder, N. D., 214, *231*
Zofall, M., 67, *74*
Zuo, Y., 13, *27*, 55, 68, 72, *75*, 135, *158*, 183, *204*
Zwartjes, C. G. M., 196, *208*

SUBJECT INDEX

Note: Page numbers followed by "*f*" indicate figures, and "*t*" indicate tables.

A

Aberrant cap-decapping protein, 170–174
Arabidopsis CSL4
  Hordum vulgare, 36
  HvRRP46, 36
  Saccharomyces cerevisiae, 34–36
  Trypanosome brucei, 34–36
Arabidopsis MTR4
  AtMTR4, 43–44
  BLAST, 42–43
  nuclear exosome, 42–43
  phosphorolytic activity, 44
  TRAMP, 43

B

Basic Local Alignment Search Tool
  (BLAST), 42–43
BLAST. *See* Basic Local Alignment Search
  Tool (BLAST)

C

Capped mRNA decapping
  Dcp2-decapping enzyme, 167–169
  description, 166–167
  mammalian rai1 homolog, 176
  Nudt16-decapping protein, 169–170
  RAI1 (*see* Retinoic acid induced 1 (Rai1))
  regulation, 176–177
  S. cerevisiae, 170–174
Carboxyl terminal domain (CTD), 144, 166,
  172, 174–175
CCR4-NOT complex
  active sites, CNOT6L and APE1,
    190–193, 194*f*
  BTG/TOB proteins, 202–203
  Caf40p protein, flies and human, 197
  Caf130 protein, yeast, 197
  catalytic domain structure, human
    CNOT6L, 190–193, 194*f*
  CNOT6 and CNOT6L, 190–193
  CNOT7/TOB complex structure,
    195–196, 196*f*

deadenylation assays, 198
DEDD family, nucleases, 195
definition, Not box, 196–197
description, 190
dominant-negative effects and
  knockdowns, 199
Drosophila, 199
in vitro deadenylase activity, 190–193,
  195–196
microRNAs, 202
"modules,", 200
Nocturnin, 194–195
Pab1p/PABPC, 203
point mutations, 199
posttranscriptional regulation, 200
PUF proteins, 201
RNAi *vs.* orthologues, 199
subunits, 190, 191*t*
TTP and AREs, 201
3' UTR, reporter RNA, 202
X-ray crystallography, 190–193
yeast composition, 190, 193*f*
Cold shock domains (CSDs), 12, 61–63,
  62*f*, 64
Crosslinking and analysis of cDNAs
  (CRAC) approach, 138–140
Cryptic unstable transcripts (CUTs)
  degradation, 16*f*, 18–19
  Mpp6p and Rrp47p, 18–19
  ncRNAs, 3
  PROMPTs, 20
  RNA polymerases, 17
  transcription termination mechanism, 19
CSDs. *See* Cold shock domains (CSDs)
CTD. *See* Carboxyl terminal domain (CTD)
CUTs. *See* Cryptic unstable transcripts
  (CUTs)

D

Dcp2-decapping enzyme, 167–169
Deadenylases
  classification, 183

Deadenylases (*Continued*)
  complex, CCR4-NOT
    (*see* CCR4-NOT complex)
  connections, 182
  intermediates, rRNA processing, 182
  modification, mRNAs, 182
  PAN (*see* Poly(A) nuclease (PAN))
  PARN (*see* Poly(A)-specific ribonuclease
    (PARN))

E

Endoribonuclease, 53–55, 61–65
Eukaryotic exosomes
  ancestors
    Archaea, 6–7
    architecture, 6, 7*f*
    unstructured RNA substrates, 6–7
  composition and localization models,
    8–11, 8*f*
  Dis3p and Rrp6p proteins, 7–8
  electron microscopy, 11
  enzymatic activities
    description, 12
    endo/exonuclease Dis3, 12–13, 12*f*
    Exo10 and Exo11, 14–15
    exonuclease Rrp6, 13
  human and yeast components, 8–11, 9*t*
  plant Exo9 complex, 11
  ratio, hRRP6, 8–11
  RNase PH domains, 7–8
Eukaryotic RNA exosome
  archaeal exosomes, 55
  bacterial and archaeal factors, 55
  11-component nuclear exosome and
    RRP44, RRP6, 70–71
  description, 55–56
  global architecture, 56
  hydrolytic endoribonuclease and
    exoribonuclease and Rrp44
    catalytic magnesium ions, 63
    mutation, 64–65
    PIN domains, 61–63, 62*f*, 64
    S1/KH-domain, 64
    stacking interactions, 63
  nuclear and cytoplasmic 3' to 5' decay,
    53–54, 54*f*
  RNase PH-like domains comprise
    amino acid, 57–59

  human exosome core, 57–59, 57*f*
  ribbon diagram, 56–57, 58*f*
  Saccharomyces cerevisiae, 57–59
  RRP44 and 10-component exosome,
    65–67
  RRP6, eukaryotic exosome, 67–70
  S1 and KH domains, 59–61
  TRAMP and SKI complexes, 71
  yeast, 54–55
EXO. *See* Exoribonuclease domain (EXO)
Exoribonuclease domain (EXO), 68–70, 69*f*
5'-3' Exoribonucleases (XRNs)
  active site, 122–124
  components, important, 126
  description, 116
  FEN, 121, 121*f*
  helix αD, 121*f*, 122
  vs. nucleases, 121*f*, 122
  Rai1/Dom3Z, 125–126
  Rat1-Rai1 complex, 124–125
  Rat1/Xrn2, 116–118
  Xrn1, 118–120
ExoRNase (XRN1)
  cellular functions, 110
  coding and noncoding RNA, 98, 99*f*
  cycloheximide, 105
  description, 98
  lithium, 109–110
  lncRNA decay, 105–107
  miRNAs-mediated decay, 101–103
  mRNA decay
    decapping enzyme, 98–100
    monophosphate, 98–100
    Xenopus oocytes, 98–100
  mRNA degradation, 104
  P-bodies, 104
  quality control, mRNA, 100–101
  rRNA and snoRNA processing, 108–109
  scavenger decapping enzyme, 109
  siRNA-mediated decay, 103
  tRNA quality control, 108
Exosome
  Mtr4p, 86*f*, 89
  nuclear, 78–79
  RNA degradation, 88
  Rrp44p/Rrp6p, 88
Exosome cofactors
  cellular RNAs, 15

Mpp6 (*see* M-phase phosphoprotein 6 (Mpp6p/hMPP6))
Mtr4, nuclear RNA metabolism, 15–17, 16*f*
NEXT (*see* Nuclear exosome targeting (NEXT) complex)
Nrd1p and Nab3p, 16*f*, 19
Rrp47p/C1D, 16*f*, 18
S. cerevisiae TRAMP complex and human counterpart, 16*f*, 17–18
SKI (*see* Superkiller (SKI) complex)

F

Fungal RNase III enzymes, RNA metabolism
description, 214
phylogenetic distribution and conservation
architecture and evolutionary relationships, 214–216, 215*f*
biochemical and structural analysis, *S. cerevisiae*, 216–217
bioinformatics analysis, 216–217
class I and IIIb enzymes, 216–217
helicase and PAZ domains, 214–216
kingdom and species, 214–216
Rnt1p and Dcr1, 216
pre-mRNA (*see* pre-mRNAs)
ribosomal processing and polymerase I transcriptional termination, 217–219
RNA polymerase II termination, 228–230
snoRNAs processing (*see* Small nucleolar RNAs (snoRNAs))
snRNAs processing (*see* Small nuclear RNAs (snRNAs))

H

Helicase and RNase D carboxy terminal (HRDC) domain, 68–69, 69*f*

I

Include long noncoding RNA (lncRNA)
bona fide regulators, 105–106
CD-CUTs, 107
DCP2/DCP1, 106–107
eukaryotic transcriptome, 99*f*, 107
XUT, 107

lncRNA. *See* Include long noncoding RNA (lncRNA)
Intron-encoded, snoRNAs
bona fide AGNN tetraloop and Rnt1p binding, 224–225
lariat structures, 224–225
RPS22B, 225–226
snR39 and snR59, 225–226

M

MicroRNAs (miRNAs)
Caenorhabditis elegans, 102–103
Drosophila S2 cells, 102
protein production, 102
RISC, 101–102
miRNAs. *See* MicroRNAs (miRNAs)
M-phase phosphoprotein 6 (Mpp6p/ hMPP6), 16*f*, 18–19
mRNA-decapping proteins, 167–170
mRNA decay
AREs, 189
cytoplasmic, PARN, 188
deadenylation, 182
Drosophila S2 cells, 200–201
5'-3' pathway, 202
mRNA degradation, 102–103, 104
Mtr4, 79, 86*f*

N

ncRNAs. *See* Noncoding RNAs (ncRNAs)
NEXT. *See* Nuclear EXosome Targeting (NEXT)
Noncoding RNAs (ncRNAs), 3, 17
stable, 82–83
unstable, 83–84
Nuclear EXosome Targeting (NEXT) complex, 16*f*, 20, 100–101, 226–227, 230
Nuclear Rat1/Xrn2 exonuclease
Arabidopsis thaliana proteins, 135
auxiliary factors, 132
catalytic properties, 134
description, 132
MET22/HAL2 deletion, 134
nuclear 5'-3' exoribonuclease activity, *S.cerevisiae*, 133
overlapping functions
cytoplasmic mRNAs, 155

Nuclear Rat1/Xrn2 exonuclease
    (*Continued*)
    deletion, *XRN1*, 154–155
    5' processing, 60S subunit rRNAs, 155
    ts-lethality, *rat1-1* mutant, 154–155
    Pol III transcription initiation, 133
    purification, Rat1/Tap1/Hke1, 133–134
    RNA surveillance (*see* RNA surveillance)
    rRNA maturation
        Arabidopsis AtXRN2 and AtXRN3,
            140
        classification, 138
        CRAC approach, 138–140
        functional analysis, 138
        ITS1 and ITS2, Rat1 substrates,
            138–140
        pre-rRNAs, 140
        and snoRNA processing, Rat1 activity,
            138, 139*f*
    Schizosaccharomyces pombe Dhp1p,
        134–135
    sequence and structural data
        classification, superfamilies, 135–136
        CR1 and CR2 regions, 136
        crystal structures, yeast, 136–137
        DOM3Z and Rai1, 137
        Rai1 binding, 137
        XRN-family enzymes, 136–137
    silencing suppressor, 155–156
    snoRNA processing (*see* Small nucleolar
        RNAs (snoRNAs))
    S. pombe Rai1, enzymatic activity, 135
    temperature-sensitive *(ts)* mutations, 133
    temperature-sensitive *rat1* mutant
        phenotype, 132–133
    torpedo model (*see* Torpedo model)
Nudt16-decapping protein, 169–170

P
PAN. *See* Poly(A) nuclease (PAN)
PARN. *See* Poly(A)-specific ribonuclease
    (PARN)
PH-domain, 32–33, 36–37, 39, 55, 56–57
Pilus-forming N-terminus (PIN) domain,
    61–63, 62*f*, 64, 66*f*
Plant exosomes and cofactors
    Arabidopsis CSL4, 34–36
    A. thaliana, 33–34

catalytically active core complex, 39–40
description, 32
duplicated genes
    PH-domain proteins, 32–33
    phylogenetic tree, 33, 34*f*
    RRP45, 33, 34*f*
functional specialization
    AtRRP4 and AtRRP41, 37–38
    CER7, 36–37
    Drosophila melanogaster, 38
    "exozyme model,", 38
    RRP45A, 37
mRNA quality control pathways, 48
nucleolar protein, 42–44
polyadenylated substrates, 44–46
RNA degradation pathways, 46–47
RRP6-like proteins, 40–41
SKI complex, 41–42
Polymerase II termination, RNA
    cotranscriptional cleavage, Rnt1p,
        228–230
    dicer enzymes, 228–230
    fail-safe, RNA Pol II, 228–230, 229*f*
    "torpedo" model, 228–230
    transcription, gene expression, 228–230
Polymerase I transcriptional termination,
    RNA
    cotranscriptional rRNA processing, 217,
        218*f*
    description, 217
    functions, 219
    RAC complex, 217–218
    Rnt1p, 217–218
    "torpedo,", 218
Poly(A) nuclease (PAN)
    C-terminal helical domain, 184
    DEDD superfamily, 3' exonucleases, 183
    dependence, PABPC, 183–184
    genetic analysis, yeast, 184–185
    inhibitory activity, 184
    miRNAs, 185
    oligomerization, Pan3p, 183
    Pabp1p-dependent activity, 183–184
    *pan2*rplain and *ccr4*rplain mutations,
        184–185
    S. cerevisiae mutant deficiency, 183
Poly(A)-specific ribonuclease (PARN)
    AREs, 189

"canonical" cap-binding proteins, 186–188
catalytic centers alignment, *S. pombe* Pop2p, 185, 187*f*
CPE, 189
HeLa cell extraction, 185
homodimeric, 185, 186*f*
intron-encoded precursors, 190
mammalian, 188
m^{7}G binding, tryptophan residues, 186–188, 188*f*
RRM domain, 185–186
T-DNA insertions, *Arabidopsis*, 188–189
Xenopus oocytes, 189
Poly(A) tails
    eukaryotic cells, 182
    PABPC, 183
    Pabp1p-dependent activity, 183–184
    RNAs, 182
    yeast PAN2/PAN3 deletions, 184–185
pre-mRNAs
    exosomes and 5'–3' exonucleases Xrn1p and Rat1p, 226–227
    fungal RNase III, 226
    intronic sequences, RPL18A and RPS22B, 226–227
    Rnt1p stem-loop structure and NMD system, 226–227
    S. pombe Pac1p overexpression, 228
    stem-loop structures, 228
    transcriptional repressor Mig2p, 227–228
    unspliced and lariat degradation, RNase III regulation, 226–227, 227*f*
Promoter upstream transcripts (PROMPTs)
    exosomal degradation, 20
    ncRNAs, 3
    S. cerevisiae CUTs, 20
PROMPTs. *See* Promoter upstream transcripts (PROMPTs)

R
Rai1/Dom3Z structure, 125–126
Rat1
    activity, Rai1, 172
    crystal structure
        and Rai1 complex, 124–125
        and Xrn2, 116–118
    lncRNAs, 107

Retinoic acid induced 1 (Rai1)
    decapping endonuclease protein, 171–172
    pyrophosphohydrolase, 170
    Rat1 activity, 172
    rRNA processing, 175–176
    structure
        Dom3Z, 125–126
        Rat1 complex, 124–125
    transcription termination and 5'-end mRNA capping quality control, 174–175
RISC. *See* RNA-induced silencing complex (RISC)
RNA degradation, 78, 83, 88
RNA degradation pathways
    cytoplasmic exosome, 46–47
    nuclear exosome, 47
    putative role, 46–47
    rRNA maturation, 47
RNA exosomes
    cellular functions
        chromatin activity, 5–6
        mRNA turnover, 5, 8*f*
        nuclear substrates, 3, 4*f*
        RNA QC, 3–5, 4*f*
    cofactors (*see* Exosome cofactors)
    components, *Saccharomyces cerevisiae*, 2
    discovery, 2
    eukaryotic, 6–15
    Rrp4p, Rrp44p and Rrp41p, 2
RNA-induced silencing complex (RISC), 101–103
RNA processing maturation, 138–140
    snoRNA, 140–141
RNA-protein particles (RNPs), 3
RNA stability, 109
RNA surveillance
    mRNA
        AREs, 151
        biogenesis, 149
        cotranscriptional pre-miRNA release, 152
        cytoplasmic degradation, 149–150
        DCP2 decapping enzyme, 152
        mammalian Xrn2 nuclease, 152
        nuclear decay pathway, 149–150

RNA surveillance (*Continued*)
    unspliced pre-mRNAs degradation, 151
    rapid tRNA decay, 152–153
    Rat1 contribution, 149, 150*f*
    rRNA and snoRNA, 153
    TERRA
        AtXRN2 and AtXRN3, 154
        lncRNA, 154
        telomeric and subtelomeric regions,
            153–154
        TRAMP complex component,
            153–154
RNPs. *See* RNA-protein particles (RNPs)
Rrp6
    Arabidopsis mutants, 41
    catalytic domains, 69–70
    crystal structures, 68–69
    domains, 68–69, 69*f*
    EXO, 68–69
    HRDC, 68–69
    nuclear exosome, 67–68
    nuclear protein, 40–41
    RRP44 and 11-component nuclear
        exosome, 70–71
RRP44
    eukaryotic 10-component exosome,
        65–67
    hydrolytic endoribonuclease and
        exoribonuclease
        exosome core, 64–65
        magnesium ions, 63
        mutation, 64–65
        PIN, 61–63, 62*f*, 64
        RNase II, 63
        S1/KH-domain, 64
        stacking interactions, 63
    RRP6, and 11-component nuclear
        exosome, 70–71

S

Saccharomyce cerevisiae
    aberrantly capped mRNAs, 171–172
    nutrient stress, 173–174
    Rai1, 172
    stress granules, 173
    uncapped RNA, 170
Saccharomyce cerevisiae TRAMP complex,
    16*f*, 17–18

S1 and KH domains
    cap subunits, 60*f*, 61
    GXNG motif, 59
    NTD, 59
    OB domains, 59–61
    Sulfolobus solfataricus, 61
    zinc ribbon domain, Csl4, 61
siRNAs. *See* Small interfering RNAs
    (siRNAs)
Small interfering RNAs (siRNAs), 103
Small nuclear RNAs (snRNAs)
    functions, snRNA 3'-end processing,
        220–221
    northern blotting, 220–221
    RNase III Rnt1p, 219–220, 220*f*
    transcripts and Pol I, II, 219–220
Small nucleolar RNAs (snoRNAs)
    5'- and 3'-end processing, 221
    box C/D and box H/ACA family, 221
    genomic organization, genes encoding,
        140
    intron-encoded (*see* Intron-encoded,
        snoRNAs)
    mature 18S rRNA, 140
    polycistronic precursors, 141
    polycistronic transcription units, 223*f*, 224
    Rat1 activity, yeast and 5'-3' exonuclease,
        vertebrates, 141
    Rat1 and Xrn1, 140
    transcribed precursors processing
        description, 222
        5'-3' exonucleases Xrn1p and Rat1p,
            222
        RNase III Rnt1p, 222, 223*f*
        snR40 transcript and U3, 222
snoRNA and rRNA processing, 108–109
snoRNAs. *See* Small nucleolar RNAs
    (snoRNAs)
snRNAs. *See* Small nuclear RNAs
    (snRNAs)
SOV. *See* Suppressor of varicose (SOV)
Superkiller (SKI) complex, 16*f*, 20–21
Suppressor of varicose (SOV), 39–40, 44

T

Torpedo model
    polymerase I, 145–147
    polymerase II

bacterial Rho DNA–RNA helicase, 144–145
CPSF-73 subunit and CPSF, 143
CTD, 144
fail-safe termination mechanism, 143
in vitro and 5'-3' degradation activity, 143–144
nascent transcript and RNA polymerases, 141
noncoding transcripts, 144–145
Rat1 function, transcription termination, 141, 142*f*
snoRNAs, 145
Xrn2 and Rat1, 142–143
yeast Rat1-cooperating proteins, 145, 146*t*
termination and RNA surveillance
cotranscriptional Drosha cleavage, 148
human Xrn2, 148
pervasive transcription, 148
site-associated RNAs, 148
transcription termination-related mechanism, 149
Transcription termination. *See* Torpedo model
TRf–Air–Mtr4 Polyadenylation (TRAMP) stimulation
Arabidopsis thaliana, 89
biochemistry and structure
enzymatic activities, 87–88
Mtr4p helicase, 85, 86*f*
RNA-binding properties, 86–87
Saccharomyces cerevisiae lysates, 85
Trf4p fragment, 85, 85*f*
canonical nuclear polyadenylation, 78, 78*f*

Drosophila melanogaster, 89
eukaryotes, 78
exosome (*see* Exosome)
Mtr4p, 79
NEXT, 90
PROMPTs, 90
proteins, 79
RNA substrate repertoire
bioinformatics analyses, 80
CUTs, 80, 84
non-protein coding, 79–80
Nrd1p-Nab3p-Sen1p complex, 80
quality control, ribosomes, 82
stable ncRNAs, 82–83
transcriptome, 83–84
yeast, 79–80, 81*t*
S. cerevisiae, 89
tRNA quality control, XRN1, 108

X
Xrn1
crystal structures
D1-D4 domains, 119, 120*f*
domains, 119–120
KlXrn1 and DmXrn1, 119
homologs, 116
K. lactis, 117*f*
Xrn2
Arabidopsis, 116
crystal structure, 116–118
silencing suppressor, 155–156
XRN4, 46–47

Z
Zinc knuckle (ZnK) proteins, 79, 85*f*

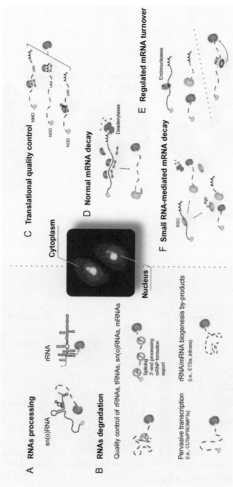

Chapter 1. Figure 1.1 (See legend in text.)

A **RNAs processing**
sn(o)RNA

rRNA

B **RNAs degradation**
Quality control of rRNAs, tRNAs, sn(o)RNAs, mRNAs

Splicing
3′-end processing
mRNP formation
export

Pervasive transcription
(i.e., CUTs/PROMPTs)

rRNA/mRNA biogenesis by-products
(i.e., ETSs, introns)

Cytoplasm

Nucleus

C **Translational quality control**
NMD
NSD
NGD

UAG

D **Normal mRNA decay**
Deadenylases

E **Regulated mRNA turnover**
Endonucleases

ARE

F **Small RNA-mediated mRNA decay**
RISC
Ago

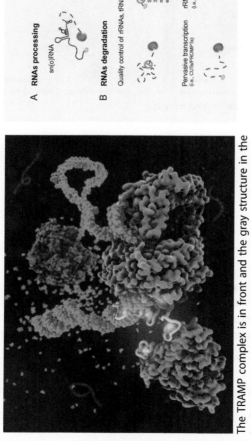

The TRAMP complex is in front and the gray structure in the back is the exosome. In yellow are zinc ions coordinated by Air2. Courtesy of Stepanka Vanacova.

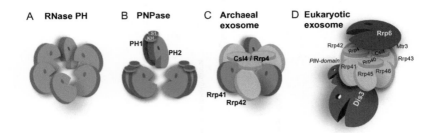

Chapter 1. Figure 1.2 (See legend in text.)

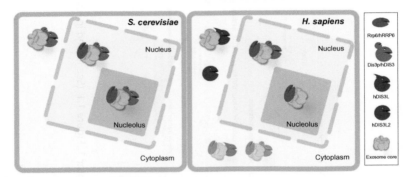

Chapter 1. Figure 1.3 (See legend in text.)

Chapter 1. Figure 1.4 (See legend in text.)

A rRNA biogenesis

 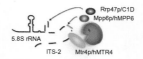

Early pre-rRNA maturation   Processing of 5.8S rRNA   Degradation of rRNA maturation by-products (5'-ETS)

B Cotranscriptional processing/degradation of unstable transcripts

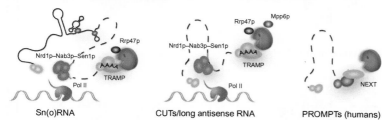

Sn(o)RNA   CUTs/long antisense RNA   PROMPTs (humans)

C Degradation of aberrant transcripts

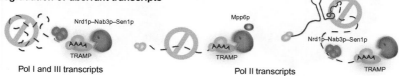

Pol I and III transcripts   Pol II transcripts

D Cytoplasmic mRNA turnover and surveillance

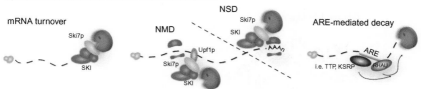

Chapter 1. Figure 1.5 (See legend in text.)

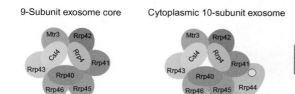

9-Subunit exosome core   Cytoplasmic 10-subunit exosome

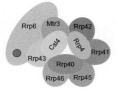 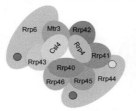

Nucleolar 10-subunit exosome   Nuclear 11-subunit exosome

Yeast	Human
Csl4	Exos1
Rrp4	Exos2
Rrp40	Exos3
Rrp41	Exos4
Rrp46	Exos5
Mtr3	Exos6
Rrp42	Exos7
Rrp43	Exos8
Rrp45	Exos9
Rrp44	Dis3
	Dis3L1
Rrp6	ExosX

Chapter 3. Figure 3.1 (See legend in text.)

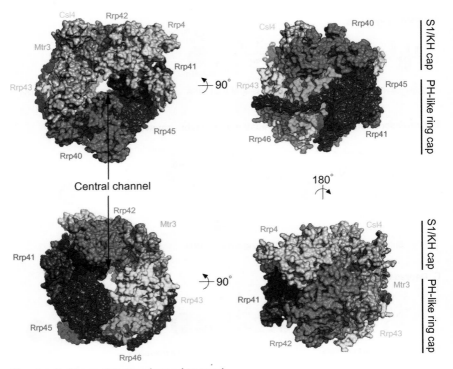

Chapter 3. Figure 3.2 (See legend in text.)

RNase PH-like ring

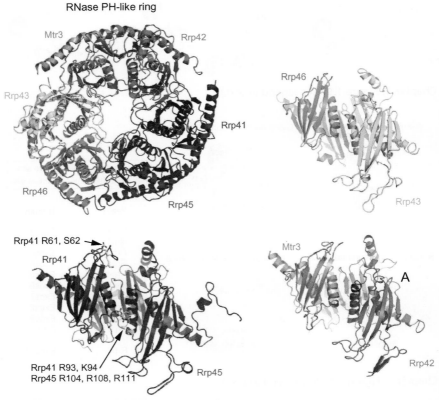

Chapter 3. Figure 3.3 (See legend in text.)

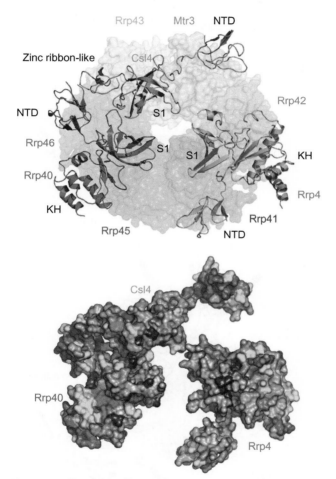

Chapter 3. Figure 3.4 (See legend in text.)

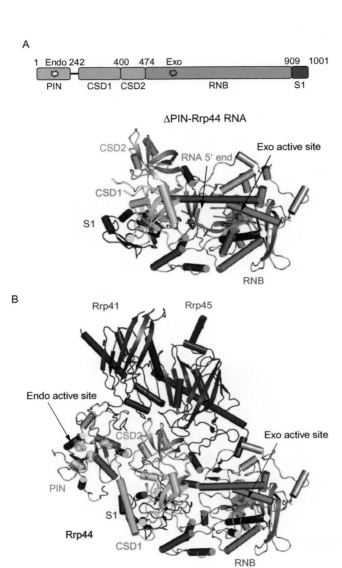

Chapter 3. Figure 3.5 (See legend in text.)

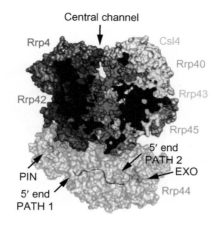

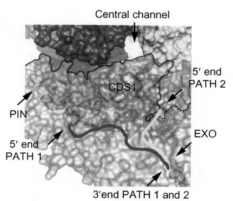

Chapter 3. Figure 3.6 (See legend in text.)

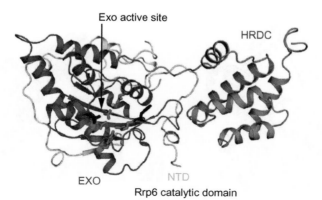

Chapter 3. Figure 3.7 (See legend in text.)

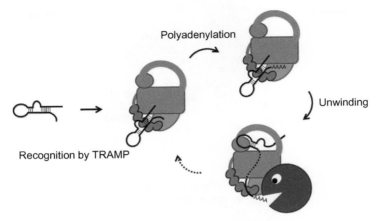

Polyadenylation

Recognition by TRAMP

Unwinding

Activation of exosome

Chapter 4. Figure 4.1 (See legend in text.)

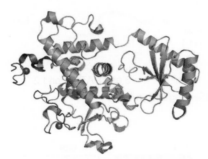

Chapter 4. Figure 4.2 (See legend in text.)

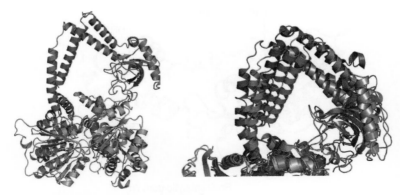

Chapter 4. Figure 4.3 (See legend in text.)

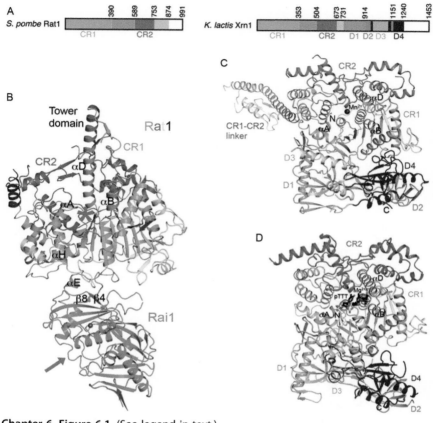

Chapter 6. Figure 6.1 (See legend in text.)

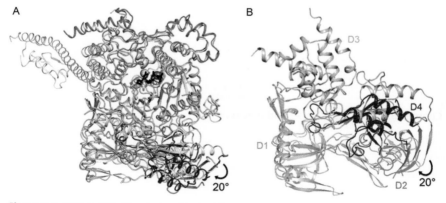

Chapter 6. Figure 6.2 (See legend in text.)

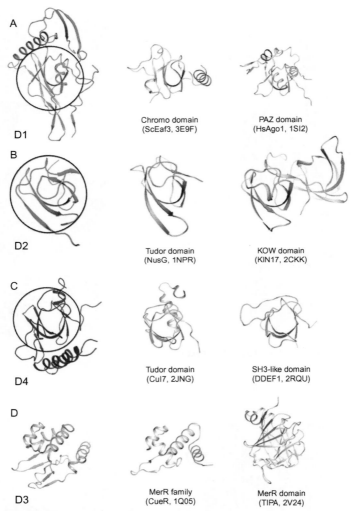

A

D1

Chromo domain
(ScEaf3, 3E9F)

PAZ domain
(HsAgo1, 1SI2)

B

D2

Tudor domain
(NusG, 1NPR)

KOW domain
(KIN17, 2CKK)

C

D4

Tudor domain
(Cul7, 2JNG)

SH3-like domain
(DDEF1, 2RQU)

D

D3

MerR family
(CueR, 1Q05)

MerR domain
(TIPA, 2V24)

Chapter 6. Figure 6.3 (See legend in text.)

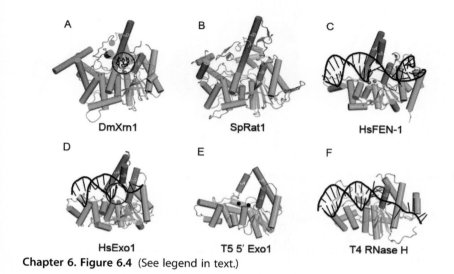

Chapter 6. Figure 6.4 (See legend in text.)

Chapter 6. Figure 6.5 (See legend in text.)

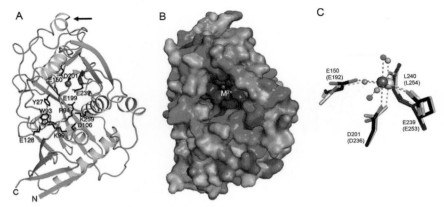

Chapter 6. Figure 6.6 (See legend in text.)

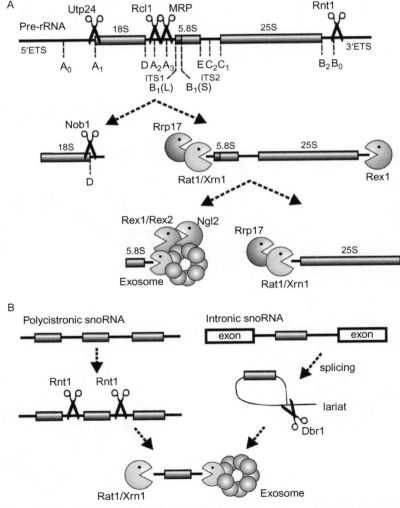

Chapter 7. Figure 7.1 (See legend in text.)

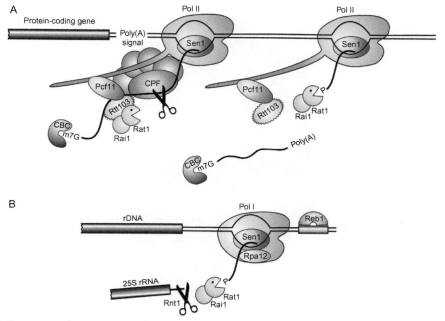

Chapter 7. Figure 7.2 (See legend in text.)

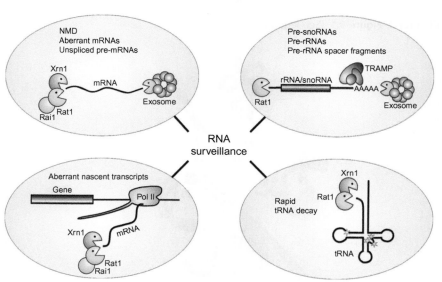

Chapter 7. Figure 7.3 (See legend in text.)

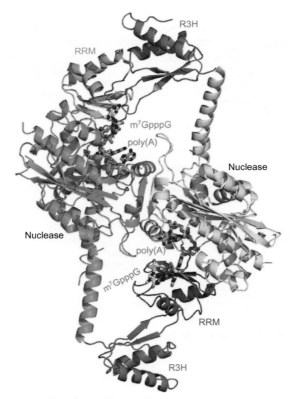

Chapter 9. Figure 9.1 (See legend in text.)

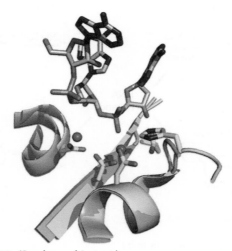

Chapter 9. Figure 9.2 (See legend in text.)

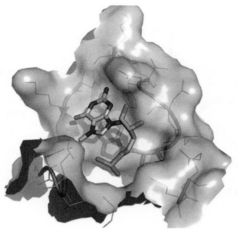

Chapter 9. Figure 9.3 (See legend in text.)

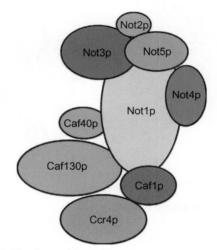

Chapter 9. Figure 9.4 (See legend in text.)

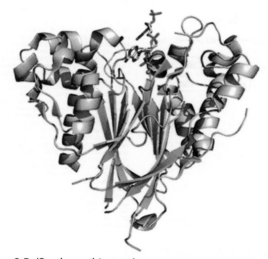

Chapter 9. Figure 9.5 (See legend in text.)

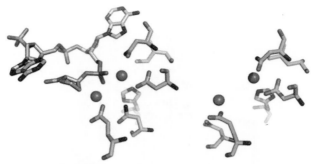

Chapter 9. Figure 9.6 (See legend in text.)

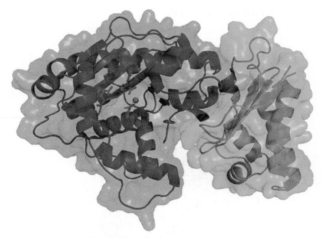

Chapter 9. Figure 9.7 (See legend in text.)

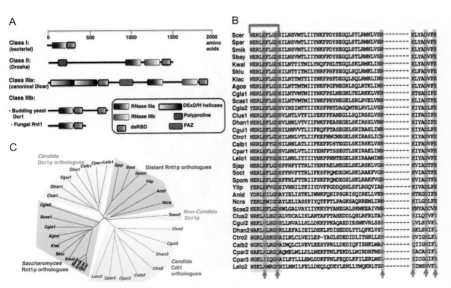

Chapter 10. Figure 10.1 (See legend in text.)

Printed and bound by CPI Group (UK) Ltd, Croydon, CR0 4YY

08/05/2025

01864957-0001